## ■ *Properties*

### Section 2.4

If both the numerator and the denominator of a given fraction are multiplied, or divided, by the same nonzero number, the resulting fraction is equal to the given fraction.

$$\frac{n}{d} = \frac{n \cdot k}{d \cdot k} \quad \text{or} \quad \frac{n}{d} = \frac{n \div k}{d \div k}$$

### Section 2.5

The sum (difference) of like fractions is a fraction with the same denominator and a numerator equal to the sum (difference) of the numerators of the original fractions.

$$\frac{a}{c} + \frac{b}{c} = \frac{a+b}{c} \quad \text{or} \quad \frac{a}{c} - \frac{b}{c} = \frac{a-b}{c}$$

### Section 2.7

The product of two fractions is a fraction whose numerator is the product of the numerators and whose denominator is the product of the denominators of the given fractions.

$$\frac{a}{b} \cdot \frac{c}{d} = \frac{a \cdot c}{b \cdot d}$$

The quotient of two fractions is a fraction equal to the product of the dividend and the reciprocal of the divisor.

$$\frac{a}{b} \div \frac{c}{d} = \frac{a}{b} \cdot \frac{d}{c} = \frac{a \cdot d}{b \cdot c}$$

### Section 3.1

The opposite of the opposite of a number is equal to the number.

$$-(-r) = r$$

### Section 3.2

#### Sum of Two Signed Numbers

If the numbers have the same sign:

1. Add their absolute values.
2. Precede the sum by the same sign.

If the numbers have opposite signs:

1. Subtract the smaller absolute value from the larger absolute value.
2. Precede the difference by the sign of the number with the larger absolute value.

If the numbers have opposite signs and the same absolute value, the sum is zero.

#### Difference of Two Signed Numbers

1. Replace the number to be subtracted with its opposite.
2. Add the resulting numbers.

$$r - s = r + (-s)$$

## Section 3.3   Product (or Quotient) of Two Signed Numbers

If the numbers have the same sign:

**1.** Multiply (divide) the absolute values.
**2.** Precede the product (quotient) by a + sign.

If the numbers have opposite signs:

**1.** Multiply (divide) the absolute values.
**2.** Precede the product (quotient) by a − sign.

## Section 3.4

If the bases are the same, the product of two numbers raised to the exponents $m$ and $n$ is equal to a power with the same base raised to the sum of the exponents.

$$b^m \cdot b^n = b^{m+n}$$

If the bases are the same, the quotient of two numbers raised to the exponents $m$ and $n$ is equal to a power with the same base raised to the difference of the exponents.

$$\frac{b^m}{b^n} = b^{m-n}$$

## Section 4.1

If a number is added to, or subtracted from, each side of an equation, the resulting equation has the same solutions as the original equation.

$$\textbf{If} \quad a = b, \quad \textbf{then} \quad a + c = b + c \quad \textbf{and} \quad a - c = b - c$$

If each side of an equation is multiplied by, or divided by, the same nonzero number, the resulting equation has the same solutions as the original equation.

$$\textbf{If} \quad a = b, \quad \textbf{then} \quad a \cdot c = b \cdot c \quad \textbf{and} \quad \frac{a}{c} = \frac{b}{c} \quad (c \neq 0)$$

## Section 4.7

In a proportion, the product of the extremes is equal to the product of the means.

$$\textbf{If} \quad \frac{a}{b} = \frac{c}{d}, \quad \textbf{then} \quad a \cdot d = b \cdot c$$

## Section 7.2

In any triangle:

$$\angle A + \angle B + \angle C = \textbf{180°}$$

In a right triangle:

$$c^2 = a^2 + b^2 \quad \textbf{Pythagorean theorem}$$

## Section 8.4

In an oblique triangle:

$$\frac{a}{\sin A} = \frac{b}{\sin B} = \frac{c}{\sin C} \quad \textbf{Law of sines}$$

## Section 8.5

$$\left.\begin{array}{l} a^2 = b^2 + c^2 - 2bc \cos A \\ b^2 = a^2 + c^2 - 2ac \cos B \\ c^2 = a^2 + b^2 - 2ab \cos C \end{array}\right\} \quad \textbf{Laws of cosines}$$

## Section 8.7

$$\frac{\textbf{Rad.}}{\pi} = \frac{\textbf{Deg.}}{180} \quad \textbf{Degree-radian conversion}$$

# Basic
# Mathematics
## for Technical Programs

### *IRVING DROOYAN*
*Los Angeles Pierce College, Emeritus*

**Prentice Hall**
*Upper Saddle River, New Jersey   Columbus, Ohio*

**Library of Congress Cataloging-in-Publication Data**

Drooyan, Irving.
   Basic mathematics for technical programs / Irving Drooyan.
     p.  cm.
   Includes index.
   ISBN 0-13-206632-7
   1. Mathematics.  I. Title
  QA39.2.D757  1997
  513′.14—dc20                                 95-42881
                                                     CIP

Editor: Stephen Helba
Production Editor: Stephen C. Robb
Design Coordinator: Jill E. Bonar
Text Designer: Susan E. Frankenberry
Production Manager: Deidra M. Schwartz
Marketing Manager: Danny Hoyt

This book was set in Times and Helvetica Bold by Bi-Comp, Inc. and was printed and bound by Quebecor Printing/Semline. The cover was printed by Phoenix Color Corp.

 © 1997 by Prentice-Hall, Inc.
Simon & Schuster/A Viacom Company
Upper Saddle River, New Jersey 07458

Earlier edition, entitled *Introduction to Technical Mathematics: A Calculator Approach,* © 1979 by John Wiley & Sons, Inc.

Printed in the United States of America

10 9 8 7 6 5 4 3 2 1

ISBN: 0-13-206632-7

Prentice-Hall International (UK) Limited, *London*
Prentice-Hall of Australia Pty. Limited, *Sydney*
Prentice-Hall Canada, Inc., *Toronto*
Prentice-Hall Hispanoamericana, S. A., *Mexico*
Prentice-Hall of India Private Limited, *New Delhi*
Prentice-Hall of Japan, Inc., *Tokyo*
Simon & Schuster Asia Pte. Ltd., *Singapore*
Editora Prentice-Hall do Brasil, Ltda., *Rio de Janeiro*

*Basic Mathematics for Technical Programs* has been prepared in an easy-to-use workbook format. Elements of algebra are introduced early with topics of arithmetic to provide a unified approach to problem solving that traditional arithmetic methods lack. Also, the use of scientific calculators allows for the efficient use of classroom time to develop mathematical concepts and problem-solving skills.

Basic arithmetic skills and problem-solving techniques are introduced in Chapters 1 through 4. The concept of percent, together with technical and consumer-oriented applications, is presented in Chapter 5. Chapter 6 provides practice with units of measurement in the United States and metric systems, using formulas related to industrial, electrical, and mechanical applications. Other technical applications are included in Chapters 7 through 9, which cover geometric figures, essentials of trigonometry, and graphing methods.

Special attention is given to mental activities that provide a balance to students' reliance on calculators. Sets of exercises in Chapters 1 through 6, which can be done orally in class, provide such activities. These exercises focus on topics that most people believe constitute "literacy" in arithmetic: multiplication and division of powers of ten, decimal equivalents of commonly used fractions and percent, estimating computations, and other topics.

Each chapter concludes with a set of review exercises that can be used as a practice test. Also, for convenient reference, a summary of definitions and properties introduced in the text are given inside the front and back covers.

Answers, including graphs, are given in the text for the odd-numbered exercises in each section and for all review exercises. The Instructors' Manual includes answers for the even-numbered exercises and a testing program.

Irving Drooyan

# CONTENTS

## 1 Fundamental Concepts      1

# 2

## Fractions                                                    39

# 5

## *Percent*                                    *153*

# 6

## Measurement

# 7

## *Geometric Figures*                                                 **225**

# 8

## Essentials of Trigonometry                                           281

# 9

## Graphing                                                                 319

# Fundamental Concepts

## 1.1  Basic Operations

As we review basic arithmetic operations we shall sometimes refer to certain kinds of numbers. As you may recall, the **natural numbers** (also called "counting numbers") are

1, 2, 3, 4, 5, and so forth

and the **whole numbers** are

0, 1, 2, 3, 4, 5, and so forth

We shall occasionally use letters of the alphabet to represent numbers. For example, a phrase such as "For any number $b$" is understood to mean that the symbol $b$ represents an unspecified number. Letters used in this manner are called **variables**.

### ■ Addition

The operation of **addition** is used to find the **sum** (or **total**) of two or more numbers. The numbers to be added are called **terms** (or **addends**).

In our work we will use a calculator instead of traditional paper and pencil methods.

**EXAMPLE 1**  $86.3 + 47.29 + 126.5 + 24 = 284.09$;   this is computed by the sequence

$$86.3 \boxed{+} 47.29 \boxed{+} 126.5 \boxed{+} 24 \boxed{=} \longrightarrow 284.09$$

An important property of addition that is useful for computations is called the **commutative property**.

*Changing the order in which numbers are added does not change the sum.*

In symbols, the commutative property is expressed as

$$a + b = b + a$$

The commutative property of addition enables us to *check* an addition by adding the same terms in reverse order. Thus the computation of Example 1 can be checked as follows:

$$24 + 126.5 + 47.29 + 86.3 = 284.09$$

## ■ Subtraction

The operation of **subtraction** is used to find the **difference** of two numbers.

**EXAMPLE 2**    $768 - 596.4 = 171.6$;   this is computed by the sequence

$$768 \boxed{-} 596.4 \boxed{=} \longrightarrow 171.6$$

The operation of subtraction can be checked by addition. Since the difference is on display, it is only necessary to check by pressing the $\boxed{+}$ key, entering the number subtracted, and then pressing the $\boxed{=}$ key.

The operation of subtraction is *not* commutative, as you can check by computing

$$28 - 17 = 11 \qquad \text{and} \qquad 17 - 28 = -11$$

The number $-11$ is a *negative* number* and is not equal to 11.

Because subtraction is not commutative, subtractions must be performed in order, from left to right.

**EXAMPLE 3**    $11.007 - 5.39 - 2.74 - 0.98 = 1.897$;   this is computed by the sequence

$$11.007 \boxed{-} 5.39 \boxed{-} 2.74 \boxed{-} 0.98 \boxed{=} \longrightarrow 1.897$$

## ■ Combined Additions and Subtractions

Computations involving both addition and subtraction are also done in order from left to right.

**EXAMPLE 4**    $79 - 12.7 - 3.11 + 16.03 = 79.22$;   this is computed by the sequence

$$79 \boxed{-} 12.7 \boxed{-} 3.11 \boxed{+} 16.03 \boxed{=} \longrightarrow 79.22$$

## ■ Multiplication

The operation of **multiplication** is used to find the **product** of two or more numbers. The numbers to be multiplied are called **factors**.

**EXAMPLE 5**    $8 \times 72 \times 6.59 = 3795.84$;   this is computed by the sequence

$$8 \boxed{\times} 72 \boxed{\times} 6.59 \boxed{=} \longrightarrow 3795.84$$

In Example 5 we used the $\times$ symbol to indicate multiplication. *Centered dots* and *parentheses* are also used to indicate multiplication. Thus the product of Example 5

---

*Negative numbers are considered in Chapter 3.

can also be written either as

$$8 \cdot 72 \cdot 6.59 \quad \text{or} \quad (8)(72)(6.59)$$

As in addition, the operation of multiplication is commutative.

> *Changing the order in which numbers are multiplied does not change the product.*

In symbols, the commutative property of multiplication is expressed as

$$a \cdot b = b \cdot a$$

The commutative property of multiplication enables us to *check* a multiplication by multiplying the factors in reverse order. Thus the computation of Example 5 can be checked by finding the product

$$(6.59)(72)(8) = 3795.84$$

## ■ Division

The operation of **division** is used to find the **quotient** of two numbers. In symbols, $n \div d$ is called the quotient of $n$ divided by $d$; the number $n$ is called the **dividend**, and $d$ is called the **divisor**. The quotient $n \div d$ can also be written in fraction form as $\frac{n}{d}$ or as $n/d$.

**EXAMPLE 6**   $417.1 \div 0.97 = \dfrac{417.1}{0.97} = 430;$   this is computed by the sequence

$$417.1 \boxed{\div} 0.97 \boxed{=} \longrightarrow 430$$

The number 417.1 is the dividend, 0.97 is the divisor, and 430 is the quotient.

The result of a division can be *checked* by a multiplication; the product of the divisor and the quotient equals the dividend. Since the quotient is on display, it is only necessary to check by pressing the $\boxed{\times}$ key, entering the divisor, and then pressing the $\boxed{=}$ key.

The operation of division is not commutative, as you can verify by computing

$$\frac{8}{2} = 8 \div 2 = 4 \quad \text{and} \quad \frac{2}{8} = 2 \div 8 = 0.25$$

Clearly, $8 \div 2$ is not equal to $2 \div 8$.

## ■ Role of Zero

Zero can be divided by any nonzero number $n$, and the quotient $\frac{0}{n}$ is always zero. However, it is not possible to divide by zero. Expressions of the form $\frac{n}{0}$ are said to be "undefined." (Most calculators will show an error signal if division by zero is attempted.) For example, $\frac{0}{18}$ equals 0, but $\frac{18}{0}$ is undefined.

## ■ Reciprocals

If $n$ is any nonzero number, then $\frac{1}{n}$, the quotient of 1 and $n$, is called the **reciprocal** of $n$. The reciprocal of a number can be found either by an ordinary division or by use of the *reciprocal key* $\boxed{1/x}$. The $\boxed{=}$ key is not needed when using the $\boxed{1/x}$ key.

### EXAMPLE 7

**(a)** The reciprocal of 4 is $\dfrac{1}{4} = 0.25$; this is computed by either of the sequences

$$1 \;\boxed{\div}\; 4 \;\boxed{=}\; \longrightarrow 0.25 \qquad \text{or} \qquad 4 \;\boxed{1/x}\; \longrightarrow 0.25$$

**(b)** The reciprocal of 0.25 is $\dfrac{1}{0.25} = 4$; this is computed by either of the sequences

$$1 \;\boxed{\div}\; 0.25 \;\boxed{=}\; \longrightarrow 4 \qquad \text{or} \qquad 0.25 \;\boxed{1/x}\; \longrightarrow 4$$

## ■ Combined Multiplications and Divisions

Computations involving both multiplications and divisions can readily be done on a calculator.

**EXAMPLE 8**   $5.75 \times 7.2 \div 0.18 = 230$; this is computed by the sequence

$$5.75 \;\boxed{\times}\; 7.2 \;\boxed{\div}\; 0.18 \;\boxed{=}\; \longrightarrow 230$$

Because a fraction bar can be used to indicate division, calculations such as that of Example 8 frequently appear in the form

$$\frac{5.75 \times 7.2}{0.18}$$

A computation that involves a quotient of products, such as

$$\frac{8 \times 19 \times 5}{3 \times 27 \times 11}$$

can be done in more than one way. A particularly efficient procedure is *first* to compute the product above the fraction bar and then divide in turn by each of the factors below the fraction bar:

$$8 \;\boxed{\times}\; 19 \;\boxed{\times}\; 5 \;\boxed{\div}\; 3 \;\boxed{\div}\; 27 \;\boxed{\div}\; 11 \;\boxed{=}\; \longrightarrow 0.8529741$$

**EXAMPLE 9**   $\dfrac{16.3 \times 86.4}{6 \times 0.9 \times 21.3} = 12.244132$; this is computed by the sequence

$$16.3 \;\boxed{\times}\; 86.4 \;\boxed{\div}\; 6 \;\boxed{\div}\; 0.9 \;\boxed{\div}\; 21.3 \;\boxed{=}\; \longrightarrow 12.24413$$

### Exercises for Section 1.1

*Compute. See Example 1.*

**1.** $6 + 9 + 8$             **2.** $8 + 14 + 6$

**3.** $7.0463 + 9.2 + 56.725$      **4.** $0.004 + 8.17 + 3.3239$

**5.** $24.1304 + 42.167 + 2.75703 + 7.79216$

**6.** 99.562 + 9.6301 + 8.54753

*Compute. See Examples 2 and 3.*

**7.** 17 − 8

**8.** 76 − 72

**9.** 503.9511 − 487.0102

**10.** 685.006 − 545.7

**11.** 15.07 − 4.885 − 0.3151

**12.** 7.8 − 3.073 − 0.4781

*Compute. See Example 4.*

**13.** 46 − 7 − 4 + 5

**14.** 12.4402 − 5.076 − 3.54 + 5.903

**15.** 691.79 − 27.928 − 1.51792 + 3.944

**16.** 60.4688 − 18.602 − 7.1194 + 9.45955

*Compute. See Example 5.*

**17.** 3 × 42

**18.** 70 × 4

**19.** 0.876 × 0.678

**20.** 879.23 × 462.714

**21.** 753.84 × 0.0475 × 3.5

**22.** 1826 × 0.0633 × 4.8

*Compute. See Example 6.*

**23.** 48 ÷ 6

**24.** 24 ÷ 6

**25.** 147.801 ÷ 8.7074

**26.** 976.669 ÷ 5.7252

**27.** 4.69012 ÷ 20.849          **28.** 8.97688 ÷ 15.368

*Find the reciprocal of each of the following numbers. See Example 7.*

**29.** 2          **30.** 3          **31.** 0.17          **32.** 0.4

*Compute. See Example 8.*

**33.** 2.64 × 18 ÷ 6.4          **34.** 11.7 × 0.94 ÷ 3.05

**35.** 99.7 × 0.305 ÷ 5.5          **36.** 84.8 × 0.29 ÷ 92

**37.** 255.5 × 0.52 ÷ 12.46          **38.** 774.9 × 0.18 ÷ 6.93

*Compute. See Example 9.*

**39.** $\dfrac{104.8 \times 1.5}{2.2 \times 36.8}$          **40.** $\dfrac{46.5 \times 7.32}{25.5 \times 9.58}$

**41.** $\dfrac{5.6 \times 8.65}{5 \times 8.59 \times 9.01}$          **42.** $\dfrac{3.15 \times 9.501}{1.5 \times 47.8 \times 0.9}$

**43.** $\dfrac{4.86 \times 0.63 \times 91.2}{54.1 \times 0.64 \times 5.84}$          **44.** $\dfrac{24.2 \times 0.133 \times 622.8}{58.7 \times 0.31 \times 458.3}$

*Find the missing dimension in each of the following drawings:*

**45.**
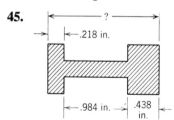

**46.**
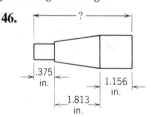

**47.**
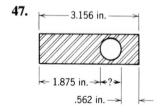

**48.**
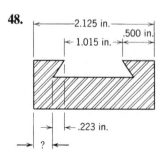

49. As shown in Figure 1.1, four pieces are cut from a steel bar. If each cut is 0.060 inches wide, and the lengths of the four pieces are 1.017 inches, 0.982 inches, 0.886 inches, and 1.201 inches, find the original length of the bar.

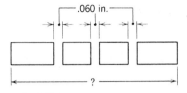

**FIGURE 1.1**

50. A wooden beam is formed by laminating strips of wood whose thicknesses are 0.51, 0.62, 1.03, 1.50, and 1.75 inches, respectively. Find the thickness of the final beam (disregard the thickness of the glue).

51. The *pitch* of the threads (distance between threads) on a screw is found by computing the reciprocal of the number of threads per inch. Find the pitch of a thread having (*a*) 16 threads per inch, and one having (*b*) 32 threads per inch.

52. The number of gallons of fresh water in a container can be found by dividing the weight in pounds by 8.35. Find the number of gallons in a container that weighs (*a*) 141.95 pounds, and another that weighs (*b*) 168.67 pounds.

## 1.2 Rounding Off

Figure 1.2 shows the names of place values for the decimal number system: *whole number places* (to the left of the decimal point) are named to ten millions; *decimal places* (to the right of the decimal point) are named to ten-thousandths. Familiarity with these place values is sufficient for most applications.

**FIGURE 1.2**

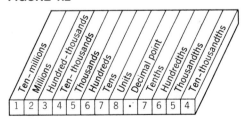

**EXAMPLE 1** Write the number 1625.3087 and underline the digit in the (*a*) tens place, (*b*) tenths place, (*c*) units place, (*d*) thousandths place, (*e*) thousands place, (*f*) ten-thousandths place.

### Solutions

(**a**) 162<u>5</u>.3087      (**b**) 1625.<u>3</u>087      (**c**) 162<u>5</u>.3087

(**d**) 1625.308<u>7</u>      (**e**) <u>1</u>625.3087      (**f**) 1625.308<u>7</u>

## ■ *Comparing Decimals*

To decide which of two decimals is greater, we compare the digits in the tenths place, in the hundredths place, and so on, until we reach a place where the digit in one number is greater than the corresponding digit in the other number.

**EXAMPLE 2**   Which number is greater: 0.024302 or 0.024299?

**Solution**   The first three digits are identical:

Because 3 is greater than 2, the number 0.024302 is greater than 0.024299.

## ■ *Rounding Off Numbers*

Results of computations are shown in calculator displays in a form called a *floating decimal point*. Frequently the results include more digits than may be needed for a particular purpose. For example, the result of a calculation that involves the cost of a machine part might appear in the display as 23.418. The actual cost, to the nearest cent, is $23.42. We say that 23.418 has been "rounded off to the nearest hundredth" (or "to the nearest cent"). Note that 23.418 is "nearer" to 23.42 than it is to 23.41.

The *flowchart* shown in Figure 1.3 indicates the rounding-off process most commonly used. In this procedure, the *round-off digit* is the last digit to be retained; the *test digit* is the digit immediately to the right of the round-off digit. For example, to round off 749.02586 to the nearest hundredth, underline the round-off digit 2 in the hundredths place, then proceed as follows:

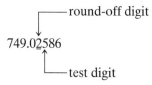

**FIGURE 1.3**

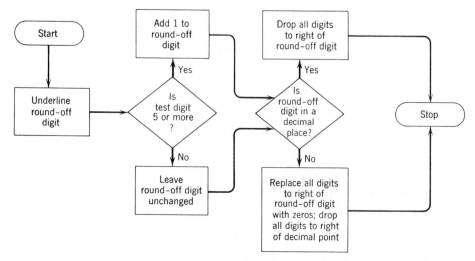

Rounding off

The test digit is 5, hence 1 is added to the round-off digit:

$$\overset{3}{749.0\underset{\displaystyle}{2}586}$$

The round-off digit is in a decimal place, so all digits to the right of the round-off digit are dropped.

$$\overset{3}{749.0\cancel{2586}} \longrightarrow 749.03$$

Thus, 249.02586 is written as 249.03, to the nearest hundredth.

**EXAMPLE 3**  In each case the round-off digit is underlined.
**(a)** Round off to the nearest ten:

$$249.0\underline{2}586 \longrightarrow 2\overset{50}{4\cancel{9.02586}} \longrightarrow 250$$

**(b)** Round off to the nearest tenth:

$$249.\underline{0}2586 \longrightarrow 249.0\cancel{2586} \longrightarrow 249.0$$

**(c)** Round off to the nearest thousandth:

$$249.02\underline{5}86 \longrightarrow 249.02\overset{6}{5\cancel{86}} \longrightarrow 249.026$$

**(d)** Round off to the nearest hundred:

$$\underline{2}49.02586 \longrightarrow 2\overset{00}{\cancel{49.02586}} \longrightarrow 200$$

Round-off instructions sometimes specify a given number of decimal places.

**EXAMPLE 4**
**(a)** Round off to two decimal places:

$$13.4\underline{9}63 \longrightarrow 13.\overset{50}{4\cancel{963}} \longrightarrow 13.50$$

**(b)** Round off to three decimal places:

$$13.49\underline{6}3 \longrightarrow 13.496\cancel{3} \longrightarrow 13.496$$

## ■ The Odd-Five Rule

In some engineering, scientific, and banking applications, a special round-off rule is used if the test digit is 5.

---

### Odd-Five Rule
If the test digit 5 is the *last* nonzero digit of a number:
  **1.** Add 1 to the round-off digit if it is odd (1, 3, 5, 7, 9).
  **2.** Retain the original round-off digit if it is even (0, 2, 4, 6, 8).

---

For example, to round off 87.035 to the nearest hundredth by the odd-five rule, first note that the test digit is 5 and it is the *last* nonzero digit. Next, because the round-off digit 3 is *odd*, add 1 to obtain 4 and drop the digit 5:

$$87.0\underline{3}5 \longrightarrow 87.0\overset{4}{\cancel{35}} \longrightarrow 87.04$$

To round off 87.025 to the nearest hundredth when the round-off digit 2 is *even*, retain the 2 and drop the 5. Thus

$$87.0\underline{2}5 \longrightarrow 87.02\cancel{5} \longrightarrow 87.02$$

**EXAMPLE 5** In each case the odd-five rule is used for rounding-off.
**(a)** To the nearest thousandth:

            ┌─odd digit; add 1 to digit

$$1.06750 \longrightarrow 1.06\underline{7}\cancel{50} \longrightarrow 1.068$$

**(b)** To the nearest hundred:

            ┌─even digit; retain digit

$$78,250 \longrightarrow 78,\underline{2}\cancel{50} \longrightarrow 78,200$$

    *Unless specified otherwise, in this book the rounding-off rule given by the flow-chart on page 8 is used.*

## Exercises for Section 1.2

*State which number in each pair is the greater. See Example 2.*

**1.** 0.06876,   0.06886      **2.** 0.73323,   0.73313

**3.** 0.89416,   0.89302      **4.** 0.91303,   0.91297

**5.** 0.58124,   0.58140      **6.** 0.49984,   0.49993

*Round off each number to the nearest (a) ten, (b) tenth, (c) hundredth, and (d) thousandth. See Example 3.*

 **7.** 14.7742     **8.** 21.6344     **9.** 76.2825     **10.** 54.6079

**11.** 69.8991    **12.** 32.8196    **13.** 45.9098    **14.** 89.8982

**15.** 70.9597    **16.** 58.9595    **17.** 19.9505    **18.** 97.9396

*Round off each number to (a) one, (b) two, and (c) three decimal places. See Example 4.*

**19.** 1.9069 **20.** 2.2591 **21.** 0.91994 **22.** 0.65232

**23.** 0.09857 **24.** 0.07579 **25.** 6.1695 **26.** 4.2945

*Round off each amount to the nearest cent (two decimal places).*

**27.** $1.285 **28.** $17.074 **29.** $8.395 **30.** $25.008

**31.** $47.291 **32.** $252.297 **33.** $57.999 **34.** $59.999

*Tax payments to the Internal Revenue Service may be rounded off to the nearest dollar. For example, $22.49 is rounded to $22, and $22.51 is rounded to $23. Round off each tax amount to the nearest dollar (to the nearest whole number).*

**35.** $38.37 **36.** $38.51 **37.** $509.62 **38.** $509.42

**39.** $600.89 **40.** $600.49 **41.** $1234.19 **42.** $1234.79

*Use the odd-five rule to round off each number (a) to the nearest tenth, and (b) to the nearest hundredth. See Example 5.*

**43.** 2.465 **44.** 6.350 **45.** 2.655 **46.** 6.045

**47.** 70.955 **48.** 58.595 **49.** 2.585 **50.** 3.555

**51.** In the fall of a recent year there were approximately 2,146,000 students enrolled in private colleges in the United States. Round this off (*a*) to the nearest hundred thousand, and (*b*) to the nearest ten thousand.

**52.** The liftoff weight of the manned spacecraft Apollo 15 was 107,142 pounds. Round this off (*a*) to the nearest ten thousand, and (*b*) to the nearest thousand.

**53.** From 1980 to 1990 the number of new passenger cars imported into the United States each year increased by 3,992,000. Round this off (*a*) to the nearest hundred thousand, and (*b*) to the nearest ten thousand.

**54.** During a recent year the railroads in the United States earned approximately $1,124,600,000 carrying transportation equipment. Round this off (*a*) to the nearest million, and (*b*) to the nearest ten million.

*Find the missing dimension in each of the following drawings, rounded off to the nearest hundredth of an inch.*

**55.**
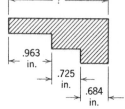

**56.**
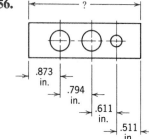

**57.**
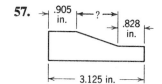

**58.**
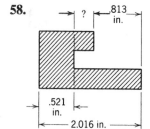

## 1.3 Exact and Approximate Numbers; Significant Digits

Any number arrived at by a counting process is an **exact number**. For example, if we say "60 students are enrolled in a class," the number 60 is exact. Exact numbers may also arise by definition. In the statement, "One hour equals 60 minutes," the number 60 is exact. On the other hand, a number that is arrived at by some process of measurement is an **approximate number**. Thus, if the distance between two cities, measured by driving from one city to the other, is given as 60 miles, this distance is an approximation to the actual distance, and the number 60 is an approximate number. The actual distance may be a little more or a little less than 60 miles.

**EXAMPLE 1**   In the statement:
**(a)** "One dime is equal to 10 cents," the number 10 is *exact*; 10 is defined to be the number of cents in a dime.
**(b)** "This sheet of postage stamps contains 10 stamps," the number 10 is *exact*; 10 is associated with the concept of counting.
**(c)** "The reading shown on a voltmeter is 10 volts," the number 10 is *approximate*; 10 is a measure of voltage.

## ■ Significant Digits

When working with approximate numbers, the term **significant digits** is used to refer to those digits in a number that have meaning. The following rules can be used to decide which digits of an approximate number are significant.

---

### Significant Digits of Approximate Numbers

1. Nonzero digits (1, 2, 3, 4, 5, 6, 7, 8, 9) are always significant.
2. Zeros that are preceded and followed by significant digits are always significant.
3. When a number has no digits on the left of the decimal point, zeros between the decimal point and the first nonzero digit are not significant.
4. Final zeros on the right of a decimal point are significant.
5. Final zeros on a whole number are assumed to be not significant unless further information is available.

---

In the examples illustrating each rule, the significant digits are underlined.

**EXAMPLE 2**
**(a)** 14.24 has four significant digits. (Rule 1)
**(b)** 0.0036 has two significant digits. (Rules 1 and 3)

**EXAMPLE 3**
**(a)** 14.0024 has six significant digits. (Rule 2)
**(b)** 7001 has four significant digits. (Rule 2)

**EXAMPLE 4**
**(a)** 8.9000 has five significant digits. (Rule 4)
**(b)** 0.03600 has four significant digits. (Rules 3 and 4)

**EXAMPLE 5**
**(a)** In the statement "Hawaii is 2400 miles west of Los Angeles," without further information on how the measurement was made, assume that 2400 has only two significant digits. (Rule 5)
**(b)** If the result of a measurement by a laser beam, which can be used to measure the distance between two points with great precision, is given as 2400 miles to the nearest mile, then the number 2400 has four significant digits.

Numbers can be rounded off to a specified number of significant digits by using the same procedure that is used to round off a number to a specified place value.

**EXAMPLE 6**
**(a)** Round off to two significant digits:

$$37.804 \longrightarrow 37.8\cancel{04} \longrightarrow 38$$

**(b)** Round off to four significant digits:

$$37.804 \longrightarrow 37.80\cancel{4} \longrightarrow 37.80$$

*EXAMPLE 7*  Use the odd-five round-off rule.
**(a)** Round off to three significant digits:

$$12.\underline{9}5 \longrightarrow 12.9\cancel{5} \longrightarrow 13.0$$

**(b)** Round off to two significant digits:

$$1\underline{2}.500 \longrightarrow 1\underline{2}.\cancel{500} \longrightarrow 12$$

## Exercises for Section 1.3

*State the number of significant digits in each number. See Examples 1 to 4.*

**1.** 8.01        **2.** 50.03        **3.** 0.0153        **4.** 0.0094

**5.** 230        **6.** 300        **7.** 230.1        **8.** 300.2

**9.** 100.0        **10.** 300.0        **11.** 5.600        **12.** 4.000

*State the number of significant digits in the number used in each statement. See Example 5.*

**13.** It is 1600 miles from here to Seattle.

**14.** The building is 150 feet high.

**15.** To the nearest pound, that machine weighs 200 pounds.

**16.** To the nearest inch, the pipe is 70 inches long.

**17.** One mile contains 5280 feet.

**18.** One ton contains 2000 pounds.

*Round off each number to (a) one, (b) two, and (c) three significant digits. See Examples 6 and 7.*

**19.** 3.981        **20.** 2.983        **21.** 5.705        **22.** 8.605

**23.** 2480.3    **24.** 3440.5    **25.** 0.01235    **26.** 0.09615

**27.** 15.025    **28.** 84.081    **29.** 104.5    **30.** 205.1

*If the total current in amperes is the sum of the individual currents, find the total current in each of the following circuits, correct to three significant digits.*

**31.**

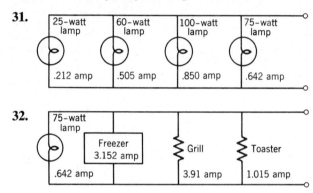

**32.**

## 1.4  Powers of a Number; Square Root

The product of two or more identical factors, such as

$$6 \cdot 6 \cdot 6 \cdot 6 \cdot 6$$

can be more concisely represented by the symbol $6^5$, read "six to the fifth power," where the number 5 indicates how many times the number 6 appears as a factor. The number 6 is called the **base**, and the number 5 is called the **exponent**. Such a form of a product is called **exponential notation** and is often referred to as *raising a number to a power*. For example, for a base $b$ (where $b$ represents any number),

$$b \cdot b = b^2 \quad \text{(read "}b\text{ to the second power" or "}b\text{ squared")}$$
$$b \cdot b \cdot b = b^3 \quad \text{(read "}b\text{ to the third power" or "}b\text{ cubed")}$$
$$b \cdot b \cdot b \cdot b = b^4 \quad \text{(read "}b\text{ to the fourth power")}$$

and, in general,

$$\overbrace{b \cdot b \cdot b \cdots b}^{n \text{ factors}} = b^n$$

### ■ The $\boxed{x^2}$ Key

Any number can be raised to the second power (squared) by multiplying the number by itself. For example, we can compute $11.2^2$ by the sequence

$$11.2 \boxed{\times} 11.2 \boxed{=} \longrightarrow 125.44$$

On the other hand, the *squaring key,* $\boxed{x^2}$, is convenient for raising a number to the second power. The $\boxed{=}$ key is not needed when using the $\boxed{x^2}$ key.

It may be necessary to use a second function key such as $\boxed{\text{2nd}}$, $\boxed{\text{SHIFT}}$, or $\boxed{\text{INV}}$ to activate the $\boxed{x^2}$ key. We will show only the $\boxed{x^2}$ key in a sequence with the understanding that you will use a second function key if necessary.

### EXAMPLE 1
**(a)** $11.2^2 = 125.44$;    this is computed by the sequence

$$11.2 \; \boxed{x^2} \longrightarrow 125.44$$

**(b)** $0.0068^2 = 0.00004624$;    this is computed by the sequence

$$0.0068 \; \boxed{x^2} \longrightarrow 0.00004624$$

## ■ *The* $\boxed{y^x}$ *Key*

The *power key,* $\boxed{y^x}$, is used to raise a number to a power.* The base is entered *before* and the exponent is entered *after* pressing the $\boxed{y^x}$ key.

### EXAMPLE 2
**(a)** $6^5 = 7776$;    this is computed by the sequence

$$6 \; \boxed{y^x} \; 5 \; \boxed{=} \longrightarrow 7776$$

**(b)** $1.7^3 = 4.913$;    this is computed by the sequence

$$1.7 \; \boxed{y^x} \; 3 \; \boxed{=} \longrightarrow 4.913$$

You may also have to use a second function key to activate the $\boxed{y^x}$ key.

## ■ *Square Root of a Number*

If a number is written as a product of two identical factors, either of the factors is called a **square root** of the number. The symbol $\sqrt{n}$ (read "the square root of *n*") is used to name a square root of a number.

### EXAMPLE 3
**(a)** $\sqrt{25} = 5$    because    $5 \cdot 5 = 25$.

**(b)** $\sqrt{81} = 9$    because    $9 \cdot 9 = 81$.

In Example 3 we used basic multiplication facts and "mental" arithmetic to find the square root of 25 and of 81. On a calculator, we use the *square root key* $\boxed{\sqrt{\phantom{x}}}$ to find the square root of a number.

### EXAMPLE 4
**(a)** $\sqrt{24.6} = 4.9598387$;    this is computed by the sequence

$$24.6 \; \boxed{\sqrt{\phantom{x}}} \longrightarrow 4.9598387$$

**(b)** $\sqrt{304.86} = 17.460241$;    this is computed by the sequence

$$304.86 \; \boxed{\sqrt{\phantom{x}}} \longrightarrow 17.460241$$

---

*Some calculators have an $\boxed{x^y}$ key.

On most calculators, as indicated in the above example, the ☐= key is not used when computing square roots.

To *check* the result of computing the square root of a number, we can multiply the square root by itself. For example, the result of Example 4(b) can be checked as

$$17.460241 \times 17.460241 = 304.86002$$

The product 304.86002 is "close" enough to 304.86 to satisfy the check. The error in the result 304.86002 is caused by the fact that 17.460241 is a rounded-off answer.

In Example 4(b), the number 17.460241 that appears in the display is actually an *approximation* for $\sqrt{304.86}$. Nevertheless, we use the "=" sign to write

$$\sqrt{304.86} = 17.460241$$

This is a practice that we follow throughout the book.

Note that if 5 is the *square root* of 25, then 25 is the *square* of 5, because 25 equals $5^2$. A number that is equal to the square of a whole number is sometimes called a "perfect square." Thus, 25 is a perfect square. You should be familiar with the perfect squares up to and including 100 and be able to state the square root of each:

| Perfect square | 1 | 4 | 9 | 16 | 25 | 36 | 49 | 64 | 81 | 100 |
|---|---|---|---|---|---|---|---|---|---|---|
| Square root | 1 | 2 | 3 | 4 | 5 | 6 | 7 | 8 | 9 | 10 |

## Exercises for Section 1.4

*Compute each square. See Example 1.*

**1.** $7.6^2$  **2.** $4.3^2$  **3.** $0.82^2$  **4.** $0.67^2$

**5.** $36.6^2$  **6.** $17.7^2$  **7.** $1.83^2$  **8.** $3.64^2$

**9.** $0.035^2$  **10.** $0.083^2$  **11.** $1.007^2$  **12.** $2.004^2$

*Compute each power. See Example 2.*

**13.** $7^3$  **14.** $5^4$  **15.** $9^5$  **16.** $8^3$

**17.** $0.4^4$  **18.** $0.2^6$  **19.** $0.09^3$  **20.** $0.05^3$

**21.** $1.06^{12}$  **22.** $1.05^{10}$  **23.** $2.12^7$  **24.** $3.04^5$

*Find each square root, without using a calculator. See Example 3.*

**25.** $\sqrt{4}$        **26.** $\sqrt{16}$        **27.** $\sqrt{36}$        **28.** $\sqrt{9}$

**29.** $\sqrt{64}$        **30.** $\sqrt{100}$        **31.** $\sqrt{900}$        **32.** $\sqrt{400}$

*Find each square root, using a calculator. See Example 4. Round off to the nearest thousandth.*

**33.** $\sqrt{9.9}$        **34.** $\sqrt{2.4}$        **35.** $\sqrt{67.6}$        **36.** $\sqrt{43.2}$

**37.** $\sqrt{500}$        **38.** $\sqrt{213}$        **39.** $\sqrt{0.68}$        **40.** $\sqrt{0.51}$

**41.** $\sqrt{0.0906}$        **42.** $\sqrt{0.0711}$        **43.** $\sqrt{1.197}$        **44.** $\sqrt{7.401}$

## 1.5  Powers of 10; Scientific Notation

In this section we will work with products in which one of the factors is a power of 10.

### ■ Multiplying and Dividing by 10, 100, 1000, and So Forth

A simple noncalculator procedure can be used to find products or quotients that involve factors 10, 100, 1000, and so forth. First verify each of the following products using a calculator.

$$4.63 \times 10 = 46.3, \quad 4.63 \times 100 = 463, \quad \text{and} \quad 4.63 \times 1000 = 4630$$

Note that in each case the decimal point in the product appears a certain number of digits *to the right* of its position in the factor 4.63. Results such as this suggest the following rules:

| To Multiply By | Move the Decimal Point |
|---|---|
| 10 | one place to the right; |
| 100 | two places to the right; |
| 1000 | three places to the right; |
| etc. | etc. |
| Add or drop zeros as needed. | |

For example,

$$568 \times 10 = 5680. = 5680;^* \qquad 0.408 \times 100 = 40.8 = 40.8;$$
$$0.0957 \times 1000 = 095.7 = 95.7; \qquad 1.625 \times 10,000 = 16250. = 16,250$$

---

*Reminder: If no decimal point appears in a number, the number is assumed to have a decimal point to the right of the units digit.

On your calculator, verify each of the following quotients:

$$\frac{39.4}{10} = 3.94, \qquad \frac{39.4}{100} = 0.394, \qquad \text{and} \qquad \frac{39.4}{1000} = 0.0394$$

Note that in each case the decimal point in the quotient appears a certain number of digits *to the left* of its position in the number 39.4. Results such as these suggest the following rules:

| To Divide By | Move the Decimal Point |
|---|---|
| 10 | one place to the left; |
| 100 | two places to the left; |
| 1000 | three places to the left; |
| etc. | etc. |
| Add or drop zeros as needed. | |

For example,

$$\frac{9.03}{10} = 0.903 = 0.903; \qquad \frac{8200}{100} = 82.00 = 82$$

$$\frac{5043.2}{1000} = 5.0432 = 5.0432; \qquad \frac{123}{10,000} = 0.0123 = 0.0123$$

## ■ Exponential Notation

Consider the following powers involving natural number exponents, which represent numbers greater than 10.

$$10^n = \overbrace{10 \times 10 \times 10 \times \cdots}^{n \text{ factors}}$$

$$\vdots$$

$$10^4 = 10 \times 10 \times 10 \times 10 = 10,000$$
$$10^3 = 10 \times 10 \times 10 \qquad = 1000$$
$$10^2 = 10 \times 10 \qquad\qquad = 100$$

It is useful to continue this pattern using "one" and "zero" in a special way to write

$$10^1 = 10$$

and

$$10^0 = 1$$

Furthermore, we use a special symbol $10^{-n}$ to write the reciprocal of a power $10^n$. The reciprocal is less than 1.

$$10^{-1} = \frac{1}{10^1} = \frac{1}{10} = 0.1$$

$$10^{-2} = \frac{1}{10^2} = \frac{1}{100} = 0.01$$

$$10^{-3} = \frac{1}{10^3} = \frac{1}{1000} = 0.001$$

$$\vdots$$

## ■ *Multiplying by Powers of 10*

Consider the product

$$6.29 \times 10^2 = 6.29 \times 100$$

and multiply 6.29 by 100 by moving the decimal point *two* places to the right. Hence,

$$6.29 \times 10^2 = 6\underset{\smile}{29}. = 629$$

Now consider the product

$$6.29 \times 10^{-2} = 6.29 \times \frac{1}{10^2} = \frac{6.29}{10^2} = \frac{6.29}{100}$$

and note that multiplication by $10^{-2}$ is equivalent to division by 100, which can be done by moving the decimal point *two places to the left*. Hence,

$$6.29 \times 10^{-2} = 0\underset{\smile}{.06}29 = 0.0629$$

Results such as these suggest the following rule:

---

### To Multiply a Number by $10^n$

**1.** Retain the original digits.
**2.** Counting from its original position, place the decimal point:
  **(a)** $n$ places to the right, if $n$ is positive;
  **(b)** $n$ places to the left, if $n$ is negative.
**3.** Add or drop zeros as needed.

---

### EXAMPLE 1

**(a)** $56.4 \times 10^3 = 56\underset{\smile}{400}. = 56{,}400$    **(b)** $56.4 \times 10^{-3} = 0.0\underset{\smile}{564} = 0.0564$

**(c)** $0.0564 \times 10^2 = 0\underset{\smile}{5.}64 = 5.64$    **(d)** $0.0564 \times 10^{-2} = 0.000564 = 0.000564$

## ■ *Numbers in Scientific Notation*

It is possible to write a given number in different factored forms in which one of the factors is a power of 10. For example,

$$3204 = 320.4 \times 10^1 \qquad 0.3204 = 3.204 \times 10^{-1}$$
$$= 32.04 \times 10^2 \qquad\qquad = 32.04 \times 10^{-2}$$
$$= 3.204 \times 10^3 \qquad\qquad = 320.4 \times 10^{-3}$$

Observe that the factored forms $3.204 \times 10^3$ and $3.204 \times 10^{-1}$ have as factors a number between 1 and 10 and a power of 10. Such factored forms or, in general, any number written in the form

$$b \times 10^n$$

where $b$ is any number between 1 and 10, and $n$ is an integer, is said to be written in **scientific notation**.

We illustrate the process of writing a number in scientific notation by the following examples.

**EXAMPLE 2**  Write each number using scientific notation.
**(a)** $8760.3 \rightarrow 8.\underset{\smile}{7603} \times 10^n$
  The decimal point is moved three places to the left. Hence,  $n = 3$  and

$$8760.3 = 8.7603 \times 10^3$$

**(b)** $0.0087603 \rightarrow 008.7603 \times 10^n$

    The decimal point is moved three places to the right. Hence, $n = -3$, and

$$0.0087603 = 8.7603 \times 10^{-3}$$

    To change a number from scientific notation to standard decimal notation, simply apply the rule for multiplying by powers of 10.

**EXAMPLE 3**   Write each number using standard decimal notation.

**(a)** $4.00621 \times 10^6 = 4006210. = 4,006,210$

**(b)** $4.00621 \times 10^{-6} = 0.00000400621 = 0.00000400621$

## ■ *Using a Calculator*

Note that the results of the products

$$37,000 \times 86,000 = 3,182,000,000$$

and

$$0.00019 \times 0.0063 = 0.000001197$$

cannot be shown in standard form on a calculator because there may be only eight digit places in the display screen. These results may appear on the screen in scientific notation as

$$\boxed{3.182 \qquad 09} \quad \text{and} \quad \boxed{1.197 \qquad -06}$$

respectively,* where the exponent on the power of 10 is displayed and the base 10 is understood. The results can now be written as

$$3.182 \times 10^9 \quad \text{and} \quad 1.197 \times 10^{-6}$$

    To enter a number in scientific notation, use the key labeled $\boxed{\text{EXP}}$ or $\boxed{\text{EE}}$. For example, to enter $6.3 \times 10^{13}$, press

$$6.3 \; \boxed{\text{EXP}} \; 13$$

The display will read

$$\boxed{6.3 \qquad 13}$$

    To enter a negative power of 10, for example, $6.3 \times 10^{-5}$, press

$$6.3 \; \boxed{\text{EXP}} \; 5 \; \boxed{\pm}$$

The display will read

$$\boxed{6.3 \qquad -5}$$

**EXAMPLE 4**   Compute

$$\frac{(4.7 \times 10^{17})(3.8 \times 10^{-8})}{5.3 \times 10^{23}}$$

to two significant digits.

---

*These numbers may appear on the screen as $\boxed{3.182^{09}}$   and   $\boxed{1.197^{-06}}$

**Solution**   Use the keying sequence

4.7 $\boxed{\text{EXP}}$ 17 $\boxed{\times}$ 3.8 $\boxed{\text{EXP}}$ 8 $\boxed{\pm}$ $\boxed{\div}$ 5.3 $\boxed{\text{EXP}}$ 23 $\boxed{=}$

The result in the display will read

$$\boxed{3.36981 \qquad -14}$$

So the result to two significant digits is   $3.4 \times 10^{-14}$.

## Exercises for Section 1.5

*Write each number in standard decimal notation. See Examples 1 and 3.*

**1.** $48.6 \times 10^5$      **2.** $54.1 \times 10^4$      **3.** $2.93 \times 10^3$      **4.** $2.48 \times 10^6$

**5.** $6.550 \times 10^7$      **6.** $4.090 \times 10^3$      **7.** $0.939 \times 10^{-6}$      **8.** $0.611 \times 10^{-2}$

**9.** $7.10 \times 10^{-4}$      **10.** $2.40 \times 10^{-5}$      **11.** $9.490 \times 10^{-3}$      **12.** $5.858 \times 10^{-6}$

*Write each number in scientific notation. See Example 2.*

**13.** 10.48          **14.** 15.01          **15.** 223.6          **16.** 241.3

**17.** 0.421          **18.** 0.933          **19.** 0.0370          **20.** 0.0770

**21.** 72,905          **22.** 69,994          **23.** 0.000174          **24.** 0.000053

*Each of the following is a number in scientific notation as displayed on a calculator. Write the number in (a) scientific notation, and (b) standard decimal form.*

**25.** $\boxed{6.7 \qquad 03}$          **26.** $\boxed{3.2 \qquad 04}$          **27.** $\boxed{2.13 \qquad 02}$

**28.** $\boxed{1.62 \qquad -02}$          **29.** $\boxed{3 \qquad -05}$          **30.** $\boxed{9 \qquad -06}$

*Each of the following measurements specifies the increase in length of a 1-inch-long bar of the given material as a result of raising the temperature of the bar 1°F. Write each measurement in standard decimal form.*

**31.** Aluminum alloy: $13.7 \times 10^{-6}$ inch      **32.** Magnesium alloy: $14.8 \times 10^{-6}$ inch

**33.** Carbon steel: $46 \times 10^{-7}$ inch

**34.** Titanium alloy: $51.5 \times 10^{-7}$ inch

**35.** Nickel base alloy: $0.79 \times 10^{-5}$ inch

**36.** Cobalt base alloy: $0.87 \times 10^{-5}$ inch

**37.** Graphite: $0.01 \times 10^{-4}$ inch

**38.** Mica: $0.018 \times 10^{-4}$ inch

*Write each measurement in Exercises 39–44 in scientific notation.*

**39.** The mass of the earth is approximately 5,980,000,000,000,000,000,000,000,000 grams.

**40.** The mass of the hydrogen atom is approximately 0.000 000 000 000 000 000 000 001 67 grams.

**41.** Light travels at a speed of 300,000,000 meters per second.

**42.** Visible blue light has a wavelength of 0.000 000 45 meters.

**43.** The average body cell of an animal has a diameter of 0.000 015 meters.

**44.** The diameter of the earth is approximately 6,450,000 meters.

*Compute. Round off answers to two significant digits. See Example 4.*

**45.** $493{,}000 \times 29{,}000$

**46.** $53{,}000 \times 8300$

**47.** $0.000073 \times 0.0091$

**48.** $0.0049 \times 0.00036$

**49.** $(7.4 \times 10^{-3})(6.4 \times 10^{-4})$

**50.** $(6.14 \times 10^{-6})(3.7 \times 10^{-5})$

**51.** $\dfrac{(6.2 \times 10^{4})(7.1 \times 10^{-3})}{3.1 \times 10^{-5}}$

**52.** $\dfrac{(1.03 \times 10^{-6})(4.5 \times 10^{3})}{1.5 \times 10^{4}}$

---

## 1.6  Grouping Symbols; Order of Operations

In this section we consider some mathematical concepts for calculations that combine two or more different operations. Since our emphasis will be on mathematical concepts, we will first use "simple" whole numbers that will not require the use of a calculator.

### ∎ *Parentheses as Grouping Symbols*

In Section 1.1 we used parentheses to indicate the factors of a product. We can also use parentheses to group numbers and operations. For example, the parentheses in the expression

$$(2 + 5) \cdot 6$$

indicate that the sum $2 + 5$ is to be multiplied by 6. The parentheses in the expression

$$2 + (5 \cdot 6)$$

indicate that the product $5 \cdot 6$ is to be added to 2. In expressions that involve parentheses, the operations within the parentheses are to be done first.

### *EXAMPLE 1*

(a) $\underbrace{(4 + 8)}_{12} \div 4$     (The addition is done first.)

$12 \div 4 = 3$

(b) $4 + \underbrace{(8 \div 4)}_{2}$     (The division is done first.)

$4 + 2 = 6$

(c) $\underbrace{(2 \cdot 9)}_{18} - 5$     (The multiplication is done first.)

$18 - 5 = 13$

(d) $2 \cdot \underbrace{(9 - 5)}_{4}$     (The subtraction is done first.)

$2 \cdot 4 = 8$

### ∎ *Fraction Bars as Grouping Symbols*

In Section 1.1 we used fraction bars to indicate quotients in fraction form. Fraction bars also serve to group two or more numbers and operations that are involved in quotients. For example, the fraction bar in the expression

$$\frac{19 - 7}{4}$$

indicates that $19 - 7$ is to be divided by 4. Given a quotient in fraction form, it is often convenient to rewrite it as a quotient using parentheses instead of a fraction bar. We refer to such a form as the **on-line** form of the mathematical expression. For example, the on-line form of $\frac{19 - 7}{4}$ is

$$(19 - 7) \div 4$$

where the parentheses are used to indicate that the difference $19 - 7$ is to be divided by 4. The subtraction is to be done first.

### EXAMPLE 2

**(a)** The on-line form of $\dfrac{6+4}{2}$ is $(6 + 4) \div 2$.

$$\underbrace{(6 + 4)}_{10} \div 2 \qquad \text{(The addition is done first.)}$$

$$10 \quad \div 2 = 5$$

**(b)** The on-line form of $\dfrac{36}{9-2-4}$ is $36 \div (9 - 2 - 4)$.

$$36 \div \underbrace{(9 - 2 - 4)}_{3} \qquad \text{(The subtraction is done first.)}$$

$$36 \div \quad 3 \quad = 12$$

## ■ Order of Operations

In the above examples, parentheses and fraction bars are used to indicate which operations are to be done first. In some expressions the order of operations is not indicated by grouping symbols. In such cases we have to make an agreement on the order of performing the operations. For example, if we do not have an agreement on a particular order of operations for the expressions $5 + 4 \cdot 6$, we can obtain either of two results:

          (1)                   (1a)

$$\underbrace{5 + 4}_{9} \cdot 6 \qquad \text{or} \qquad 5 + \underbrace{4 \cdot 6}_{24}$$

$$9 \quad \cdot 6 = 54 \qquad\qquad 5 + \quad 24 \quad = 29$$

As another example, note that if we don't have an agreement on an order of operations for $2 \cdot 5^2$, we can obtain either

          (2)                   (2a)

$$\underbrace{2 \cdot 5}^{2} \qquad \text{or} \qquad 2 \cdot \underbrace{5^2}$$

$$10^2 = 100 \qquad\qquad 2 \cdot \quad 25 \quad = 50$$

Because we want a collection of symbols to represent exactly one number, we shall agree that operations are to be performed as follows.

---

### Order of Operations

   **1.** Operations inside parentheses or above or below a fraction bar.
   **2.** Raising to powers.
   **3.** Multiplications and divisions in the order in which they appear, left to right.
   **4.** Additions and subtractions in the order in which they appear, left to right.

---

Applying these rules to the above expression, $5 + 4 \cdot 6$, the multiplication is done first, and the correct answer is 29, as in procedure (1a). In the expression $2 \cdot 5^2$, the power is computed first and the correct answer is 50, as in procedure (2a).

Note that *it is not always correct to do the operations in the exact order in which the numbers and symbols appear, left to right*! The following examples show how the order of operations rules are used in various computations.

### EXAMPLE 3
**(a)** $(3 \cdot 2)^2$                      **(b)** $3 \cdot 2^2$

#### Solutions
**(a)** Do the multiplication in the parentheses first, and then raise the product to the second power.

$$(3 \cdot 2)^2 = 6^2 = 36$$

**(b)** Compute the power $2^2$ first, and then multiply by 3.

$$3 \cdot 2^2 = 3 \cdot 4 = 12$$

### EXAMPLE 4
**(a)** $6 \cdot (5 - 2)^2$                **(b)** $6 \cdot 5 - 2^2$

#### Solutions
**(a)** Do the subtraction in the parentheses first, and then raise the difference to the second power and multiply the result by 6.

$$6 \cdot (5 - 2)^2 = 6 \cdot 3^2 = 6 \cdot 9$$
$$= 54$$

**(b)** First compute the power $2^2$. Next, find the product $6 \cdot 5$ and compute the resulting difference.

$$6 \cdot 5 - 2^2 = 6 \cdot 5 - 4 = 30 - 4$$
$$= 26$$

It is sometimes helpful to use parentheses to group products or quotients in expressions that involve combined operations.

### EXAMPLE 5    **(a)** $3 \cdot 2 + 7 - 5$      **(b)** $3 + 2 \cdot 7 - 5$

#### Solutions
**(a)** Group the product $3 \cdot 2$ in parentheses; do the multiplication first.

$$\underbrace{(3 \cdot 2)}_{6} + 7 - 5$$
$$6 \quad + 7 - 5 = 8$$

**(b)** Group the product $2 \cdot 7$ in parentheses; do the multiplication first.

$$3 + \underbrace{(2 \cdot 7)}_{14} - 5$$
$$3 + \quad 14 \quad - 5 = 12$$

### EXAMPLE 6    $8 + \dfrac{15}{3}$

**Solution**    The on-line form is $8 + (15 \div 3)$.

$$8 + \underbrace{(15 \div 3)}_{5}$$
$$8 + \quad 5 \quad = 13$$

If more than one set of grouping symbols are needed in an expression, we use *brackets*, [   ], in the same manner as parentheses.

**EXAMPLE 7** $\dfrac{2 + 3 \cdot 4}{7}$

**Solution** The on-line form is $[2 + (3 \cdot 4)] \div 7$, where the parentheses around $(3 \cdot 4)$ indicate that the multiplication is done first.

$$[2 + \underbrace{(3 \cdot 4)}] \div 7$$
$$\underbrace{[2 + \quad 12 \quad]} \div 7$$
$$14 \qquad \div 7 = 2$$

## ■ Using a Calculator

In the preceding sections we used a calculator to compute arithmetic expressions that only involved sums and differences and expressions that only involved products and quotients. The keystroke sequences were simple and direct. Computations that involve combined operations, which are governed by the order of operations specified on page 25, may require the use of parentheses.

**EXAMPLE 8** Compute $(4.1 + 3.7) \times 6.2$.

**Solution** Use the sequence

$$\boxed{(}\ 4.1 + 3.7\ \boxed{)}\ \boxed{\times}\ 6.2 \longrightarrow 48.36$$

The result is 48.4 to the nearest tenth.

**EXAMPLE 9** Compute $\dfrac{13.7 - 8.9}{1.5}$.

**Solution** The on-line form is $(13.7 - 8.9) \div 1.5$. Use the sequence

$$\boxed{(}\ 13.7 - 8.9\ \boxed{)}\ \boxed{\div}\ 1.5 \longrightarrow 3.2$$

Sometimes, as in the next example, parentheses are not necessary because the operations are written in a way that agrees with the correct order of operations.

**EXAMPLE 10** Compute $2.6 \times 3.5 + 14.1^3$.

**Solution** Use the sequence

$$2.6\ \boxed{\times}\ 3.5\ \boxed{+}\ 14.1\ \boxed{y^x}\ 3 \longrightarrow 2812.321$$

The result is 2812.3 to the nearest tenth.

## Exercises for Section 1.6

*Each exercise 1 to 60 should be completed without a calculator.*

*For each expression, state which operation is to be performed first, and then do the computation. See Example 1.*

**1.** $5 + (6 \cdot 7)$      **2.** $(5 + 6) \cdot 7$      **3.** $8 \cdot (2 + 3)$

**4.** $(8 \cdot 2) + 3$      **5.** $36 \div (12 - 3)$      **6.** $(36 \div 12) - 3$

**7.** $15 - (9 \div 3)$     **8.** $(15 - 9) \div 3$     **9.** $(3 \cdot 7) - 5$

**10.** $3 \cdot (7 - 5)$     **11.** $(48 \div 12) \cdot 4$     **12.** $48 \div (12 \cdot 4)$

*Use parentheses to write each expression in on-line form, state which operation is performed first, and then do the computation. See Example 2.*

**13.** $\dfrac{60}{5 + 7}$     **14.** $\dfrac{36}{8 - 4}$     **15.** $\dfrac{12}{6 - 2}$

**16.** $\dfrac{21}{4 + 3}$     **17.** $\dfrac{19 - 3}{4}$     **18.** $\dfrac{36 - 4}{8}$

**19.** $\dfrac{8 + 12}{4}$     **20.** $\dfrac{15 + 9}{6}$     **21.** $\dfrac{16}{12 - 8}$

**22.** $\dfrac{20}{10 - 6}$     **23.** $\dfrac{30}{6 + 9}$     **24.** $\dfrac{30}{3 + 7}$

*Perform the indicated computations. See Examples 3, 4, and 5.*

**25.** $(2 \cdot 4)^2$     **26.** $(4 \cdot 3)^2$     **27.** $2 \cdot 4^2$

**28.** $4 \cdot 3^2$     **29.** $(5 \cdot 2)^2$     **30.** $(3 \cdot 5)^2$

**31.** $5 \cdot 2^2$     **32.** $3 \cdot 5^2$     **33.** $3 \cdot 5 + 2$

**34.** $6 \cdot 7 + 4$     **35.** $3 + 5 \cdot 2 - 6$     **36.** $6 + 7 \cdot 4 - 8$

**37.** $20 - 4 \cdot 3 + 1$     **38.** $10 - 2 \cdot 6 + 4$     **39.** $6 \cdot (8 - 3)^2$

**40.** $7 \cdot (9 - 5)^2$     **41.** $6 \cdot 8 - 3^2$     **42.** $7 \cdot 9 - 5^2$

*Write the on-line form; perform the indicated operations. See Examples 6 and 7.*

**43.** $3 + \dfrac{6}{2}$      **44.** $2 + \dfrac{8}{2}$      **45.** $10 - \dfrac{12}{4}$

**46.** $9 - \dfrac{16}{4}$      **47.** $\dfrac{15}{3} + 8$      **48.** $\dfrac{20}{5} + 5$

**49.** $\dfrac{16}{2} - 3$      **50.** $\dfrac{24}{2} - 6$      **51.** $\dfrac{2 + 2 \cdot 5}{3}$

**52.** $\dfrac{3 + 2 \cdot 6}{5}$      **53.** $\dfrac{20 - 2 \cdot 4}{4}$      **54.** $\dfrac{18 - 2 \cdot 3}{6}$

**55.** $\dfrac{5 \cdot 4 - 8}{3}$      **56.** $\dfrac{3 \cdot 5 - 6}{3}$      **57.** $\dfrac{16}{2 + 3 \cdot 2}$

**58.** $\dfrac{32}{4 + 3 \cdot 4}$      **59.** $\dfrac{18}{5 \cdot 2 - 4}$      **60.** $\dfrac{28}{4 \cdot 4 - 2}$

*Use your calculator to compute the following exercises. Round off answers to the nearest tenth. See Examples 8–10.*

**61.** $(9.1 + 4.3) \times 2.9$      **62.** $(14.7 - 8.3) \times 1.7$

**63.** $3.8\,(8.1 - 6.3)$      **64.** $5.8\,(4.9 + 7.3)$

**65.** $\dfrac{4.9 + 3.9}{4.4}$      **66.** $\dfrac{17.9 - 6.1}{0.59}$

**67.** $\dfrac{4.2}{7.6 - 6.4}$      **68.** $\dfrac{8.2}{5.9 - 3.2}$

**69.** $(4.7 \times 2.1) - (3.8 \times 1.3)$      **70.** $(8.1 \times 3.4) + (1.3 \times 4.7)$

**71.** $3.3 \times 2.5 + 1.6^2$                        **72.** $4.2 \times 6.3 - 2.1^2$

**73.** $1.3^3 + (2.1 \times 4.7)$                      **74.** $(2.4)^3 - (3.7 \times 2.1)$

*Round off answers to the same number of decimal places as the given data.*

**75.** Compute the average of 14.8, 17.9, 15.3, and 15.5. (Hint: Find the sum of the numbers and divide by the number of addends.)

**76.** Compute the average of 8.204, 7.931, and 10.002. (See hint for Exercise 75.)

**77.** A wooden beam is formed by laminating strips of wood whose thicknesses are 0.69, 0.73, 1.05, and 1.43 inches. Find the average thickness of the strips of wood.

**78.** A machinist selected six samples of ball bearings from a large production run. The diameter measurements were 0.372, 0.369, 0.381, 0.362, 0.371, and 0.372. What is the average diameter of the six samples?

## 1.7  Mental Calculations

In the preceding sections of this chapter we considered methods for doing computations on calculators. As you develop your skill in using a calculator, you should also learn some basic arithmetic facts and how to do simple computations mentally.

It is equally important that you be able to estimate answers as a rough check on computations you make using a calculator.

### ■ Estimating Sums and Differences

One method that can be used to estimate an answer involves the use of rounded-off numbers. It is usually convenient to use rounded-off numbers that consist of only *one* nonzero digit.

**EXAMPLE 1**  Estimate each sum.
**(a)** $83 + 35 + 27$     **(b)** $867 + 147 + 558$     **(c)** $\$74.23 + \$29.50 + \$19.63$

**Solutions**
**(a)** Round off each number to the nearest ten and mentally add the rounded-off numbers:

$$80 + 40 + 30 = 150$$

The estimated sum is 150.
**(b)** Round off each number to the nearest hundred and mentally add the rounded-off numbers:

$$900 + 100 + 600 = 1600$$

The estimated sum is 1600.
**(c)** Round off each number to the nearest ten dollars and mentally add the rounded-off numbers:

$$\$70 + \$30 + \$20 = \$120$$

The estimated sum is $120.

**EXAMPLE 2**  Estimate each difference.

**(a)** $94 - 39$     **(b)** $406 - 297$     **(c)** $\$337.11 - \$142.60$

**Solutions**

**(a)** Round off each number to the nearest ten and mentally subtract the rounded-off numbers:

$$90 - 40 = 50$$

The estimated difference is 50.

**(b)** Round off each number to the nearest hundred and mentally subtract the rounded-off numbers:

$$400 - 300 = 100$$

The estimated difference is 100.

**(c)** Round off each number to the nearest hundred dollars and mentally subtract the rounded-off numbers:

$$\$300 - \$100 = \$200$$

The estimated difference is $200.

## ■ Estimating Products and Quotients

Rounded-off numbers, together with scientific notation, can be used to estimate products and quotients. As an example, let us estimate the product

$$237 \times 26 \times 39$$

First, we round off each number to the nearest unit, or 10, or 100, and so forth, so that *only the first digit to the left is a nonzero digit,* to obtain

$$200 \times 30 \times 40$$

Next, we write each number in scientific notation and compute as follows:

$$200 \times 30 \times 40 = 2 \times 10^2 \times 3 \times 10^1 \times 4 \times 10^1$$
$$= (2 \times 3 \times 4) \times (10^2 \times 10^1 \times 10^1)$$
$$= 24 \times 10^4 = 240,000$$

The estimated product is 240,000. By using a calculator, the actual product is found to be 240,318.

In the following examples, note how the powers of 10 are grouped separately from the other numbers and are then replaced by a single power of 10.

**EXAMPLE 3**  Estimate the quotient $\dfrac{864.3}{41.07}$; then use your calculator to find the result.

**Solution**  Round off each number and write in scientific notation.

$$\frac{900}{40} = \frac{9 \times 10^2}{4 \times 10^1} = \frac{9}{4} \times \frac{10^2}{10^1}$$
$$= 2.25 \times 10^1 = 22.5$$

The estimated quotient is 22.5. Using your calculator,

$$\frac{864.3}{41.07} = 21.044558$$

(Note that 2, rather than 2.5, can be used as an approximation for $\frac{9}{4}$, in which case the estimated quotient would be $2 \times 10^1$, or 20.)

**EXAMPLE 4**  Estimate the computation $\frac{215.2 \times 67.5}{44.1}$; then use your calculator to find the result.

**Solution**  Round off each number and write in scientific notation.

$$\frac{200 \times 70}{40} = \frac{2 \times 10^2 \times 7 \times 10^1}{4 \times 10^1} = \frac{2 \times 7}{4} \times \frac{10^2 \times 10^1}{10^1}$$
$$= 3.5 \times 10^2 = 350$$

The estimated result is 350. Using your calculator,

$$\frac{215.2 \times 67.5}{44.1} = 329.38776$$

(If 3 is used, instead of 3.5, as an approximation for $\frac{2 \times 7}{4}$, the estimated result would be $3 \times 10^2$, or 300.)

## Exercises for Section 1.7

*Perform all calculations mentally.*

*Compute.*

**1.** $4 + 7$      **2.** $8 + 6$      **3.** $9 + 8$      **4.** $11 + 12$

**5.** $15 - 6$      **6.** $13 - 5$      **7.** $21 - 8$      **8.** $19 - 13$

**9.** $6 \times 7$      **10.** $8 \times 5$      **11.** $9 \times 8$      **12.** $9 \times 9$

**13.** $15 \div 5$      **14.** $24 \div 3$      **15.** $36 \div 9$      **16.** $21 \div 7$

**17.** $6 + 9 - 4$      **18.** $7 + 12 - 5$      **19.** $(8 + 10) \div 6$

**20.** $(18 - 3) \div 5$      **21.** $6 \times 4 \div 8$      **22.** $8 \times 6 \div 12$

*Specify each square root.*

**23.** $\sqrt{9}$  **24.** $\sqrt{49}$  **25.** $\sqrt{25}$  **26.** $\sqrt{36}$

**27.** $\sqrt{16}$  **28.** $\sqrt{81}$  **29.** $\sqrt{64}$  **30.** $\sqrt{100}$

*Specify each power.*

**31.** $2^2$  **32.** $6^2$  **33.** $8^2$  **34.** $9^2$

**35.** $2^3$  **36.** $3^3$  **37.** $4^3$  **38.** $10^3$

*Compute.*

**39.** $1{,}342 - 41 + 41$  **40.** $762 + 21 - 21$  **41.** $484 + 621 - 484$

**42.** $76 + 124 - 76$  **43.** $6.1 + 12.2 - 6.1$  **44.** $8.4 + 62 - 8.4$

*Multiply each number by* (a) *10,* (b) *100, and* (c) *10,000.*

**45.** 8.34  **46.** 9.17  **47.** 52.62  **48.** 73.11

**49.** 0.912  **50.** 0.786  **51.** 0.034  **52.** 0.062

*Divide each number by* (a) *10,* (b) *100, and* (c) *10,000.*

**53.** 701  **54.** 520  **55.** 2.625  **56.** 1.137

**57.** 0.219  **58.** 0.687  **59.** 0.043  **60.** 0.026

*Use rounded-off numbers to estimate each computation. See Examples 1 and 2. Use a calculator to find the result.*

**61.** $28 + 34 + 48$  **62.** $64 + 97 + 83$  **63.** $54.1 + 20.3 + 65.2$

**64.** 38.9 + 43.1 + 10.8     **65.** 962 + 843 + 784     **66.** 527 + 688 + 789

**67.** 981 − 783          **68.** 622 − 428

**69.** 1126 − 713          **70.** 3917 − 1009

**71.** 356.20 + 248.19 − 109.37     **72.** 66.66 + 44.44 − 22.22

*Use scientific notation to estimate each computation. See Examples 3 and 4. Use a calculator to find the result.*

**73.** $98.7 \times 99.5$        **74.** $68.6 \times 198$        **75.** $382 \times 109.5 \times 7$

**76.** $50.9 \times 699 \times 89$     **77.** $0.0163 \times 0.298$     **78.** $0.483 \times 0.068$

**79.** $\dfrac{49.3 \times 710.2}{51.1}$        **80.** $\dfrac{0.0613 \times 0.0049}{0.0294}$

■ *Review Exercises*

**Section 1.1**

**1.** Find the reciprocal of 0.56.

**2.** Find the reciprocal of 56.

*Compute.*

**3.** $75.01 - 45.88 - 3.511$    **4.** $68.102 - 4.608 - 17.947 + 4.9555$

**5.** $812.6 \times 0.3036 \times 8.4$    **6.** $688.06 \div 46.232$

**7.** $48.8 \times 0.92 \div 4.57$    **8.** $\dfrac{8.64 \times 0.74 \times 0.658}{0.997 \times 74.8 \times 0.13}$

## Sections 1.2 and 1.3

*State which number in each pair is the greater.*

**9.** $0.44651, 0.44660$    **10.** $3.9424, 3.9431$

*Round off each number to the nearest* (a) *ten,* (b) *tenth,* (c) *hundredth, and* (d) *thousandth.*

**11.** $32.8195$    **12.** $45.2685$

*Round off to the nearest cent.*

**13.** $\$2.3749$    **14.** $\$3.115$

*Use the odd-five rule to round off each number* (a) *to the nearest tenth, and* (b) *to the nearest hundredth.*

**15.** $3.655$    **16.** $23.850$

## Section 1.4

*Compute each power.*

**17.** $1.32^2$    **18.** $0.27^2$    **19.** $0.9^5$    **20.** $4.03^4$

*Find each square root.*

**21.** $\sqrt{49}$      **22.** $\sqrt{81}$      **23.** $\sqrt{2.07}$      **24.** $\sqrt{19.3}$

## Section 1.5

*Write each number in standard decimal notation.*

**25.** $8.42 \times 10^4$      **26.** $1.45 \times 10^{-6}$

*Write each number in scientific notation.*

**27.** $0.00123$      **28.** $37,500$

*Compute.*

**29.** $\dfrac{(6.7 \times 10^{-5})\,(4.8 \times 10^7)}{1.6 \times 10^3}$      **30.** $\dfrac{(4.1 \times 10^3)\,(7.8 \times 10^{-6})}{3.9 \times 10^{-2}}$

**31.** $\dfrac{(5.8 \times 10^4)\,(2.1 \times 10^{-6})}{2.9 \times 10^{-3}}$      **32.** $\dfrac{(6.4 \times 10^{-5})\,(1.7 \times 10^4)}{3.2 \times 10^3}$

## Section 1.6

*Exercises 33 to 40 should be worked mentally, without a calculator. State which operation is to be performed first, and then do the computation.*

**33.** $3 + 4 \cdot 7$      **34.** $(16 - 4) \div 4$

*Use parentheses to write the on-line form, state which operation is performed first, and then do the computation.*

**35.** $\dfrac{40}{5 \times 4}$      **36.** $\dfrac{28}{5 + 9}$

*Perform the indicated computations.*

**37.** $5 + 7 \cdot 6 - 4$      **38.** $8 \cdot (12 - 9)^2$

*Write the on-line form, and perform the indicated computations.*

**39.** $\dfrac{6 + 2 \cdot 6}{9}$      **40.** $\dfrac{24}{5 \cdot 3 - 7}$

*Compute. Round off answers to two significant digits.*

**41.** $5.4 \times (2.8 + 10.3)$          **42.** $(9.4 \times .8) - 3.56$

**43.** $2.3 \times 1.2^3$          **44.** $(2.3 \times 1.2)^3$

**45.** $(4.5 \times 4.1) + 7 \times 8.3$          **46.** $(6.1 + 2.8) \times (8.9 - 1.6)$

**47.** $3.41^5 + 3.8 \times 7.6$          **48.** $\dfrac{2.9 - 1.4}{4.1 + 6.7}$

## Section 1.7

*Perform all calculations mentally.*

**49.** $3 \cdot 2^2$          **50.** $2 \cdot 3^2$          **51.** $\dfrac{6^2}{12}$          **52.** $\dfrac{8^2}{16}$

**53.** $\dfrac{8 + 2^3}{4}$          **54.** $\dfrac{12 - 3^2}{3}$          **55.** $4^2 + 3^2$          **56.** $5^2 + 12^2$

**57.** $4\sqrt{9}$          **58.** $6\sqrt{49}$          **59.** $\dfrac{3\sqrt{64}}{12}$          **60.** $\dfrac{4\sqrt{25}}{10}$

**61.** $\sqrt{12 + 13}$          **62.** $\sqrt{45 - 9}$          **63.** $\sqrt{4^2 + 3^2}$          **64.** $\sqrt{5^2 + 12^2}$

*Multiply each number by* (a) *10,* (b) *100, and* (c) *1000.*

**65.** 6.8          **66.** 12.2          **67.** 0.43          **68.** 0.08

*Divide each number by* (a) *10,* (b) *100, and* (c) *1000.*

**69.** 439          **70.** 28          **71.** 1.4          **72.** 0.81

# Fractions

In Chapter 1 we showed instructions for using your calculator. In the remainder of this book we will show calculator sequences *only* in cases where new techniques are being considered. However, you should continue to use your calculator to check all computations that appear in the examples, whether or not the calculator sequences are shown.

In the first three sections of this chapter we consider decimal forms of fractions. In the remaining sections we consider fractions using the fraction bar. For computations with fractions in either form, we need the concept of *factoring*.

## 2.1 Factoring

When a number such as 75 is written as the product of $3 \cdot 25$, we say that $3 \cdot 25$ is a *factored form* of 75. The process of changing a number to a factored form is called **factoring**.

If one whole number is divided by a second whole number so that the quotient is a whole number, then the second whole number is an *exact divisor* of the first. Thus, because

$$75 \div 3 = 25$$

and 25 is a whole number, 3 is an exact divisor of 75.

An exact divisor of a whole number is also called a *factor* of the number. For example, 25 and 3 are two of the factors of 75.

### EXAMPLE 1
**(a)** 15 is a factor of 75 because $\ 75 \div 15 = 5,\ $ and 5 is a whole number.

**(b)** 4 is not a factor of 75 because $\ 75 \div 4 = 18.75,\ $ and 18.75 is not a whole number.

## ■ *Completely Factored Form*

A whole number greater than 1 that has no exact divisor except for itself and 1 is called a **prime number**. The first 10 prime numbers are

$$2, 3, 5, 7, 11, 13, 17, 19, 23, 29$$

Whole numbers greater than 1 that are not prime numbers are called **composite numbers**. For example, 15 is a composite number because it is exactly divisible by 3 and by 5. When a composite number is written as a product of prime numbers, it is said to be *completely factored*, and this product is called the **completely factored form** of the number. Such factored forms can sometimes be found by inspection, together with a knowledge of fundamental multiplication facts.

### *EXAMPLE 2*
(a) $21 = 3 \cdot 7$     (b) $25 = 5 \cdot 5$     (c) $30 = 2 \cdot 3 \cdot 5$
(d) $4 \cdot 5$ is not the completely factored form of 20 because 4 is not a prime number.

The following rules can be used to decide whether or not a given number can be exactly divided by 2 or 3 or 5 and can be helpful in writing the completely factored form of some numbers.

---

### *Divisibility Rules*

1. A whole number is divisible by 2 if the last digit to the right is an even digit (0, 2, 4, 6, 8).
2. A whole number is divisible by 3 if the sum of its digits is divisible by 3.
3. A whole number is divisible by 5 if the last digit to the right is 0 or 5.

---

### *EXAMPLE 3*
(a) 36 is divisible by 2 because the last digit to the right is an even digit, 6.
(b) 36 is divisible by 3 because   $3 + 6 = 9$, and 9 is divisible by 3.
(c) 35 is divisible by 5 because the last digit to the right is 5.

The completely factored form of a number may include the same factor more than once, in which case it may be convenient to use exponents.

### *EXAMPLE 4*
(a) $25 = 5 \cdot 5 = 5^2$       (b) $63 = 3 \cdot 3 \cdot 7 = 3^2 \cdot 7$

If a number $N$ cannot be readily factored by inspection or by using the divisibility rules for 2, 3, and 5, we can try each prime factor in turn (more than once, if necessary) to see if it is an exact divisor. The process ends when $N$ is completely factored or when we find that $N$ is prime.

### *EXAMPLE 5*   Write 308 in completely factored form.

**Solution**   First, note that 308 is divisible by 2 because the last digit, 8, is an even digit. Thus,

$$308 \div 2 = 154 \quad \text{or} \quad 308 = 2 \cdot 154$$

Next, note that 154 is divisible by 2. Thus,

$$154 \div 2 = 77 \quad \text{or} \quad 154 = 2 \cdot 77$$

Hence

$$308 = 2 \cdot 154$$
$$= 2 \cdot 2 \cdot 77$$

Now, observe that 77 is not divisible by 2, or 3, or 5. Divide 77 by the next greatest prime number, 7:

$$77 \div 7 = 11 \qquad \text{or} \qquad 77 = 7 \cdot 11$$

Hence,

$$308 = 2 \cdot 154$$
$$= 2 \cdot 2 \cdot 77$$
$$= 2 \cdot 2 \cdot 7 \cdot 11$$

Because 2, 7, and 11 are prime numbers, the completely factored form of 308 is $2 \cdot 2 \cdot 7 \cdot 11$ (or $2^2 \cdot 7 \cdot 11$), and because multiplication is commutative, any arrangement of the factors 2, 2, 7, and 11 is correct.

When trying to factor a number $N$, or to show that $N$ is prime, we need only try to divide $N$ by each number in the list of prime numbers until we have tried the *greatest* prime number that is less than or equal to the square root of $N$. We shall refer to this number as the *greatest prime trial divisor* of $N$.

**EXAMPLE 6**   Show that 127 is a prime number

**Solution**   $\sqrt{127} = 11.269428$.   The greatest prime number less than or equal to 11.269428 is 11. Hence, 11 is the greatest prime trial divisor of 127, and only the prime numbers, 2, 3, 5, 7, and 11 need to be tried. Note that 127 is not divisible by 2, or by 3, or by 5. Try 7 and 11.

$$127 \div 7 = 18.142857 \qquad 127 \div 11 = 11.545455$$

Observe that neither 7 nor 11 is a factor of 127, and we can conclude that 127 is a prime number.

**EXAMPLE 7**   Write 2667 in completely factored form.

**Solution**   First, note that 2667 is not divisible by 2 but is divisible by 3.

$$2667 \div 3 = 889 \qquad \text{or} \qquad 2667 = 3 \cdot 889$$

Next, note that 889 is not divisible by either 3 or 5. Try 7.

$$889 \div 7 = 127 \qquad \text{or} \qquad 889 = 7 \cdot 127$$

Thus,

$$2667 = 3 \cdot 889$$
$$= 3 \cdot 7 \cdot 127$$

From Example 6, the number 127 is prime. Hence the completely factored form of 2667 is $3 \cdot 7 \cdot 127$.

## *Exercises for Section 2.1*

*Determine if the first number is a factor of the second number. See Example 1.*

**1.** 4;   244      **2.** 9;   342      **3.** 3;   142      **4.** 8;   196

**5.** 79;   395     **6.** 12;   408     **7.** 21;   455     **8.** 19;   493

**9.** 43;   752     **10.** 37;   902    **11.** 29;   696    **12.** 41;   615

*By inspection, write each number in completely factored form. See Example 2.*

**13.** 14      **14.** 22      **15.** 33      **16.** 34

**17.** 10      **18.** 38      **19.** 39      **20.** 26

**21.** 70      **22.** 42      **23.** 65      **24.** 51

*By inspection and the rules used in Example 3, determine if each number is divisible by 2, 3, or 5.*

**25.** 120      **26.** 130      **27.** 72       **28.** 82

**29.** 111      **30.** 114      **31.** 201      **32.** 405

**33.** 225      **34.** 126      **35.** 20,712    **36.** 17,235

*Write each number in completely factored form, using exponents where suitable. See Examples 4 and 5.*

**37.** 4       **38.** 8       **39.** 27       **40.** 9

**41.** 49      **42.** 18      **43.** 116      **44.** 245

**45.** 540     **46.** 720     **47.** 756      **48.** 504

*Show that each number is prime. See Example 6.*

**49.** 211   **50.** 757   **51.** 839   **52.** 821

*Write each number in completely factored form. See Examples 6 and 7.*

**53.** 254   **54.** 253   **55.** 164   **56.** 185

**57.** 188   **58.** 358   **59.** 393   **60.** 660

**61.** 415   **62.** 446   **63.** 746   **64.** 799

*In each of the following exercises, only one number is prime. Which number is it?*

**65.** (*a*) 2187   (*b*) 1409   (*c*) 9305   (*d*) 2114

**66.** (*a*) 2043   (*b*) 5245   (*c*) 1663   (*d*) 3002

## 2.2 Decimal Equivalents

In Section 1.1 we stated that a quotient can be written in fraction form as $\frac{n}{d}$. In such fractions $n$ is the **numerator** and $d$ is the **denominator**; these numbers are the **terms** of the fraction.

As you may recall, fractions with denominators such as 10, 100, 1000, and so forth can be written in decimal form. For example,

$$\frac{3}{10} = 0.3 \qquad \frac{27}{100} = 0.27 \qquad \frac{531}{1000} = 0.531$$

Because a fraction represents a quotient, fractions with denominators other than 10, 100, 1000, and so forth can also be written in decimal form by actually doing the division.

**EXAMPLE 1** Use your calculator to verify that

**(a)** $\frac{3}{5} = 3 \div 5 = 0.6$   **(b)** $\frac{5}{6} = 5 \div 6 = 0.83333333$

In Example 1*a*, note that no digits appear to the right of the digit 6. When the decimal form of a quotient has a last digit, as in 0.6, it is called a **terminating decimal**. In Example 1*b*, we cannot tell from the display whether 0.83333333 is a terminating decimal, because eight-digit calculators display, at most, eight decimal places. In fact, the quotient $\frac{5}{6}$ is equal to

$$0.83333333 \cdots$$

where the symbol "$\cdots$" indicates that there is no last digit. Such a decimal is called a **nonterminating decimal**.

***EXAMPLE 2***   Write each fraction as a decimal.

**(a)** $\frac{7}{16} = 0.4375$        (under floating decimal point)

   $= 0.4$          (to the nearest tenth)
   $= 0.44$         (to the nearest hundredth)
   $= 0.438$        (to the nearest thousandth)

**(b)** $\frac{11}{12} = 0.91666667*$   (under floating decimal point)

   $= 0.9$          (to the nearest tenth)
   $= 0.92$         (to the nearest hundredth)
   $= 0.917$        (to the nearest thousandth)

***EXAMPLE 3***

**(a)** $\frac{9}{8} = 1.125$          (under floating decimal point)

   $= 1.1$          (to one decimal place)
   $= 1.13$         (to two decimal places)

**(b)** $\frac{17}{6} = 2.8333333$     (under floating decimal point)

   $= 2.8$          (to one decimal place)
   $= 2.83$         (to two decimal places)

## ■ Mixed Numbers

The sum of a whole number and a fraction is a **mixed number**. For example, $11 + \frac{3}{4}$ is a mixed number and is customarily written as $11\frac{3}{4}$. A mixed number can be changed to decimal form by replacing the fraction part by its decimal equivalent.

***EXAMPLE 4***

**(a)** $11\frac{3}{4} = 11.75;$   because $\frac{3}{4} = 0.75$

**(b)** $16\frac{2}{3} = 16.66666667;$   because $\frac{2}{3} = 0.66666667$

## ■ Operations with Decimal Equivalents

The methods introduced in Chapter 1 for computing with decimals can be applied to calculations involving fractions and mixed numbers, if decimal equivalents are used.

---

*Note that we use the "=" sign between the fraction $\frac{11}{12}$ and the *approximation* 0.91666667 that appears in the display.

**EXAMPLE 5**

(a) $3\dfrac{1}{2} + 5\dfrac{7}{8} + 7\dfrac{11}{16} = 3.5 + 5.875 + 7.6875$

$$= 17.0625$$

(b) $23\dfrac{5}{32} - 19\dfrac{3}{4} = 23.15625 - 19.75$

$$= 3.40625$$

(c) $2\dfrac{5}{6} \div 3\dfrac{15}{16} = 2.8333333 \div 3.9365$

$$= 0.71957671$$

Combined operations with fractions can also be done by using decimal equivalents.

**EXAMPLE 6**  $2 \times 13\dfrac{5}{8} + 5 \times 14\dfrac{3}{8} + 7 \times 15\dfrac{9}{16}$

**Solution**  Replace each mixed number by its decimal equivalent, and group each product in parentheses.

$$(2 \times 13.625) + (5 \times 14.375) + (7 \times 15.5625) = 208.0625$$

The result is 208.06 to the nearest hundredth.

**EXAMPLE 7**  $\dfrac{\dfrac{22}{7} \times 11\dfrac{7}{16} \times \left(5\dfrac{3}{8}\right)^2}{1728}$

**Solution**  Write the on-line form of the quotient, and replace each fraction and mixed number by its decimal equivalent.

$$\dfrac{22}{7} \times 11\dfrac{7}{16} \times \left(5\dfrac{3}{8}\right)^2 \div 1728 = 3.1428571 \times 11.4375 \times 5.375^2 \div 1728$$

$$= 0.6009923$$

The result is 0.60, to the nearest hundredth.

## Exercises for Section 2.2

*Write the decimal equivalent of each fraction rounded off to the nearest thousandth. See Examples 1, 2, and 3.*

**1.** $\dfrac{7}{8}$    **2.** $\dfrac{11}{16}$    **3.** $\dfrac{5}{12}$    **4.** $\dfrac{1}{12}$

**5.** $\dfrac{19}{32}$    **6.** $\dfrac{31}{64}$    **7.** $\dfrac{11}{6}$    **8.** $\dfrac{23}{15}$

**9.** $\dfrac{37}{21}$    **10.** $\dfrac{43}{36}$    **11.** $\dfrac{41}{18}$    **12.** $\dfrac{91}{45}$

*Write the decimal equivalent of each number. See Example 4.*

**13.** $3\dfrac{9}{16}$     **14.** $7\dfrac{4}{5}$     **15.** $4\dfrac{3}{8}$     **16.** $16\dfrac{1}{8}$

**17.** $5\dfrac{5}{6}$     **18.** $4\dfrac{6}{7}$     **19.** $13\dfrac{7}{9}$     **20.** $63\dfrac{7}{11}$

*Use the methods of this section to perform the following computations. Round off answers to the nearest thousandth. See Example 5.*

**21.** $3\dfrac{1}{4} + 6\dfrac{2}{5} + 2\dfrac{1}{16}$     **22.** $2\dfrac{3}{5} + 7\dfrac{3}{4} + 5\dfrac{3}{16}$     **23.** $7\dfrac{2}{3} + 3\dfrac{7}{8} + 9\dfrac{7}{16}$

**24.** $2\dfrac{1}{6} + 4\dfrac{9}{16} + 8\dfrac{3}{8}$     **25.** $74\dfrac{9}{16} - 28\dfrac{3}{4}$     **26.** $46\dfrac{7}{8} - 44\dfrac{2}{5}$

**27.** $19\dfrac{3}{8} - 4\dfrac{1}{3}$     **28.** $49\dfrac{5}{6} - 13\dfrac{7}{16}$     **29.** $12\dfrac{3}{4} \times 8\dfrac{5}{8}$

**30.** $2\dfrac{7}{8} \times 76\dfrac{9}{16}$     **31.** $3\dfrac{1}{3} \times 0.21\dfrac{3}{8}$     **32.** $16\dfrac{5}{12} \times 0.34\dfrac{7}{8}$

**33.** $8\dfrac{3}{4} \div 15\dfrac{9}{16}$     **34.** $9\dfrac{3}{8} \div 17\dfrac{1}{2}$     **35.** $75\dfrac{2}{3} \div 29\dfrac{7}{8}$

**36.** $31\dfrac{1}{7} \div 4\dfrac{4}{5}$

*Round off the answers to the following exercises to the nearest hundredth. See Examples 6 and 7.*

**37.** $14 \times 3\frac{5}{8} + 2 \times 7\frac{5}{32} + 5 \times 6\frac{3}{4}$

**38.** $5 \times 4\frac{3}{32} + 18 \times 6\frac{1}{3} + 7 \times 5\frac{3}{16}$

**39.** $44 \times 13\frac{1}{2} + 4\frac{2}{3} + 8\frac{3}{5} + 84 \times 4\frac{5}{8}$

**40.** $30\frac{3}{8} \times 2\frac{1}{4} + 6 \times 16\frac{2}{3} - 16 \times 9\frac{7}{8}$

**41.** $56\frac{7}{16} \times 8\frac{1}{3} + 7\frac{3}{4} + 18\frac{1}{2} - 8\frac{3}{16} \times 2\frac{1}{32}$

**42.** $32\frac{5}{8} \times 7\frac{9}{16} + 3\frac{5}{6} \times 21\frac{1}{64} - 5\frac{7}{8} \times 15\frac{1}{2}$

**43.** $\dfrac{3\frac{1}{7} \times \left(5\frac{1}{4}\right)^2 \times 51\frac{1}{2}}{360}$

**44.** $\dfrac{4\frac{1}{3} \times 17\frac{2}{5} \times \left(9\frac{3}{8}\right)^2}{144}$

## 2.3  Reducing and Building Fractions

In the first sections of this chapter we used decimal equivalents for computations with fractions. In the remaining sections we consider the more traditional methods for such calculations. Although calculators are not always appropriate when using these methods, you will find that they can sometimes be used. Answers to exercises in these sections will, in general, be left in fraction form.

### ■ Equal Fractions

The following test can be used to determine whether or not two fractions are equal:

**Equal Fractions Test**

If the product of the numerator of the first fraction and the denominator of the second is equal to the product of the denominator of the first fraction and the numerator of the second, then the fractions are equal.

In symbols, the equal fractions test is given by

$$\frac{a}{b} = \frac{c}{d} \quad \text{if} \quad a \cdot d = b \cdot c$$

where $b \neq 0$ and $d \neq 0$. (The symbol $\neq$ is read "is not equal to.")

**EXAMPLE 1**

**(a)** $\dfrac{1}{2} = \dfrac{2}{4}$  because $1 \cdot 4 = 2 \cdot 2$;      $(4 = 4)$.

**(b)** $\dfrac{1}{2} = \dfrac{3}{6}$  because $1 \cdot 6 = 2 \cdot 3$;      $(6 = 6)$.

**(c)** $\dfrac{1}{2} \neq \dfrac{3}{8}$  because $1 \cdot 8 \neq 2 \cdot 3$;      $(8 \neq 6)$.

From Examples 1*a* and 1*b*, note that

$$\frac{1}{2} = \frac{2}{4} = \frac{3}{6}$$

which indicates that fractions can be written in different forms. Now, note that

$$\frac{1 \cdot 2}{2 \cdot 2} = \frac{2}{4} \quad \text{and} \quad \frac{1 \cdot 3}{2 \cdot 3} = \frac{3}{6}$$

and so

$$\frac{1}{2} = \frac{1 \cdot 2}{2 \cdot 2} = \frac{1 \cdot 3}{2 \cdot 3}$$

Results such as this suggest the following rule, which enables us to change the form of a fraction.

---

**Fundamental Principle of Fractions**

If both the numerator and the denominator of a given fraction are multiplied or divided by the same nonzero number, the resulting fraction is equal to the given fraction.

---

This principle is expressed in symbols as

$$\frac{n}{d} = \frac{n \cdot k}{d \cdot k} \quad \text{or} \quad \frac{n}{d} = \frac{n \div k}{d \div k}$$

where $k \neq 0$.

**EXAMPLE 2**

**(a)** $\dfrac{1}{4} = \dfrac{1 \cdot 9}{4 \cdot 9} = \dfrac{9}{36}$;      $\dfrac{1}{4}$ and $\dfrac{9}{36}$ are equal fractions.

**(b)** $\dfrac{12}{36} = \dfrac{12 \div 4}{36 \div 4} = \dfrac{3}{9}$;      $\dfrac{12}{36}$ and $\dfrac{3}{9}$ are equal fractions.

■ **Reducing Fractions**

If each of two whole numbers has the same whole number as a factor, that factor is called a *common factor* (or *common divisor*) of the two numbers. Thus, in Example 2*b* above, 4 is a common factor of 12 and 36. A fraction in which the numerator and denominator do not have any common factors is said to be in **lowest terms**. To change (or reduce) a fraction to lowest terms, we can factor the numerator and the denominator and "divide out" common factors. Slash bars (/) are frequently used on common factors to indicate which factors are being divided out.

### EXAMPLE 3

(a) $\dfrac{6}{10} = \dfrac{\cancel{2} \cdot 3}{\cancel{2} \cdot 5} = \dfrac{3}{5}$      (b) $\dfrac{9}{30} = \dfrac{\cancel{3} \cdot 3}{2 \cdot \cancel{3} \cdot 5} = \dfrac{3}{10}$

We sometimes need to divide out more than one common factor when reducing a fraction to lowest terms.

### EXAMPLE 4

(a) $\dfrac{12}{32} = \dfrac{\cancel{2} \cdot \cancel{2} \cdot 3}{\cancel{2} \cdot \cancel{2} \cdot 2 \cdot 2 \cdot 2} = \dfrac{3}{8}$      (b) $\dfrac{72}{360} = \dfrac{\cancel{2} \cdot \cancel{2} \cdot \cancel{2} \cdot \cancel{3} \cdot \cancel{3}}{\cancel{2} \cdot \cancel{2} \cdot \cancel{2} \cdot \cancel{3} \cdot \cancel{3} \cdot 5} = \dfrac{1}{5}$

Note that if all the factors in the numerator of a fraction are divided out, as in Example 4b, the factor 1 must be written in the numerator of the reduced fraction.

## ■ Building Fractions

In Example 2a we saw that the fundamental principle of fractions can be used to write

$$\frac{1}{4} = \frac{1 \cdot 9}{4 \cdot 9} = \frac{9}{36}$$

Note that each term of $\dfrac{9}{36}$ is greater than the corresponding term of $\dfrac{1}{4}$. We say that the fraction $\dfrac{1}{4}$ has been *raised*, or *built*, *to higher terms* and we refer to the factor 9 as the *building factor*. Most often, we need to build a fraction to higher terms with the requirement that the resulting fraction will have a specified denominator. For example, to build the fraction $\dfrac{3}{4}$ to an equal fraction with the denominator 64, we need to find a building factor $k$ such that

$$\frac{3 \cdot k}{4 \cdot k} = \frac{?}{64}$$

A comparison of the denominators indicates that we want a number $k$ such that $4 \cdot k$ will be equal to 64. By inspection, or by dividing 64 by 4, we obtain the building factor $k = 16$. Thus,

$$\frac{3}{4} = \frac{3 \cdot 16}{4 \cdot 16} = \frac{48}{64}$$

---

### To Raise a Fraction to Higher Terms

  **1.** Obtain the building factor by inspection or by dividing the new denominator by the given denominator.
  **2.** Multiply the numerator and denominator of the given fraction by the building factor.

---

**EXAMPLE 5**   Build the fraction $\dfrac{5}{8}$ to a fraction with denominator 32.

### Solution

**1.** The new denominator is 32, the original denominator is 8. Thus the building factor equals $32 \div 8 = 4$.

**2.** Hence, $\dfrac{5}{8} = \dfrac{5 \cdot 4}{8 \cdot 4} = \dfrac{20}{32}$.

The fundamental principle of fractions can also be used to change a whole number to a fraction with a specified denominator.

### EXAMPLE 6

(a) $1 = \dfrac{?}{8}$     (b) $7 = \dfrac{?}{8}$

### Solutions

(a) $1 = \dfrac{1}{1} = \dfrac{1 \cdot 8}{1 \cdot 8} = \dfrac{8}{8}$     (b) $7 = \dfrac{7}{1} = \dfrac{7 \cdot 8}{1 \cdot 8} = \dfrac{56}{8}$

## Exercises for Section 2.3

*Use the method of Example 1 to show whether the fractions in each pair are or are not equal.*

1. $\dfrac{5}{8}, \dfrac{35}{56}$     2. $\dfrac{1}{3}, \dfrac{18}{52}$     3. $\dfrac{17}{32}, \dfrac{52}{96}$

4. $\dfrac{3}{4}, \dfrac{49}{64}$     5. $\dfrac{7}{11}, \dfrac{329}{507}$     6. $\dfrac{25}{53}, \dfrac{325}{689}$

*Reduce each fraction to lowest terms. See Examples 2b, 3, and 4.*

7. $\dfrac{8}{14}$     8. $\dfrac{22}{33}$     9. $\dfrac{30}{34}$     10. $\dfrac{12}{15}$

11. $\dfrac{25}{35}$     12. $\dfrac{15}{25}$     13. $\dfrac{35}{85}$     14. $\dfrac{48}{72}$

15. $\dfrac{15}{48}$     16. $\dfrac{40}{65}$     17. $\dfrac{12}{42}$     18. $\dfrac{18}{54}$

19. $\dfrac{24}{32}$     20. $\dfrac{16}{64}$     21. $\dfrac{27}{81}$     22. $\dfrac{18}{27}$

**23.** $\dfrac{16}{44}$    **24.** $\dfrac{24}{40}$    **25.** $\dfrac{36}{60}$    **26.** $\dfrac{42}{72}$

**27.** $\dfrac{25}{125}$    **28.** $\dfrac{30}{150}$    **29.** $\dfrac{120}{280}$    **30.** $\dfrac{150}{250}$

*Build each fraction or whole number to an equal fraction with the given denominator.*
*See Examples 2a, 5, and 6.*

**31.** $\dfrac{1}{2} = \dfrac{?}{24}$    **32.** $\dfrac{2}{3} = \dfrac{?}{21}$    **33.** $\dfrac{3}{5} = \dfrac{?}{35}$    **34.** $\dfrac{4}{7} = \dfrac{?}{56}$

**35.** $\dfrac{5}{8} = \dfrac{?}{40}$    **36.** $\dfrac{5}{9} = \dfrac{?}{54}$    **37.** $\dfrac{7}{12} = \dfrac{?}{36}$    **38.** $\dfrac{5}{32} = \dfrac{?}{128}$

**39.** $\dfrac{11}{16} = \dfrac{?}{80}$    **40.** $\dfrac{3}{64} = \dfrac{?}{256}$    **41.** $\dfrac{9}{16} = \dfrac{?}{144}$    **42.** $\dfrac{7}{24} = \dfrac{?}{96}$

**43.** $\dfrac{3}{8} = \dfrac{?}{128}$    **44.** $\dfrac{5}{6} = \dfrac{?}{120}$    **45.** $\dfrac{2}{7} = \dfrac{?}{105}$    **46.** $\dfrac{2}{15} = \dfrac{?}{180}$

**47.** $\dfrac{5}{18} = \dfrac{?}{108}$    **48.** $\dfrac{5}{16} = \dfrac{?}{128}$    **49.** $\dfrac{1}{9} = \dfrac{?}{135}$    **50.** $\dfrac{1}{13} = \dfrac{?}{390}$

**51.** $1 = \dfrac{?}{12}$    **52.** $4 = \dfrac{?}{16}$    **53.** $5 = \dfrac{?}{16}$    **54.** $6 = \dfrac{?}{20}$

## 2.4 Addition; Subtraction

Fractions with the same denominators are called **like fractions**. Like fractions can be added or subtracted by using the following rule.

> ### To Add or Subtract Like Fractions
> The sum (difference) of like fractions is a fraction with the same denominator and a numerator equal to the sum (difference) of the numerators of the original fractions.

In symbols, we have

$$\frac{a}{c} + \frac{b}{c} = \frac{a+b}{c} \qquad \text{and} \qquad \frac{a}{c} - \frac{b}{c} = \frac{a-b}{c}$$

where $c \neq 0$.

### EXAMPLE 1

**(a)** $\dfrac{3}{8} + \dfrac{1}{8} = \dfrac{3+1}{8}$            **(b)** $\dfrac{3}{8} - \dfrac{1}{8} = \dfrac{3-1}{8}$

$\qquad\qquad = \dfrac{4}{8} = \dfrac{1}{2}$              $\qquad\qquad = \dfrac{2}{8} = \dfrac{1}{4}$

Note that, as in Example 1, we follow the customary practice of writing fraction answers in lowest terms.

### ◼ Least Common Denominator

The number that is the denominator of a set of like fractions is the **common denominator** of the fractions. The smallest number exactly divisible by each of the denominators of a set of unlike fractions is called the **least common denominator (LCD)** of the fractions. For example, the LCD of $\dfrac{1}{2}$ and $\dfrac{2}{3}$ is 6, because 6 is the smallest number that is exactly divisible by 2 and by 3.

The LCD of a set of unlike fractions can be found by the following procedure:

> ### To Find the LCD of a Set of Unlike Fractions
> **1.** Express each denominator in completely factored form.
> **2.** Write a product whose factors are each of the different prime factors that occur in any of the denominators, and include each of these factors the greatest number of times that it appears in any one of the given denominators.

**EXAMPLE 2   (a)** Find the LCD for $\dfrac{2}{3}$ and $\dfrac{3}{5}$.   **(b)** Write $\dfrac{2}{3}$ and $\dfrac{3}{5}$ as like fractions, with their LCD as the common denominator.

### Solutions

**(a)** Each denominator is prime. The LCD must contain each of the factors 3 and 5 once. Hence the LCD is

$$3 \cdot 5 = 15$$

**(b)** $\dfrac{2}{3} = \dfrac{?}{15}$   and   $\dfrac{3}{5} = \dfrac{?}{15}$.   Because

$$15 \div 3 = 5 \qquad \text{and} \qquad 15 \div 5 = 3$$

the respective building factors are 5 and 3. Use the fundamental principle of fractions.

$$\frac{2}{3} = \frac{2 \cdot 5}{3 \cdot 5} = \frac{10}{15} \quad \text{and} \quad \frac{3}{5} = \frac{3 \cdot 3}{5 \cdot 3} = \frac{9}{15}$$

**EXAMPLE 3**   **(a)** Find the LCD for $\frac{5}{6}$ and $\frac{7}{15}$.   **(b)** Write $\frac{5}{6}$ and $\frac{7}{15}$ as like fractions, with their LCD as the common denominator.

**Solutions**
**(a)** Factor each denominator.
The LCD must contain each of the factors 2, 3, and 5 once. The LCD is 30.

**(b)** $\frac{5}{6} = \frac{?}{30}$ and $\frac{7}{15} = \frac{?}{30}$. Because

$$30 \div 6 = 5 \quad \text{and} \quad 30 \div 15 = 2$$

the respective building factors are 5 and 2. Use the fundamental principle of fractions.

$$\frac{5}{6} = \frac{5 \cdot 5}{6 \cdot 5} = \frac{25}{30} \quad \text{and} \quad \frac{7}{15} = \frac{7 \cdot 2}{15 \cdot 2} = \frac{14}{30}$$

**EXAMPLE 4**   **(a)** Find the LCD for $\frac{3}{4}$, $\frac{5}{8}$, and $\frac{7}{10}$.   **(b)** Write $\frac{3}{4}$, $\frac{5}{8}$, and $\frac{7}{10}$ as like fractions, with their LCD as the common denominator.

**Solutions**
**(a)** Factor each denominator.
The LCD must contain the factor 2 three times (because 8 contains 2 as a factor three times), and the factor 5 once. The LCD is 40.

**(b)** $\frac{3}{4} = \frac{?}{40}$, $\frac{5}{8} = \frac{?}{40}$, and $\frac{7}{10} = \frac{?}{40}$. Because

$$40 \div 4 = 10, \quad 40 \div 8 = 5, \quad \text{and} \quad 40 \div 10 = 4$$

the respective building factors are 10, 5, and 4. Use the fundamental principle of fractions.

$$\frac{3}{4} = \frac{3 \cdot 10}{4 \cdot 10} = \frac{30}{40}, \quad \frac{5}{8} = \frac{5 \cdot 5}{8 \cdot 5} = \frac{25}{40}, \quad \text{and} \quad \frac{7}{10} = \frac{7 \cdot 4}{10 \cdot 4} = \frac{28}{40}$$

## ■ Sums and Differences of Unlike Fractions

To add or subtract unlike fractions, we first use the fundamental principle of fractions to change the given fractions to a set of like fractions. In general, the following procedure is used.

---

### To Add or Subtract Unlike Fractions

1. Find the LCD of the fractions.
2. Change the given fractions to like fractions with the LCD as the common denominator.
3. Find the sum or difference of the resulting fractions.
4. Reduce the answer to lowest terms.

---

**EXAMPLE 5** **(a)** $\dfrac{7}{8} + \dfrac{5}{16}$     **(b)** $\dfrac{7}{8} - \dfrac{5}{16}$

**Solutions**   In each case the LCD is 16. Use the fundamental principle of fractions to change the given fractions to like fractions with denominator 16.

**(a)** $\dfrac{7}{8} + \dfrac{5}{16} = \dfrac{7 \cdot 2}{8 \cdot 2} + \dfrac{5}{16}$          **(b)** $\dfrac{7}{8} - \dfrac{5}{16} = \dfrac{7 \cdot 2}{8 \cdot 2} - \dfrac{5}{16}$

$\qquad\qquad = \dfrac{14}{16} + \dfrac{5}{16}$                       $= \dfrac{14}{16} - \dfrac{5}{16}$

$\qquad\qquad = \dfrac{14 + 5}{16} = \dfrac{19}{16}$                     $= \dfrac{14 - 5}{16} = \dfrac{9}{16}$

Often, fewer details are needed in a solution than are shown in the above example. In the following example, some of the details are omitted.

**EXAMPLE 6** $\quad \dfrac{1}{2} + \dfrac{31}{32} - \dfrac{15}{64} = \dfrac{32}{64} + \dfrac{62}{64} - \dfrac{15}{64}$

$$= \dfrac{32 + 62 - 15}{64} = \dfrac{79}{64}$$

## ■ Comparing Fractions

In a set of fractions with the same denominator the fraction with the smallest numerator is the smallest fraction. If the fractions are unlike fractions, we can first change them to like fractions in order to determine the smallest fraction.

**EXAMPLE 7**

**(a)** $\dfrac{61}{64}, \ \dfrac{63}{64}$

Because 61 is less than 63, the fraction $\dfrac{61}{64}$ is less than $\dfrac{63}{64}$.

**(b)** $\dfrac{5}{6}, \ \dfrac{23}{30}$

From Example 3*b*, $\dfrac{5}{6} = \dfrac{25}{30}$. Because 23 is less than 25, the fraction $\dfrac{23}{30}$ is less than the fraction $\dfrac{25}{30}$. Thus, $\dfrac{23}{30}$ is less than $\dfrac{5}{6}$.

Another way to compare fractions is to change each fraction to its decimal equivalent and then compare the results, as in Section 1.2.

**EXAMPLE 8**

$$\dfrac{5}{6} = 0.83333333 \quad \text{and} \quad \dfrac{23}{30} = 0.76666667$$

Because 0.76666667 is less than 0.83333333, the fraction $\dfrac{23}{30}$ is less than $\dfrac{5}{6}$.

## Exercises for Section 2.4

*Unless otherwise noted, express fractional answers in lowest terms. Find each sum or difference. See Example 1.*

1. $\dfrac{2}{7}+\dfrac{3}{7}$

2. $\dfrac{1}{9}+\dfrac{4}{9}$

3. $\dfrac{1}{16}+\dfrac{5}{16}$

4. $\dfrac{3}{32}+\dfrac{5}{32}$

5. $\dfrac{6}{7}-\dfrac{2}{7}$

6. $\dfrac{3}{4}-\dfrac{1}{4}$

7. $\dfrac{9}{64}-\dfrac{5}{64}$

8. $\dfrac{17}{32}-\dfrac{3}{32}$

9. $\dfrac{1}{16}+\dfrac{5}{16}+\dfrac{7}{16}$

10. $\dfrac{1}{15}+\dfrac{7}{15}+\dfrac{4}{15}$

11. $\dfrac{27}{32}-\dfrac{11}{32}-\dfrac{13}{32}$

12. $\dfrac{23}{25}-\dfrac{7}{25}-\dfrac{6}{25}$

13. $\dfrac{15}{16}-\dfrac{9}{16}+\dfrac{2}{16}$

14. $\dfrac{11}{20}+\dfrac{7}{20}-\dfrac{4}{20}$

*Write each group of fractions as a group of like fractions with the LCD as the common denominator. See Examples 2, 3, and 4.*

15. $\dfrac{3}{8},\ \dfrac{3}{4}$

16. $\dfrac{3}{5},\ \dfrac{5}{6}$

17. $\dfrac{7}{16},\ \dfrac{1}{6}$

18. $\dfrac{3}{8},\ \dfrac{1}{20}$

19. $\dfrac{5}{12},\ \dfrac{3}{32}$

20. $\dfrac{3}{16},\ \dfrac{5}{24}$

21. $\dfrac{1}{4},\ \dfrac{2}{3},\ \dfrac{1}{16}$

22. $\dfrac{1}{12},\ \dfrac{3}{4},\ \dfrac{7}{16}$

23. $\dfrac{5}{18},\ \dfrac{1}{9},\ \dfrac{1}{12}$

**24.** $\frac{5}{6}, \ \frac{1}{18}, \ \frac{3}{4}$   **25.** $\frac{5}{1}, \ \frac{2}{15}, \ \frac{3}{20}$   **26.** $\frac{1}{12}, \ \frac{1}{10}, \ \frac{4}{1}$

*Find each sum or difference. See Examples 5 and 6.*

**27.** $\frac{2}{3} + \frac{3}{4}$   **28.** $\frac{1}{3} + \frac{2}{5}$   **29.** $\frac{5}{8} + \frac{3}{10}$

**30.** $\frac{3}{8} + \frac{1}{20}$   **31.** $\frac{5}{6} - \frac{3}{5}$   **32.** $\frac{3}{4} + \frac{2}{3}$

**33.** $\frac{3}{8} - \frac{1}{20}$   **34.** $\frac{5}{24} - \frac{3}{16}$   **35.** $\frac{3}{4} + \frac{5}{8} - \frac{3}{16}$

**36.** $\frac{1}{2} + \frac{3}{4} - \frac{3}{32}$   **37.** $\frac{27}{32} - \frac{5}{64} - \frac{1}{8}$   **38.** $\frac{7}{8} - \frac{3}{16} - \frac{5}{32}$

**39.** $\frac{27}{16} - \frac{5}{4} + \frac{5}{32}$   **40.** $\frac{15}{8} + \frac{7}{16} + \frac{1}{4}$   **41.** $3 + \frac{11}{12} - \frac{5}{16}$

**42.** $6 - \frac{1}{3} + \frac{5}{32}$   **43.** $2 - \frac{5}{8} - \frac{1}{2}$   **44.** $3 - \frac{5}{16} - \frac{3}{10}$

*Write each set of fractions in order, from the least to the greatest. See Example 7.*

**45.** $\frac{5}{8}, \frac{3}{8}$

**46.** $\frac{7}{16}, \frac{5}{16}$

**47.** $\frac{1}{4}, \frac{1}{8}, \frac{5}{16}$

**48.** $\frac{1}{8}, \frac{3}{32}, \frac{5}{64}$

**49.** $\frac{9}{16}, \frac{15}{32}, \frac{7}{8}, \frac{37}{64}$

**50.** $\frac{1}{8}, \frac{7}{16}, \frac{19}{32}, \frac{37}{64}$

*Certain types of wrenches, such as open-end wrenches, are referred to by the opening size. For each problem, arrange the wrenches in order, starting with the smallest.*

**51.** $\frac{1}{2}$ inch, $\frac{3}{8}$ inch, $\frac{9}{16}$ inch

**52.** $\frac{7}{16}$ inch, $\frac{1}{4}$ inch, $\frac{5}{32}$ inch

**53.** $\frac{7}{8}$ inch, $\frac{3}{4}$ inch, 1 inch

**54.** $\frac{11}{16}$ inch, $\frac{7}{32}$ inch, $\frac{3}{8}$ inch

**55.** $\frac{1}{2}$ inch, $\frac{7}{16}$ inch, $\frac{5}{8}$ inch

**56.** $\frac{9}{32}$ inch, $\frac{5}{16}$ inch, $\frac{1}{4}$ inch

*Find the missing dimension in each of the following drawings.*

**57.**
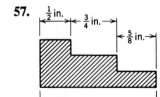
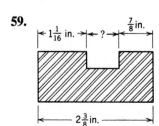

**58.**
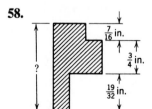

**59.**
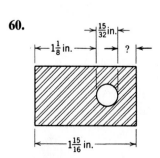

**60.**

## 2.5 Computations with Mixed Numbers

If the numerator of a fraction is greater than or equal to the denominator, the fraction is an **improper fraction**. For example, $\dfrac{79}{64}$ is such a fraction. When necessary, mixed numbers can be written as improper fractions. For example,

$$3\frac{5}{6} = 3 + \frac{5}{6} = \frac{3}{1} + \frac{5}{6}$$

$$= \frac{3 \cdot 6}{1 \cdot 6} + \frac{5}{6}$$

$$= \frac{3 \cdot 6 + 5}{6} = \frac{23}{6}$$

The above example suggests a rapid method for computing the numerator of the desired improper fraction:

(whole number part) $\times$ (original denominator) + (original numerator)

**EXAMPLE 1**   $13\dfrac{11}{64} = \dfrac{13 \cdot 64 + 11}{64} = \dfrac{843}{64}$

An improper fraction can be changed to a mixed number by a division process.

**EXAMPLE 2**   We can change $\dfrac{23}{4}$ to a mixed number by dividing the numerator 23 by the denominator 4

$$
\begin{array}{r}
5 \quad \longleftarrow \quad \text{quotient} \\
4\overline{)23} \\
\underline{20} \\
3 \quad \longleftarrow \quad \text{remainder}
\end{array}
$$

to obtain the quotient 5 and the remainder 3. From this result we determine that

$$\frac{23}{4} = 5 + \frac{3}{4} = 5\frac{3}{4}$$

### ■ Sums and Differences

An efficient procedure for adding or subtracting mixed numbers involves separating the whole number and fractional parts.

**EXAMPLE 3**   $12\dfrac{3}{8} + 5\dfrac{7}{8} - 4\dfrac{1}{8} = (12 + 5 - 4) + \left( \dfrac{3}{8} + \dfrac{7}{8} - \dfrac{1}{8} \right)$

$$= 13 + \frac{9}{8}$$

$$= 13 + 1 + \frac{1}{8} = 14\frac{1}{8}$$

**EXAMPLE 4**   $19\dfrac{3}{8} - 7 - 5\dfrac{5}{16} = 19\dfrac{6}{16} - 7 - 5\dfrac{5}{16}$

$$= (19 - 7 - 5) + \left( \frac{6}{16} - \frac{5}{16} \right)$$

$$= 7 + \frac{1}{16} = 7\frac{1}{16}$$

In problems such as $8\frac{1}{5} - 3\frac{4}{5}$, where the fractional part of the second number is greater than the fractional part of the first number, we borrow 1 from the whole number part of the first number, change it to a fraction with an appropriate denominator, such as $\frac{5}{5}$ or $\frac{3}{3}$, and add it to the fractional part. Thus, to subtract $3\frac{4}{5}$ from $8\frac{1}{5}$, we first have

$$8\frac{1}{5} = 7 + 1\frac{1}{5} = 7\frac{6}{5}$$

Now, using $7\frac{6}{5}$ in place of $8\frac{1}{5}$,

$$8\frac{1}{5} - 3\frac{4}{5} = 7\frac{6}{5} - 3\frac{4}{5} = (7-3) + \left(\frac{6}{5} - \frac{4}{5}\right)$$

$$= 4 + \frac{2}{5} = 4\frac{2}{5}$$

**EXAMPLE 5**  $26\frac{1}{6} - 14\frac{3}{4}$

**Solution**   The LCD of the fractional parts is 12. Hence,

$$26\frac{1}{6} - 14\frac{3}{4} = 26\frac{2}{12} - 14\frac{9}{12}$$

Because $\frac{9}{12}$ is greater than $\frac{2}{12}$, change $26\frac{2}{12}$ as follows:

$$26\frac{2}{12} = 25 + 1\frac{2}{12} = 25\frac{14}{12}$$

Substitute $25\frac{14}{12}$ for $26\frac{2}{12}$.

$$25\frac{14}{12} - 14\frac{9}{12} = (25 - 14) + \left(\frac{14}{12} - \frac{9}{12}\right) = 11\frac{5}{12}$$

**EXAMPLE 6**   $18 - 9\frac{5}{8}$

**Solution**   Substitute $17\frac{8}{8}$ for 18.

$$17\frac{8}{8} - 9\frac{5}{8} = (17 - 9) + \left(\frac{8}{8} - \frac{5}{8}\right) = 8\frac{3}{8}$$

### Exercises for Section 2.5

*Change each mixed number to an improper fraction. See Example 1.*

**1.** $12\frac{4}{5}$       **2.** $10\frac{1}{4}$       **3.** $21\frac{5}{16}$

**4.** $30\frac{3}{16}$      **5.** $29\frac{11}{32}$      **6.** $42\frac{27}{32}$

*Change each improper fraction to a mixed number. See Example 2.*

**7.** $\frac{43}{8}$      **8.** $\frac{37}{8}$      **9.** $\frac{107}{16}$

**10.** $\frac{123}{16}$      **11.** $\frac{317}{33}$      **12.** $\frac{361}{41}$

*Find each sum or difference. (Reduce proper fraction answers to lowest terms; change improper fraction answers to mixed numbers.) See Examples 3 and 4.*

**13.** $6\frac{5}{8} + 5\frac{3}{8}$          **14.** $5\frac{3}{4} + 4\frac{1}{4}$          **15.** $3\frac{7}{16} + 7\frac{11}{16}$

**16.** $3\frac{5}{8} + 1\frac{5}{8}$          **17.** $5\frac{3}{5} + 2\frac{7}{8}$          **18.** $6\frac{1}{3} + 3\frac{1}{4}$

**19.** $10\frac{5}{6} - 3\frac{1}{4}$          **20.** $6\frac{2}{3} - 4\frac{2}{5}$          **21.** $7\frac{1}{2} - 4\frac{1}{8}$

**22.** $3\frac{5}{8} - 1\frac{3}{16}$          **23.** $7\frac{4}{5} - 3\frac{1}{3}$          **24.** $8\frac{6}{7} - 3\frac{1}{2}$

**25.** $9\dfrac{1}{8} + 2\dfrac{7}{8} - 4\dfrac{3}{8}$      **26.** $2\dfrac{11}{16} + 8\dfrac{9}{16} - 3\dfrac{7}{16}$      **27.** $7\dfrac{5}{12} - 1\dfrac{1}{12} + 2\dfrac{7}{12}$

**28.** $8\dfrac{4}{7} - 4\dfrac{3}{7} + 5\dfrac{2}{7}$      **29.** $1\dfrac{1}{15} + 2\dfrac{2}{3} - 1\dfrac{1}{12}$      **30.** $3\dfrac{5}{6} - 2\dfrac{5}{27} + 2\dfrac{1}{18}$

**31.** $12\dfrac{25}{32} - 5\dfrac{3}{8} + 2\dfrac{1}{4}$      **32.** $25\dfrac{3}{8} - 4\dfrac{3}{16} + 2\dfrac{1}{4}$

*See Examples 5 and 6.*

**33.** $9\dfrac{3}{8} - 2\dfrac{1}{2}$      **34.** $8\dfrac{7}{16} - 4\dfrac{3}{4}$      **35.** $7\dfrac{1}{4} - 3\dfrac{2}{3}$

**36.** $12\dfrac{1}{6} - 8\dfrac{3}{4}$      **37.** $8 - 3\dfrac{1}{4}$      **38.** $6 - 2\dfrac{2}{3}$

*Some lumberyards sell lumber in even lengths, such as 8 feet, 10 feet, 12 feet, and so forth. What is the shortest single board from which a carpenter could cut the pieces with the given lengths? (Assume no waste in cutting.)*

**39.** Three pieces, each $3\dfrac{1}{2}$ feet long.

**40.** Two pieces $2\dfrac{1}{4}$ feet long and one piece $3\dfrac{3}{4}$ feet long.

**41.** Three pieces of lengths $2\frac{1}{2}, 3\frac{1}{4}$, and $5\frac{3}{4}$ feet, respectively.

**42.** Three pieces of lengths $4\frac{3}{8}, 2\frac{1}{2}$, and $6\frac{1}{6}$ feet, respectively.

*Find the indicated dimension in the drawings in Exercises 43 to 46 (Figures 2.1 and 2.2). All dimensions are in feet.*

**43.** Dimension A.

**44.** Dimension B.

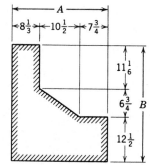

**FIGURE 2.1**

**45.** Dimension C.

**46.** Dimension D.

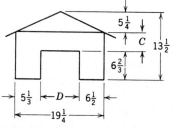

**FIGURE 2.2**

**47.** Find the length of the cap screw shown in Figure 2.3.

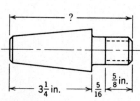

**FIGURE 2.3**

**48.** Find the length of the tapered shaft shown in Figure 2.4.

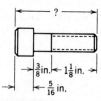

**FIGURE 2.4**

## 2.6  *Multiplication; Division*

### ■ *Multiplying Fractions*

In this section we review methods for finding products and quotients of fractions.

Two or more fractions can be multiplied by using the following rule:

---

**To Multiply Fractions**

The product of two or more fractions is a fraction whose numerator is the product of the numerators and whose denominator is the product of the denominators of the given fractions.

---

In symbols, we have

$$\frac{a}{b} \cdot \frac{c}{d} = \frac{a \cdot c}{b \cdot d}$$

where  $b \neq 0$  and  $d \neq 0$.

**EXAMPLE 1**  (a) $\dfrac{2}{3} \cdot \dfrac{4}{5} = \dfrac{2 \cdot 4}{3 \cdot 5}$           (b) $\dfrac{4}{7} \cdot \dfrac{3}{2} = \dfrac{4 \cdot 3}{7 \cdot 2}$

$= \dfrac{8}{15}$                $= \dfrac{12}{14} = \dfrac{6}{7}$

In example 1b, note that *after* the product $\dfrac{12}{14}$ was obtained, it was reduced to lowest terms. It is usually easier to write all numerators and denominators in completely factored form and divide out any factors that are common factors of the numerators and denominators *before* multiplying. The following steps provide an efficient method for finding products of fractions in which some numerators and denominators may have common factors:

1. Write all numerators and denominators in completely factored form.
2. Divide out any factors that are common factors of the numerator and denominator. Write "1" in place of each factor that has been divided out.
3. Multiply the remaining factors.

**EXAMPLE 2**   $\dfrac{2}{3} \cdot \dfrac{15}{16} \cdot \dfrac{3}{5} = \dfrac{\overset{1}{\cancel{2}}}{\underset{1}{\cancel{3}}} \cdot \dfrac{\overset{1}{\cancel{3}} \cdot \overset{1}{\cancel{5}}}{\underset{1}{\cancel{2} \cdot 2 \cdot 2 \cdot 2}} \cdot \dfrac{3}{\underset{1}{\cancel{5}}} = \dfrac{3}{8}$

In actual practice, we do not always show the completely factored form of numerators and denominators. Instead, if we recognize *any* common factors, we "divide out" accordingly. For example, the computation of Example 2 may appear as

$$\dfrac{2}{3} \cdot \dfrac{15}{16} \cdot \dfrac{3}{5} = \dfrac{\overset{1}{\cancel{2}}}{\underset{1}{\cancel{3}}} \cdot \dfrac{\overset{\overset{1}{\cancel{5}}}{\cancel{15}}}{\underset{8}{\cancel{16}}} \cdot \dfrac{3}{\underset{1}{\cancel{5}}} = \dfrac{3}{8}$$

The word "of" is often used to indicate multiplication.

**EXAMPLE 3**  $\dfrac{3}{4}$ of $80 = \dfrac{3}{4} \times 80 = \dfrac{3}{\cancel{4}} \times \dfrac{\overset{20}{\cancel{80}}}{1} = 60$

If one or more of the factors of a product are mixed numbers, we change the mixed numbers to improper fractions before doing the calculations.

**EXAMPLE 4**

**(a)** $4\dfrac{1}{2} \cdot 7\dfrac{5}{8} = \dfrac{9}{2} \cdot \dfrac{61}{8} = \dfrac{549}{16} = 34\dfrac{5}{16}$        **(b)** $16 \times 3\dfrac{3}{32} = \dfrac{\overset{1}{\cancel{16}}}{1} \times \dfrac{99}{\underset{2}{\cancel{32}}} = \dfrac{99}{2} = 49\dfrac{1}{2}$

## ■ Reciprocals of Numbers

Recall from Section 1.1 that the reciprocal of a nonzero number $n$ is the number $\dfrac{1}{n}$. More generally, if the product of two numbers is 1, each of the numbers is the *reciprocal* of the other. For example, because

$$3 \cdot \dfrac{1}{3} = \dfrac{3}{3} = 1$$

$\dfrac{1}{3}$ is the reciprocal of 3, and 3 is the reciprocal of $\dfrac{1}{3}$. Also, because

$$\dfrac{3}{4} \cdot \dfrac{4}{3} = \dfrac{12}{12} = 1$$

$\dfrac{4}{3}$ is the reciprocal of $\dfrac{3}{4}$, and $\dfrac{3}{4}$ is the reciprocal of $\dfrac{4}{3}$. Results such as these indicate that the reciprocal of $\dfrac{n}{d}$ is $\dfrac{d}{n}$.

**EXAMPLE 5**

**(a)** The reciprocal of $\dfrac{17}{64}$ is $\dfrac{64}{17}$.        **(b)** The reciprocal of 7 is $\dfrac{1}{7}$.

The reciprocal of a mixed number can be obtained by first changing the mixed number to an improper fraction.

**EXAMPLE 6**  Because  $9\dfrac{3}{16} = \dfrac{147}{16}$,  the reciprocal of $9\dfrac{3}{16}$ is $\dfrac{16}{147}$.

Fractions can be divided by using the following rule.

---

**To Divide Fractions**
Invert the divisor and multiply.

---

In symbols, we have

$$\dfrac{a}{b} \div \dfrac{c}{d} = \dfrac{a}{b} \cdot \dfrac{d}{c} = \dfrac{a \cdot d}{b \cdot c}$$

where  $b \neq 0$,  $c \neq 0$,  and  $d \neq 0$.

**EXAMPLE 7**
$$\frac{5}{8} \div \frac{7}{12} = \frac{5}{\overset{}{\underset{2}{8}}} \cdot \frac{\overset{3}{12}}{7} = \frac{15}{14} = 1\frac{1}{14}$$

**EXAMPLE 8**
$$4\frac{1}{2} \div 3\frac{3}{8} = \frac{9}{2} \div \frac{27}{8} = \frac{9}{2} \cdot \frac{8}{27}$$
$$= \frac{\overset{1}{\cancel{9}}}{\underset{1}{\cancel{2}}} \cdot \frac{\overset{4}{\cancel{8}}}{\underset{3}{\cancel{27}}} = \frac{4}{3} = 1\frac{1}{3}$$

**EXAMPLE 9**
$$2\frac{3}{4} \div 5 = \frac{11}{4} \div 5$$
$$= \frac{11}{4} \cdot \frac{1}{5} = \frac{11}{20}$$

## Exercises for Section 2.6

*Reduce proper fraction answers to lowest terms; change improper fraction answers to mixed numbers.*

*Compute. See Examples 1 and 2.*

**1.** $\dfrac{3}{16} \cdot \dfrac{4}{9}$      **2.** $\dfrac{7}{12} \cdot \dfrac{3}{14}$      **3.** $\dfrac{6}{25} \cdot \dfrac{5}{9}$

**4.** $\dfrac{14}{15} \cdot \dfrac{3}{4}$      **5.** $\dfrac{2}{3} \cdot \dfrac{5}{7} \cdot \dfrac{3}{4}$      **6.** $\dfrac{5}{24} \cdot \dfrac{3}{16} \cdot \dfrac{8}{15}$

**7.** $\dfrac{9}{20} \cdot \dfrac{5}{18} \cdot \dfrac{11}{12}$      **8.** $\dfrac{9}{16} \cdot \dfrac{5}{24} \cdot \dfrac{8}{27}$

*Compute. See Example 3.*

**9.** $\dfrac{2}{3}$ of 60      **10.** $\dfrac{5}{6}$ of 84      **11.** $\dfrac{3}{8}$ of 96

**12.** $\dfrac{5}{16}$ of 80      **13.** $\dfrac{3}{4}$ of 38      **14.** $\dfrac{5}{12}$ of 78

*Compute. See Example 4.*

**15.** $7\frac{1}{3} \cdot 2\frac{1}{4}$     **16.** $3\frac{3}{7} \cdot 2\frac{1}{12}$     **17.** $19\frac{3}{32} \cdot 28$

**18.** $18 \cdot 15\frac{3}{4}$     **19.** $4\frac{2}{3} \cdot \frac{6}{7} \cdot 2\frac{1}{4}$     **20.** $1\frac{3}{4} \cdot \frac{3}{8} \cdot 2\frac{2}{3}$

*Write the reciprocal of each number. See Examples 5 and 6.*

**21.** 5     **22.** 9     **23.** $\frac{3}{5}$

**24.** $\frac{2}{3}$     **25.** $3\frac{1}{2}$     **26.** $4\frac{2}{3}$

*See Examples 7, 8, and 9 for Exercises 27 to 40.*

**27.** $\frac{3}{4} \div \frac{5}{8}$     **28.** $\frac{15}{32} \div \frac{25}{48}$     **29.** $\frac{4}{5} \div 6$

**30.** $\frac{5}{6} \div 10$     **31.** $12 \div \frac{8}{13}$     **32.** $16 \div \frac{12}{25}$

**33.** $3\frac{3}{4} \div 1\frac{7}{8}$     **34.** $5\frac{5}{6} \div 2\frac{5}{8}$     **35.** $8 \div 3\frac{1}{5}$

**36.** $6 \div 4\frac{2}{7}$     **37.** $5\frac{1}{2} \div \frac{3}{4}$     **38.** $3\frac{1}{3} \div \frac{3}{5}$

**39.** $7\frac{1}{5} \div 4$     **40.** $6\frac{1}{8} \div 7$

**41.** The distance a screw will advance when turned can be found by multiplying the pitch by the number of turns. Find the distance that each screw will advance.

|                 | a              | b              | c             | d              |
|-----------------|----------------|----------------|---------------|----------------|
| Pitch (inches)  | $\frac{1}{16}$ | $\frac{1}{32}$ | $\frac{1}{8}$ | $\frac{1}{12}$ |
| Number of turns | 6              | 20             | $7\frac{1}{2}$ | $3\frac{1}{2}$ |

**42.** The distance that a car will travel can be found by multiplying the car's mileage (in miles per gallon) by the number of gallons of gasoline used. Find the distance that each car will travel.

|         | a               | b               | c               | d               |
|---------|-----------------|-----------------|-----------------|-----------------|
| Gallons | 8               | 10              | $20\frac{1}{2}$ | $16\frac{1}{4}$ |
| Mileage | $15\frac{1}{2}$ | $16\frac{1}{4}$ | $14\frac{1}{4}$ | $18\frac{3}{4}$ |

**43.** The current (amperes) in an electric circuit can be found by dividing the voltage (volts) by the resistance (ohms). Find the current in each circuit.

|            | a              | b              | c               | d               |
|------------|----------------|----------------|-----------------|-----------------|
| Voltage    | 21             | 18             | $18\frac{3}{4}$ | $21\frac{1}{4}$ |
| Resistance | $3\frac{1}{2}$ | $2\frac{1}{4}$ | 15              | $1\frac{1}{4}$  |

**44.** The diametral pitch of a spur gear is found by dividing the number of teeth in the gear by the pitch diameter (in inches). Find the diametral pitch of each gear.

|                  | a              | b              | c               | d              |
|------------------|----------------|----------------|-----------------|----------------|
| Number of teeth  | 16             | 40             | 75              | 50             |
| Pitch diameter   | $4\frac{1}{2}$ | $5\frac{1}{8}$ | $12\frac{1}{2}$ | $8\frac{1}{3}$ |

## 2.7  Mental Calculations

The following exercises will improve your skills in working with fractions.

### Exercises for Section 2.7

*Do the following exercises mentally.*

*Write each number in completely factored form.*

**1.** 4        **2.** 12        **3.** 18        **4.** 15

**5.** 16        **6.** 27        **7.** 30        **8.** 38

*Reduce each fraction to lowest terms.*

**9.** $\dfrac{2}{4}$        **10.** $\dfrac{3}{6}$        **11.** $\dfrac{6}{9}$        **12.** $\dfrac{8}{12}$

**13.** $\dfrac{10}{15}$        **14.** $\dfrac{12}{16}$        **15.** $\dfrac{21}{35}$        **16.** $\dfrac{18}{27}$

*Specify the missing numerator.*

**17.** $\dfrac{1}{3} = \dfrac{?}{12}$        **18.** $\dfrac{3}{4} = \dfrac{?}{24}$        **19.** $\dfrac{3}{5} = \dfrac{?}{30}$

**20.** $\dfrac{5}{6} = \dfrac{?}{18}$        **21.** $\dfrac{4}{7} = \dfrac{?}{49}$        **22.** $\dfrac{5}{12} = \dfrac{?}{48}$

*Specify the least common denominator for each set of fractions.*

**23.** $\dfrac{1}{2}, \dfrac{2}{3}, \dfrac{1}{6}$        **24.** $\dfrac{3}{4}, \dfrac{1}{6}, \dfrac{5}{12}$        **25.** $\dfrac{1}{3}, \dfrac{1}{6}, \dfrac{5}{18}$        **26.** $\dfrac{1}{2}, \dfrac{3}{8}, \dfrac{7}{16}$

*Find each sum. Reduce fractions to lowest terms.*

**27.** $\dfrac{1}{4} + \dfrac{1}{4}$          **28.** $\dfrac{1}{2} + \dfrac{1}{2}$          **29.** $\dfrac{2}{4} + \dfrac{3}{4}$

**30.** $1\dfrac{1}{3} + \dfrac{2}{3}$          **31.** $5\dfrac{1}{4} + \dfrac{1}{4}$          **32.** $3\dfrac{1}{8} + 2\dfrac{3}{8}$

**33.** $3\dfrac{1}{4} + 2\dfrac{1}{4} + 1\dfrac{3}{4}$      **34.** $\dfrac{1}{8} + 5 + 3\dfrac{3}{8}$      **35.** $2\dfrac{11}{16} + \dfrac{1}{16} + 3\dfrac{3}{16}$

**36.** $\dfrac{2}{3} + \dfrac{2}{9}$          **37.** $\dfrac{1}{4} + \dfrac{3}{8}$          **38.** $\dfrac{1}{2} + \dfrac{7}{16}$

*Change each mixed number to an improper fraction.*

**39.** $1\dfrac{1}{4}$      **40.** $2\dfrac{1}{2}$      **41.** $3\dfrac{1}{3}$      **42.** $2\dfrac{5}{8}$

**43.** $4\dfrac{2}{5}$      **44.** $5\dfrac{3}{4}$      **45.** $5\dfrac{3}{7}$      **46.** $2\dfrac{7}{16}$

**47.** $9\dfrac{3}{10}$      **48.** $3\dfrac{7}{10}$      **49.** $5\dfrac{17}{100}$      **50.** $3\dfrac{93}{100}$

*Change each improper fraction to a mixed number.*

**51.** $\dfrac{3}{2}$      **52.** $\dfrac{19}{4}$      **53.** $\dfrac{19}{8}$      **54.** $\dfrac{7}{3}$

**55.** $\dfrac{27}{10}$      **56.** $\dfrac{9}{4}$      **57.** $\dfrac{27}{5}$      **58.** $\dfrac{14}{3}$

*Find each difference. Reduce fractions to lowest terms.*

**59.** $\dfrac{3}{4} - \dfrac{1}{4}$          **60.** $\dfrac{5}{6} - \dfrac{3}{6}$          **61.** $\dfrac{6}{7} - \dfrac{1}{7}$

**62.** $2\frac{7}{8} - 1\frac{1}{8}$          **63.** $3\frac{11}{16} - 2\frac{1}{16}$          **64.** $4\frac{9}{32} - 4\frac{7}{32}$

**65.** $\frac{4}{9} - \frac{1}{3}$          **66.** $\frac{9}{16} - \frac{3}{8}$          **67.** $\frac{17}{20} - \frac{1}{2}$

**68.** $\frac{7}{8} - \frac{5}{8} + \frac{1}{2}$          **69.** $\frac{7}{8} + \frac{5}{8} - \frac{1}{2}$          **70.** $\frac{13}{20} - \frac{3}{20} - \frac{5}{10}$

*Find each product. Reduce fractions to lowest terms.*

**71.** $\frac{1}{2} \times \frac{1}{2}$          **72.** $\frac{1}{2} \times \frac{2}{3}$          **73.** $\frac{3}{5} \times \frac{5}{3}$

**74.** $\frac{2}{3} \times 9$          **75.** $\frac{5}{8}$ of 16          **76.** $\frac{3}{4}$ of 12

**77.** $\frac{3}{5} \times 5 \times \frac{2}{3}$          **78.** $7 \times \frac{2}{7} \times \frac{1}{2}$          **79.** $\frac{1}{2} \times \frac{2}{3} \times 3$

*Give the reciprocal of each number.*

**80.** $\frac{4}{5}$      **81.** 4      **82.** $\frac{1}{5}$      **83.** $\frac{3}{2}$      **84.** 1

*Find each quotient. Reduce fractions to lowest terms.*

**85.** $\frac{4}{5} \div 4$      **86.** $\frac{1}{8} \div \frac{1}{4}$      **87.** $4 \div \frac{1}{3}$      **88.** $12 \div \frac{2}{3}$

■ **Review Exercises**

**Section 2.1**

*Determine whether the first number is or is not a factor of the second number.*

**1.** 26;  936      **2.** 48;  884

*Write each number in completely factored form.*

**3.** 42    **4.** 66    **5.** 407    **6.** 428

## Section 2.2

*Write the decimal equivalent of each fraction.*

**7.** $\dfrac{11}{32}$    **8.** $\dfrac{22}{19}$    **9.** $2\dfrac{5}{16}$    **10.** $93\dfrac{3}{8}$

*Use decimal equivalents to perform the following computations. Round off answers to the nearest thousandth.*

**11.** $3\dfrac{2}{3} + 2\dfrac{7}{16} + 7\dfrac{3}{4}$    **12.** $27\dfrac{3}{16} - 16\dfrac{1}{32}$

**13.** $7 \times 2\dfrac{3}{8} + 3 \times 6\dfrac{3}{4}$    **14.** $58\dfrac{1}{5} \div 89\dfrac{5}{12}$

## Section 2.3

*State whether the fractions in each pair are or are not equal.*

**15.** $\dfrac{5}{16}, \ \dfrac{25}{80}$    **16.** $\dfrac{15}{32}, \ \dfrac{31}{64}$

*Reduce each fraction to lowest terms.*

**17.** $\dfrac{22}{72}$    **18.** $\dfrac{24}{36}$    **19.** $\dfrac{75}{105}$    **20.** $\dfrac{72}{128}$

*Build each fraction to an equal fraction with the given denominator.*

**21.** $\dfrac{3}{8} = \dfrac{?}{32}$     **22.** $\dfrac{5}{12} = \dfrac{?}{180}$

## Section 2.4

*Write each group of fractions as like fractions, with the LCD as the common denominator.*

**23.** $\dfrac{2}{5}, \ \dfrac{3}{7}$     **24.** $\dfrac{3}{4}, \ \dfrac{5}{12}$     **25.** $\dfrac{2}{3}, \ \dfrac{5}{6}, \ \dfrac{2}{15}$     **26.** $\dfrac{4}{1}, \ \dfrac{5}{18}, \ \dfrac{7}{24}$

*Write each set of fractions in order, from the smallest to the greatest.*

**27.** $\dfrac{6}{13}, \ \dfrac{5}{13}, \ \dfrac{8}{13}$     **28.** $\dfrac{2}{5}, \ \dfrac{1}{7}, \ \dfrac{9}{25}$

*Find each sum.*

**29.** $\dfrac{5}{24} + \dfrac{7}{24}$     **30.** $\dfrac{3}{20} + \dfrac{5}{12} + \dfrac{1}{6}$

*Compute each of the following.*

**31.** $\dfrac{25}{36} - \dfrac{11}{36} - \dfrac{7}{36}$     **32.** $\dfrac{27}{32} - \dfrac{3}{8} - \dfrac{1}{4}$

## Section 2.5

**33.** Change $6\dfrac{3}{8}$ to an improper fraction.

**34.** Change $\dfrac{137}{15}$ to a mixed number.

*Compute each of the following.*

**35.** $3\dfrac{2}{7} + 6\dfrac{3}{7}$     **36.** $15\dfrac{5}{16} + 2\dfrac{3}{8} + 4\dfrac{3}{4}$     **37.** $9\dfrac{25}{32} - 2\dfrac{5}{32}$

**38.** $15\dfrac{5}{6} - 3\dfrac{1}{4}$     **39.** $18 - 6\dfrac{5}{6}$     **40.** $5\dfrac{2}{3} + 3\dfrac{1}{5} - 4\dfrac{1}{6}$

**Section 2.6**

*Find each product.*

**41.** $\dfrac{3}{4} \times \dfrac{2}{5}$     **42.** $\dfrac{6}{25} \times \dfrac{5}{12}$     **43.** $\dfrac{3}{7} \times \dfrac{14}{15} \times \dfrac{5}{8}$     **44.** $16 \times 3\dfrac{1}{4}$

*Write the reciprocal of each number.*

**45.** $\dfrac{3}{7}$     **46.** $2\dfrac{3}{5}$

*Compute each of the following.*

**47.** $\dfrac{15}{28} \div \dfrac{3}{7}$     **48.** $\dfrac{7}{8} \div 14$     **49.** $3\dfrac{1}{2} \div 1\dfrac{3}{4}$     **50.** $2\dfrac{3}{16} \div 2\dfrac{1}{2}$

# *Signed Numbers*

## *3.1 Graphical Representation; Absolute Value*

Numbers greater than zero, which are used for most applications, are called **positive numbers**. However, sometimes it is helpful to use numbers that are less than zero to express quantities such as temperatures below zero, distances below sea level, overdrawn bank accounts, and so forth. Such numbers are called **negative numbers** and are represented by numbers preceded by minus signs. For example, a temperature of 18 degrees below zero can be represented by $-18°$, and a distance of 282 feet below sea level can be represented by $-282$ feet. Positive numbers are sometimes preceded by a plus sign. For instance, 18 degrees above zero can be represented by $+18°$, and 282 feet above sea level can be represented by $+282$ feet.

Positive and negative numbers, together, are called **signed numbers**. Although zero (0) is not usually preceded by any sign, it may be considered either positive or negative and included as a signed number. The numbers

$$\ldots -3, -2, -1, 0, +1, +2, +3 \ldots$$

are called **integers**.

### ■ *Number Lines*

You may recall that a number line, such as that shown in Figure 3.1, is used to associate (pair) numbers with points on a line.

**FIGURE 3.1**

The point associated with a number is called the **graph** of the number; the number is called the **coordinate** of the point. The point associated with zero is called the **origin**. Positive numbers are associated with points to the right of the origin, and negative numbers are associated with points to the left of the origin.

**EXAMPLE 1**   The numbers −7, −4, +2, +4, and +6 can be graphed as shown in Figure 3.2.

**FIGURE 3.2**

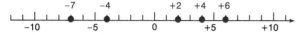

## ■ *Opposite of a Signed Number*

Given any nonzero number, the number that results from changing only the sign is the opposite of the number. For example, the opposite of +7 is −7 and the opposite of −12 is +12.

A phrase such as "For any signed number *r*" means that *r* represents any number—positive, negative, or zero; the symbol −*r* means the opposite of the number *r*.

**EXAMPLE 2**
**(a)** If *r* represents +3, then −*r*, the opposite of +3, is −3.
**(b)** If *r* represents −3, then −*r*, the opposite of −3, is +3.

Example 2 illustrates the fact that a variable preceded by a minus sign can represent either a positive number or a negative number.

The phrase "the opposite of −3 is +3" can be written in symbols as

$$-(-3) = +3$$

In general, we have the following property. For any signed number *r*,

$$-(-r) = r$$

Note that we have now used the minus sign in three ways:

**1.** To indicate a subtraction: +8 − (+4)
**2.** To indicate a negative number: −4
**3.** To indicate the opposite of a number: −(+4) or −(−4)

## ■ *Greater Than; Less Than*

Referring to a number line, it is traditional to use the terms "number" and "point" interchangeably. For example, in Figure 3.3, we may simply refer to −9, −2, +3, and +8 on the number line to mean the points associated with these numbers. This idea is particularly convenient when we consider the relative *order* of two signed numbers. Observe in Figure 3.3 that +8 is to the right of +3. We say that +8 is greater than +3 and that +3 is less than +8. Consistent with this notion for positive numbers, we consider all signed numbers to *increase* to the right and to *decrease* to the left on a number line. Thus, +3 is greater than −2, and −2 is less than +3. Also −2 is greater than −9, and −9 is less than −2. More generally, if a number *c* is to the right of a number *b*, as in Figure 3.4, then *c is greater than b*, and we write

$$c > b$$

Also, *b is less than c*, and we write

$$b < c$$

**FIGURE 3.3**

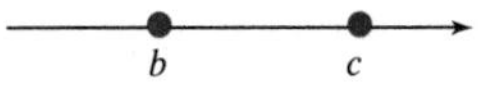

**FIGURE 3.4**

**EXAMPLE 3**  From Figure 3.3:
(a) $+8 > +3$  and  $+3 < +8$
(b) $+3 > -2$  and  $-2 < +3$
(c) $-2 > -9$  and  $-9 < -2$

**EXAMPLE 4**  For each pair of numbers, replace the comma with either $<$ or $>$.
(a) $+4.6, +3$   (b) $-4.6, -3$   (c) $+4.6, -3$

**Solutions**  You may either visualize the numbers on a number line or actually draw a number line in order to compare each pair of numbers. See Figure 3.5.

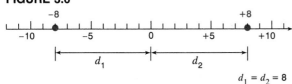

**FIGURE 3.5**

(a) $+4.6$ is to the right of $+3$;  hence,  $+4.6 > +3$
(b) $-4.6$ is to the left of $-3$;  hence,  $-4.6 < -3$
(c) $+4.6$ is to the right of $-3$;  hence,  $+4.6 > -3$

### ∎ Absolute Value of a Number

In Figure 3.6, observe that although $+8$ and $-8$ are in opposite directions from the origin, each of the two numbers is the same distance (eight units) from the origin. The positive number 8 that specifies the distance between $+8$ and the origin, or $-8$ and the origin, is called the **absolute value** of $+8$ and $-8$. A symbol involving two vertical bars is used to name the absolute value of a number. For example,

$$|+8| = +8 = 8 \quad \text{and} \quad |-8| = +8 = 8$$

**FIGURE 3.6**

More generally, the absolute value of a number $r$, represented by $|r|$, is the positive number (or zero) that specifies the distance between $r$ and the origin. It follows that *the absolute value of any signed number is either positive or zero.*

**EXAMPLE 5**
(a) $|+4.53| = 4.53$   (b) $|-4.53| = 4.53$   (c) $|0| = 0$

**EXAMPLE 6**  Specify the number in each pair that has the lesser absolute value.
(a) $-28; +15$   (b) $-19; +34$

**Solutions**
(a) $|-28| = 28$,  $|+15| = 15$,  and  $15 < 28$.  Hence, $+15$ has the lesser absolute value.
(b) $|-19| = 19$,  $|+34| = 34$,  and  $19 < 34$.  Hence, $-19$ has the lesser absolute value.

Note that we have written positive numbers with or without the $(+)$ sign. We will continue to do so as convenient.

## Exercises for Section 3.1

*Graph each set of numbers on a number line. See Example 1.*

**1.** −8, −3, 0, 1, 5      **2.** −6, −4, −2, 2, 4

**3.** −5, −1, 4, 6, 8      **4.** −9, −1, 0, 2, 6

**5.** −10, −5, 0, 5, 10      **6.** −8, −4, 0, 4, 8

*Write the opposite of each number. See Example 2.*

**7.** +4      **8.** +2      **9.** +5      **10.** +8

**11.** −6      **12.** −9      **13.** −10      **14.** −1

**15.** +7      **16.** +12      **17.** −8      **18.** −13

*Referring to the given number line, replace each comma with either < or >. See Example 3.*

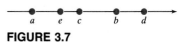

**FIGURE 3.7**

**19.** *a, b*      **20.** *e, a*      **21.** *c, d*

**22.** *b, c*      **23.** *d, e*      **24.** *a, d*

*For each pair of numbers, replace the comma with either < or >. See Example 4.*

**25.** −15, +8      **26.** −12, +5      **27.** +8, +17

**28.** +4, +7      **29.** −0.9, −1.2      **30.** −0.2, −1.5

**31.** −10, −7   **32.** −13, −1   **33.** +1.1, +0.3

**34.** +1.5, +0.6   **35.** +1, −10   **36.** +2, −5

*Write the indicated absolute value. See Example 5.*

**37.** $|+3.2|$   **38.** $|+5|$   **39.** $|-2.15|$

**40.** $|-8.3|$   **41.** $|-6.4|$   **42.** $|-7.02|$

*Specify the number in each pair that has the lesser absolute value. See Example 6.*

**43.** −8; −6   **44.** −5; −2   **45.** +4; −6

**46.** +5; −7   **47.** −2.4; +3.5   **48.** −5.6; +7.4

## 3.2 Addition; Subtraction

In this section we consider methods for finding sums and differences of signed numbers. Computations in the first part of this section should be done mentally.

### ■ Adding Signed Numbers

The sum of two signed numbers is obtained by the following rule:

---

**To Add Two Signed Numbers**

If the numbers have the *same sign*:
  **1.** Add their absolute values.
  **2.** Precede the sum by the same sign.

If the numbers have *opposite signs*:
  **1.** Subtract the smaller absolute value from the larger absolute value.
  **2.** Precede the difference by the sign of the number with the larger absolute value.

---

   It may be helpful to view this rule using the number line. Positive numbers are viewed as moving to the right and negative numbers to the left as illustrated in the following example.

**EXAMPLE 1**  Add.
**(a)** $(+7) + (+5)$   **(b)** $(-7) + (-5)$   **(c)** $(+7) + (-5)$   **(d)** $(-7) + (+5)$

## Solutions

**(a)** The numbers have the same sign (both +). Add the absolute values $(7 + 5)$ to obtain 12, and precede this sum with a (+) sign.

$$(+7) + (+5) = +(7 + 5) = +12$$

**FIGURE 3.8**

**(b)** The numbers have the same sign (both −). Add the absolute values $(7 + 5)$ to obtain 12, and precede this sum with a (−) sign.

$$(-7) + (-5) = -(7 + 5) = -12$$

**FIGURE 3.9**

**(c)** The numbers have opposite signs, and the absolute value of −5 is less than the absolute value of +7. Subtract the smaller absolute value from the larger absolute value $(7 - 5)$ to obtain 2. Because +7 has a larger absolute value than −5, precede the difference by a (+) sign.

$$(+7) + (-5) = +(7 - 5) = +2$$

**FIGURE 3.10**

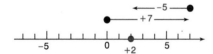

**(d)** The numbers have opposite signs, and the absolute value of +5 is less than the absolute value of −7. Subtract the smaller absolute value from the larger absolute value $(7 - 5)$ to obtain 2. Because −7 has a larger absolute value than +5, precede the difference by a (−) sign.

$$(-7) + (+5) = -(7 - 5) = -2$$

**FIGURE 3.11**

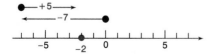

Now consider the sums

$$(+7) + (-7) = 0 \quad \text{and} \quad (+12) + (-12) = 0$$

In both cases we obtained zero. In general, the sum of any number and its opposite is zero. In symbols, this useful concept is stated as follows. For any signed number $r$,

$$(+r) + (-r) = 0$$

It is not customary to enclose all signed numbers in parentheses, except in those cases where confusion may arise. For example, we write $7 + 5$ rather than $(+7) + (+5)$; however, we write $7 + (-5)$ rather than the confusing notation $7 + -5$.

**EXAMPLE 2** Using only necessary (+) signs and parentheses, the sums in Example 1 appear as follows:

(a) $7 + 5 = 12$   (b) $-7 + (-5) = -12$   (c) $7 + (-5) = 2$   (d) $-7 + 5 = -2$

## ■ Subtracting Signed Numbers

The difference of two signed numbers can be obtained by first changing the difference to a sum.

---

### To Subtract Two Signed Numbers

**1.** Replace the number to be subtracted with its opposite.
**2.** Add the resulting numbers.

---

In symbols, this rule can be written as follows. For all signed numbers $r$ and $s$,

$$r - s = r + (-s)$$

**EXAMPLE 3** Find each difference.
(a) $7 - 5$      (b) $-7 - 5$      (c) $7 - (-5)$      (d) $-7 - (-5)$

**Solutions**   In Example 3a and b, replace 5 by $-5$ and then add the two numbers.
(a) $7 - 5 = 7 + (-5)$          (b) $-7 - 5 = -7 + (-5)$
　　　$= +(7 - 5) = 2$                  $= -(7 + 5) = -12$

In Example 3c and d, replace $-5$ by 5 and then add the two numbers.

(c) $7 - (-5) = 7 + 5$          (d) $-7 - (-5) = -7 + 5$
　　　　　$= 12$                          $= -(7 - 5) = -2$

## ■ Calculator Additions and Subtractions; the $\boxed{+/_-}$ Key

Scientific calculators have a *sign change key*, $\boxed{+/_-}$, that can be used to change the sign of a number in the display. This enables us to enter a negative number directly into the display. The $\boxed{+/_-}$ key is pressed *after* the absolute value of a number has been entered.

**EXAMPLE 4** Enter each number in the display.
(a) $-23.18$      (b) $-0.00475$

**Solutions**
(a) $23.18 \boxed{+/_-} \longrightarrow -23.18$      (b) $0.00475 \boxed{+/_-} \longrightarrow -0.00475$

The sign change key enables us to use a calculator to find sums and differences of signed numbers.

**EXAMPLE 5**   Compute.

**(a)** $10.8 + (-12.2) - (-4.9) + 13.6 = 17.1$;   this is computed by

$$10.8 \boxed{+} 12.2 \boxed{^+/_-} \boxed{-} 4.9 \boxed{^+/_-} \boxed{+} 13.6 \boxed{=} \longrightarrow 17.1$$

**(b)** $-99.13 + (-7.56) - 8.45 - (-18.03) = -97.11$;   this is computed by

$$99.13 \boxed{^+/_-} \boxed{+} 7.56 \boxed{^+/_-} \boxed{-} 8.45 \boxed{-} 18.03 \boxed{^+/_-} \boxed{=} \longrightarrow -97.11$$

## Exercises for Section 3.2

*Rewrite each expression as a sum, using only necessary (+) signs and parentheses, and then compute mentally. See Examples 1 and 2.*

**1.** $(+8) + (+6)$        **2.** $(+7) + (+2)$        **3.** $(+3) + (+3)$

**4.** $(+4) + (+5)$        **5.** $(-6) + (-7)$        **6.** $(-5) + (-3)$

**7.** $(-8) + (-1)$        **8.** $(-9) + (-2)$        **9.** $(+7) + (-4)$

**10.** $(+9) + (-3)$       **11.** $(+4) + (-8)$       **12.** $(+6) + (-9)$

**13.** $(-5) + (+9)$       **14.** $(-3) + (-8)$       **15.** $(-4) + (+6)$

**16.** $(-7) + (+3)$       **17.** $(-6) + (+6)$       **18.** $(+8) + (-8)$

*Rewrite each expression, without using parentheses, and compute mentally. See Examples 2 and 3.*

**19.** $(-5) + (-2)$       **20.** $(-4) + (-3)$       **21.** $(-7) - (-4)$

**22.** $(-6) - (-1)$       **23.** $(-9) - (-4)$       **24.** $(-8) - (+2)$

**25.** $9 - (-2)$          **26.** $10 - (-4)$         **27.** $-10 + (-3)$

**28.** $-12 + (-5)$        **29.** $-6 - (-4)$         **30.** $-7 - (-6)$

*Compute, using a calculator. See Examples 4 and 5.*

**31.** $-14.3 - 11.9$  **32.** $-21.39 - 16.3$  **33.** $6.3 + (-2.1)$

**34.** $17 + (-5.5)$  **35.** $0.74 - (-0.12)$  **36.** $1.34 - (-1.66)$

**37.** $6.87 + 6.39 + (-5.76)$  **38.** $88.9 + 9.4 + (-4.96)$

**39.** $46.7 + (-4.63) - (-3.36)$  **40.** $9.97 + (-9.3) - (-6.66)$

**41.** $-5.46 - (-8.49) + (-3.6)$  **42.** $(-8.5) + (-13.2) + (-6.43)$

**43.** $7.8 + (-4.47) - (-25.1) - 8.36$  **44.** $83.61 + (-41.34) - (-3.81) - 60.95$

**45.** $-47.74 - 13.04 + (-8.89) + 61.13$  **46.** $-88.56 - 94.23 + (-86.56) + 84.62$

*Find the average of each set of temperature readings to the nearest tenth of a degree.*

**47.** $5°, \ -4°, \ -1°, \ 8°$  **48.** $38°, \ -23°, \ 35°, \ -17°$

**49.** $-18°, \ 16°, \ 20°, \ -3°, \ -11°$  **50.** $-28°, \ 22°, \ 39°, \ -47°, \ 4°$

**51.** $-6°, \ -5°, \ -1°, \ 7°, \ -3°$  **52.** $-4°, \ -1°, \ -7°, \ 9°, \ -7°$

## 3.3 Multiplication; Division; Combined Operations

The rules for finding products and quotients of signed numbers are very much alike.

> ### To Multiply or Divide Two Signed Numbers
> If the numbers have the *same sign*:
> 1. Multiply (divide) the absolute values.
> 2. Precede the product (quotient) by a $+$ sign.
>
> If the numbers have *opposite signs*:
> 1. Multiply (divide) the absolute values.
> 2. Precede the product (quotient) by a $-$ sign.

You may find it helpful to remember the following sign patterns that are determined by the above rules.

$$(+) \cdot (+) = (+) \qquad (-) \cdot (-) = (+) \qquad (+) \cdot (-) = (-) \qquad (-) \cdot (+) = (-)$$

$$\frac{(+)}{(+)} = (+) \qquad\qquad \frac{(-)}{(-)} = (+) \qquad\qquad \frac{(+)}{(-)} = (-) \qquad\qquad \frac{(-)}{(+)} = (-)$$

**EXAMPLE 1**
**(a)** $8 \cdot 2 = +(8 \cdot 2)$
$\qquad = 16$
**(b)** $-8 \cdot (-2) = +(8 \cdot 2)$
$\qquad\qquad\quad = 16$
**(c)** $8 \cdot (-2) = -(8 \cdot 2)$
$\qquad\qquad = -16$
**(d)** $-8 \cdot 2 = -(8 \cdot 2)$
$\qquad\qquad = -16$

**EXAMPLE 2**
**(a)** $\dfrac{8}{2} = +\dfrac{8}{2} = 4$
**(b)** $\dfrac{-8}{-2} = +\dfrac{8}{2} = 4$
**(c)** $\dfrac{8}{-2} = -\dfrac{8}{2} = -4$
**(d)** $\dfrac{-8}{2} = -\dfrac{8}{2} = -4$

As in Section 3.2, the $\boxed{+/_-}$ key can be used to enter negative numbers in the display. Thus, when computing products or quotients of signed numbers, the sign of the final result is given by the calculator.

**EXAMPLE 3**  Compute:

$$\frac{5.061(-3.12)}{-4.7}$$

**Solution**   The on-line form is $5.061 \times (-3.12) \div (-4.7)$. Use the sequence

$$5.061 \;\boxed{\times}\; 3.12 \;\boxed{+/_-}\; \boxed{\div}\; 4.7 \;\boxed{+/_-}\; \boxed{=} \longrightarrow 3.359642$$

**EXAMPLE 4**  Compute:

$$\frac{-5.4(6.28)(-9.13)}{-7.01}$$

**Solution**   The on-line form is $-5.4 \times 6.28 \times (-9.13) \div (-7.01)$. Use the sequence

$$5.4 \;\boxed{+/_-}\; \boxed{\times}\; 6.28 \;\boxed{\times}\; 9.13 \;\boxed{+/_-}\; \boxed{\div}\; 7.01 \;\boxed{+/_-}\; \boxed{=} \longrightarrow -44.16784$$

## ▪ Powers

Some calculators will not accept a negative number as a base when using the $\boxed{y^x}$ key to raise a negative number to a power. In that case, the *sign* of the resulting power must be determined separately; the *absolute value* of the power is computed on the calculator. To decide upon the sign, we use the following rule:

---
**To Find the Sign of $b^n$**

If $b$ is a negative number and $n$ is a natural number, then

   **1.** $b^n$ is a positive number if $n$ is even $(2, 4, 6, \ldots)$.
   **2.** $b^n$ is a negative number if $n$ is odd $(1, 3, 5, \ldots)$.
---

**EXAMPLE 5**  Compute each power: **(a)** $(-5)^4$  **(b)** $(-5)^3$

**Solutions**
**(a)**
$$5 \boxed{y^x} \, 4 \boxed{=} \longrightarrow 625$$

Because 4 is an even number, $(-5)^4$ is a positive number. Hence,
$$(-5)^4 = 625$$

**(b)**
$$5 \boxed{y^x} \, 3 \boxed{=} \longrightarrow 125$$

Because 3 is an odd number, $(-5)^3$ is a negative number. Hence,
$$(-5)^3 = -125$$

## ■ Combined Operations

As we noted in Section 1.6, computations that involve combined operations, which are governed by the specified order of operations on page 25, require the use of the parentheses keys on a calculator.

**EXAMPLE 6**  Compute:
$$\frac{6.3^2 - 4.1(3.6)}{2.1}$$

**Solution**  The on-line form is $[6.3^2 - 4.1(3.6)] \div 2.1$. Use the sequence
$$\boxed{(} \, 6.3 \, \boxed{x^2} \, \boxed{-} \, 4.1 \, \boxed{\times} \, 3.6 \, \boxed{)} \, \boxed{\div} \, 2.1 \longrightarrow 11.871429$$

The result is 11.9 to the nearest tenth.

**EXAMPLE 7**  Compute:
$$\frac{(-3.1)^3 + 8.4}{2.3}$$

**Solution**  The on-line form is $[(-3.1)^3 + 8.4] \div 2.3$. Use the sequence
$$\boxed{(} \, 3.1 \, \boxed{+/_-} \, \boxed{y^x} \, 3 \, \boxed{+} \, 8.4 \, \boxed{)} \, \boxed{\div} \, 2.3 \longrightarrow -9.3004348$$

The result is $-9.3$ to the nearest tenth.

## Exercises for Section 3.3

*Compute mentally. See Examples 1 and 2.*

**1.** $3 \cdot (-4)$  **2.** $2 \cdot (-4)$  **3.** $(-2) \cdot 5$  **4.** $(-3) \cdot 6$

**5.** $(-5) \cdot (-3)$  **6.** $(-7) \cdot (-2)$  **7.** $\dfrac{12}{-2}$  **8.** $\dfrac{14}{-7}$

**9.** $\dfrac{-16}{-4}$  **10.** $\dfrac{-18}{-9}$  **11.** $\dfrac{-6}{3}$  **12.** $\dfrac{-10}{5}$

*Compute. Round off to no more than three decimal places.*

**13.** $-43 \times (-9.3)$        **14.** $-62 \times (-5.7)$        **15.** $-1.81 \times (-0.39)$

**16.** $-6.2 \times (-3.3)$        **17.** $-1.6 \times (3.45)$        **18.** $1.5 \times (-3.8)$

**19.** $\dfrac{-5.2}{-7}$        **20.** $\dfrac{-6.3}{-2.9}$        **21.** $\dfrac{-0.356}{8.91}$

**22.** $\dfrac{-0.51}{15.1}$        **23.** $\dfrac{1.32}{-2.2}$        **24.** $\dfrac{6.6}{-5.17}$

*Compute. See Examples 3 and 4. Round off to three significant digits.*

**25.** $\dfrac{6.165 \times 0.93}{-1.73}$        **26.** $\dfrac{3.28 \times 6.158}{-0.87}$

**27.** $\dfrac{-9.21(-5.9)}{-0.72}$        **28.** $\dfrac{-64.9(-0.384)}{-37.8}$

**29.** $\dfrac{19.2(-5.9)}{-0.83}$        **30.** $\dfrac{1.19(-76.3)}{-20.6}$

**31.** $\dfrac{-1.59 \times 43.8 \times (-2.56)}{71.79}$        **32.** $\dfrac{-14.9 \times 56.8 \times (-0.094)}{25.29}$

**33.** $\dfrac{9.36(-3.49)(-7.15)}{-49.1}$        **34.** $\dfrac{74.8(-0.812)(-5.1)}{-27.5}$

**35.** $\dfrac{-41.5(-4.63)(-.125)}{-0.954}$        **36.** $\dfrac{-1.85(-0.47)(-3.43)}{-0.373}$

*Compute each power. Write answers in floating decimal point. See Example 5.*

**37.** $(-2)^5$        **38.** $(-3)^7$        **39.** $(-1.21)^8$        **40.** $(-5.37)^6$

**41.** $(-11.3)^4$     **42.** $(-13.2)^4$     **43.** $(-0.26)^3$     **44.** $(-0.125)^5$

*Compute. See Examples 6 and 7. Round off to the nearest hundredth.*

**45.** $\dfrac{12.5 - 6(-6.5)}{8.4}$

**46.** $\dfrac{16.4 - 3(-9.1)}{14}$

**47.** $\dfrac{8(-8.9) + 1.6}{1.9}$

**48.** $\dfrac{1.9(-5.9) + 2.5}{6.2}$

**49.** $7.9^2 - 0.8(-4.2)(2.5)$

**50.** $9.6^2 - 0.9(6.9)(-.8)$

**51.** $(-3.4)^2 + 6.9(-3.9)(4.4)$

**52.** $(-6.2)^2 + 2.8(-7.6)(3.1)$

**53.** $\dfrac{8.4(-0.85)(-5.2) + (-9.8)^2}{-8.8}$

**54.** $\dfrac{7.2(-0.82)(-8.1) + (-7.7)^2}{-5.7}$

## 3.4 Integer Exponents; Laws of Exponents

Recall from Section 1.5 that we gave a special meaning to powers of 10 with negative exponents:

$$10^{-1} = \frac{1}{10}, \quad 10^{-2} = \frac{1}{10^2}, \quad 10^{-3} = \frac{1}{10^3}, \quad \cdots$$

A power of $10^{-n}$ is just the reciprocal of $10^n$. We also use negative exponents with other bases in the same way.

### ■ Negative Integer Exponents

For example, we can use $3^{-1}$ to represent $\dfrac{1}{3^1}$ and $3^{-5}$ to represent $\dfrac{1}{3^5}$. More generally, the meaning of a power with a negative exponent is given by the following rule. If $b \neq 0$ and $n$ is any integer, then

$$b^{-n} = \frac{1}{b^n} \quad \text{and} \quad b^n = \frac{1}{b^{-n}}$$

A number with a negative exponent is sometimes said to be "a number raised to a negative power."

### EXAMPLE 1

**(a)** $4^{-2} = \dfrac{1}{4^2}$     **(b)** $4^2 = \dfrac{1}{4^{-2}}$     **(c)** $\dfrac{1}{10^{-3}} = 10^3$     **(d)** $\dfrac{1}{10^3} = 10^{-3}$

The $\boxed{y^x}$ key, together with the $\boxed{^{+/}\!_-}$ key, can be used to obtain the decimal form of a number raised to a negative integer exponent.

**EXAMPLE 2**  $5^{-3} = 0.008$;  this is computed by

$$5 \boxed{y^x} 3 \boxed{^{+/}\!_-} \boxed{=} \longrightarrow 0.008$$

## ■ First Law of Exponents

Consider the product $6^3 \cdot 6^2$, where the bases are the same, and note that the base 6 is first used as a factor *three* times and then *two* more times, for a total of *five* times. That is,

$$6^3 \cdot 6^2 = \overbrace{6 \cdot 6 \cdot 6} \cdot \overbrace{6 \cdot 6}$$
$$= 6^{3+2} = 6^5$$

This result can be verified by observing that

$$6^3 \cdot 6^2 = 216 \cdot 36 = 7776 \quad \text{and} \quad 6^5 = 7776$$

The preceding example suggests the following property of powers called the **first law of exponents**.

$$b^m \cdot b^n = b^{m+n}$$

**EXAMPLE 3**  Use the first law of exponents to write each product as a single power. Evaluate the result.
**(a)** $4^2 \cdot 4^5$      **(b)** $0.3^3 \times 0.3^2$

**Solutions**
**(a)** $4^2 \cdot 4^5 = 4^{2+5}$                    **(b)** $0.3^3 \times 0.3^2 = 0.3^{3+2}$
$\quad\quad = 4^7; \quad 4^7 = 16{,}384$                    $\quad\quad\quad = 0.3^5; \quad 0.3^5 = 0.00243$

The first law of exponents applies to *all* integer exponents—positive, negative, and zero.

**Example 4**
**(a)** $3^{-5} \cdot 3^7 = 3^{-5+7}$      **(b)** $3^2 \cdot 3^{-3} = 3^{2-3}$      **(c)** $2^3 \cdot 2^0 = 2^{3+0}$
$\quad\quad = 3^2$                         $\quad\quad\quad = 3^{-1}$                         $\quad\quad\quad = 2^3$

## ■ Second Law of Exponents

Consider the quotient

$$\frac{4^5}{4^2} = \frac{4 \cdot 4 \cdot 4 \cdot 4 \cdot 4}{4 \cdot 4}$$

By the fundamental principle of fractions,

$$\frac{4 \cdot 4 \cdot 4 \cdot \cancel{4} \cdot \cancel{4}}{\cancel{4} \cdot \cancel{4}} = 4 \cdot 4 \cdot 4 = 4^3$$

Note that the same result can be obtained by *subtracting* exponents, as follows:

$$\frac{4^5}{4^2} = 4^{5-2} = 4^3$$

Results such as these suggest the following **second law of exponents**.

$$\frac{b^m}{b^n} = b^{m-n}$$

**EXAMPLE 5**  Use the second law of exponents to write each quotient as a single power; evaluate the result.

(a) $\dfrac{3^6}{3^4}$      (b) $\dfrac{3^4}{3^{-2}}$

**Solutions**

(a) $\dfrac{3^6}{3^4} = 3^{6-4} = 3^2;$      $3^2 = 9$      (b) $\dfrac{3^4}{3^{-2}} = 3^{4-(-2)} = 3^{4+2} = 3^6;$      $3^6 = 729$

**EXAMPLE 6**  Use the second law of exponents to write each quotient as a single power; evaluate the result.

(a) $\dfrac{3^4}{3^6}$      (b) $\dfrac{3^{-2}}{3^4}$

**Solutions**

(a) $\dfrac{3^4}{3^6} = 3^{4-6} = 3^{-2};$   $3^{-2} = 0.11111111$      (b) $\dfrac{3^{-2}}{3^4} = 3^{-2-4} = 3^{-6};$   $3^{-6} = 0.00137174$

The first and the second laws of exponents can be applied to products and quotients involving more than two factors.

**EXAMPLE 7**  Use the first and/or the second law of exponents to write each expression as a single power; evaluate the result.

(a) $\dfrac{2^3 \cdot 2^{-5} \cdot 2^4}{2 \cdot 2^2 \cdot 2^{-6}}$      (b) $\dfrac{10^2 \cdot 10 \cdot 10^{-4}}{10^{-1} \cdot 10^5}$

**Solutions**

(a) $\dfrac{2^3 \cdot 2^{-5} \cdot 2^4}{2 \cdot 2^2 \cdot 2^{-6}} = \dfrac{2^{3-5+4}}{2^{1+2-6}} = \dfrac{2^2}{2^{-3}}$      (b) $\dfrac{10^2 \cdot 10 \cdot 10^{-4}}{10^{-1} \cdot 10^5} = \dfrac{10^{2+1-4}}{10^{-1+5}} = \dfrac{10^{-1}}{10^4}$

$\qquad\qquad = 2^{2-(-3)} = 2^5;$      $\qquad\qquad\qquad = 10^{-1-4} = 10^{-5};$

$\quad 2^5 = 32$      $\qquad\qquad 10^{-5} = 0.00001$

## Exercises for Section 3.4

*Write each expression using a positive exponent. See Example 1a and c.*

**1.** $2^{-5}$      **2.** $4^{-6}$      **3.** $5^{-2}$      **4.** $8^{-3}$

**5.** $\dfrac{1}{2^{-5}}$      **6.** $\dfrac{1}{4^{-6}}$      **7.** $\dfrac{1}{5^{-2}}$      **8.** $\dfrac{1}{8^{-3}}$

*Write each expression using a negative exponent. See Example 1b and d.*

**9.** $4^6$     **10.** $2^5$     **11.** $8^3$     **12.** $5^2$

**13.** $\dfrac{1}{4^6}$     **14.** $\dfrac{1}{2^5}$     **15.** $\dfrac{1}{8^3}$     **16.** $\dfrac{1}{5^2}$

*Compute. See Example 2. Round off to three significant digits.*

**17.** $2^{-5}$         **18.** $3^{-2}$         **19.** $4^{-2}$         **20.** $6^{-3}$

**21.** $(1.2)^{-3}$     **22.** $(3.1)^{-4}$     **23.** $(0.67)^{-3}$     **24.** $(0.92)^{-5}$

Because $b^{-n} = \dfrac{1}{b^n}$, the reciprocal key $\boxed{1/x}$ can be used, together with the $\boxed{y^x}$ key, to raise a number to a negative power.

*Use the $\boxed{1/x}$ key to compute each power. Round off to three significant digits.*

**EXAMPLE 8**   $6^{-3} = 0.00463$ to three significant digits, as computed by

$$6 \; \boxed{y^x} \; 3 \; \boxed{=} \; \boxed{1/x} \longrightarrow 0.0046296$$

**25.** $3^{-3}$         **26.** $2^{-6}$         **27.** $6^{-4}$         **28.** $4^{-3}$

**29.** $(1.3)^{-5}$     **30.** $(2.1)^{-3}$     **31.** $(0.29)^{-3}$     **32.** $(0.76)^{-4}$

*Write each expression as a single power; evaluate the result. See Examples 3 to 6.*
*Write answers in floating decimal point.*

**33.** $5^{-4} \cdot 5^7$             **34.** $6^{-3} \cdot 6^5$             **35.** $4^5 \cdot 4^{-2}$

**36.** $2^7 \cdot 2^{-4}$             **37.** $3^{-2} \cdot 3^{-1}$             **38.** $5^{-2} \cdot 5^{-2}$

**39.** $(1.2)^{-5} \times (1.2)^3$     **40.** $(0.78)^{-8} \times (0.78)^5$     **41.** $\dfrac{6^5}{6^2}$

**42.** $\dfrac{6^{10}}{6^7}$  **43.** $\dfrac{(0.45)^7}{(0.45)^3}$  **44.** $\dfrac{(1.5)^8}{(1.5)^6}$

**45.** $\dfrac{2^3}{2^{-2}}$  **46.** $\dfrac{2^5}{2^{-1}}$  **47.** $\dfrac{5^{-1}}{5^2}$

**48.** $\dfrac{5^{-2}}{5^3}$  **49.** $\dfrac{(6.2)^2}{(6.2)^6}$  **50.** $\dfrac{(0.76)^4}{(0.76)^7}$

*Write each expression as a single power; evaluate the result. See Example 7. Write answers in floating decimal point.*

**51.** $\dfrac{3^2 \cdot 3^{-4} \cdot 3^5}{3^{-1} \cdot 3^3}$  **52.** $\dfrac{4^5 \cdot 4^{-3} \cdot 4}{4^2 \cdot 4^{-2}}$

**53.** $\dfrac{(1.2)^3 \cdot (1.2)^6 \cdot (1.2)^{-5}}{(1.2) \cdot (1.2)^6 \cdot (0.2)^{-5}}$  **54.** $\dfrac{(0.7)^4 \cdot (0.7)^5 \cdot (0.7)^{-6}}{(0.7)^3 \cdot (0.7) \cdot (0.7)^{-2}}$

**55.** $\dfrac{5 \cdot 5^{-3} \cdot 5^5}{5^{-2} \cdot 5^4 \cdot 5^2}$  **56.** $\dfrac{8^{-4} \cdot 8 \cdot 8^6}{8^3 \cdot 8^{-1} \cdot 8^4}$

**57.** $\dfrac{10^3 \cdot 10^{-4} \cdot 10^{-2}}{10^{-8} \cdot 10^2}$  **58.** $\dfrac{10 \cdot 10^{-2} \cdot 10^5}{10^3 \cdot 10^{-4}}$

*Compute each power. Write answers in floating decimal point.*

**EXAMPLE 9**
**(a)** $(-2)^{-3}$    **(b)** $(-3)^{-6}$

**Solutions**
**(a)** $(-2)^{-3} = -0.125$,  by noting that a negative number raised to an odd power is negative, and using the sequence

$$2 \boxed{y^x} \, 3 \boxed{+/_-} \boxed{=} \longrightarrow 0.125$$

Note that 2 and *not* $-2$ is used as the base. (Some calculators will not accept a negative base.)

**(b)** $(-3)^{-6} = 0.00137174$,  by noting that a negative number raised to an even power is positive, and using the sequence

$$3 \boxed{y^x} 6 \boxed{+/_-} \boxed{=} \longrightarrow 0.00137174$$

**59.** $(-3)^{-4}$          **60.** $(-4)^{-6}$          **61.** $(-1.4)^{-5}$          **62.** $(-2.6)^{-3}$

**63.** $(-2.31)^{-3}$          **64.** $(-1.46)^{-3}$          **65.** $(-0.39)^{-2}$          **66.** $(-0.29)^{-4}$

## 3.5   Mental Calculations

In this section, mental practice on the basic operations with signed numbers is provided, together with practice in using the two laws of exponents.

### Exercises for Section 3.5

*Each exercise in this section is to be done mentally. Compute.*

**1.** $(+100) + (+100)$          **2.** $(+200) + (-100)$          **3.** $(+70) + (-40)$

**4.** $(-60) + (+30)$          **5.** $(-20) + (+50)$          **6.** $(+20) + (+50)$

**7.** $(-10) - (-40)$          **8.** $(-200) + (-30)$          **9.** $(+10) - (+30)$

**10.** $(-10) - (+30)$          **11.** $(-100) - (+200)$          **12.** $(+100) - (+200)$

**13.** $(-50) - (-60)$          **14.** $(-40) - (-30)$          **15.** $(+200) - (-50)$

**16.** $(+300) - (-100)$          **17.** $7 \cdot (-10)$          **18.** $(-8) \cdot (-100)$

**19.** $(-30) \cdot (+5)$          **20.** $(-40) \cdot (+6)$          **21.** $(-10) \cdot (-20)$

**22.** $8 \cdot (-20)$          **23.** $\dfrac{100 - (-100)}{2}$          **24.** $\dfrac{50 - (-50)}{10}$

**25.** $\dfrac{(-60) + (-40)}{-5}$    **26.** $\dfrac{(-30) + (-50)}{-4}$

**27.** $3^{-2}$    **28.** $2^{-3}$    **29.** $12 \cdot 2^{-1}$    **30.** $6 \cdot 3^{-1}$

**31.** $\dfrac{1}{2^{-1}}$    **32.** $\dfrac{1}{3^{-1}}$    **33.** $10^3 \times 10^{-2}$    **34.** $10^5 \times 10^{-3}$

**35.** $\dfrac{10^2}{10^{-1}}$    **36.** $\dfrac{10^{-1}}{10^2}$    **37.** $\dfrac{10^{-3} \times 10^6}{10^2}$    **38.** $\dfrac{10^{-5} \times 10^7}{10}$

**39.** $\dfrac{10^2 \times 10^{-3} \times 10}{10^{-2}}$    **40.** $\dfrac{10^{-3} \times 10^3 \times 10^{-1}}{10^{-2}}$

## ■ *Review Exercises*

### Section 3.1

**1.** Graph the numbers $-7, -4, -1, 0, 5$ on a number line.

**2.** Write the negatives of $+3$ and $-6$.

*For Exercises 3 and 4 refer to the number line and replace each comma with $<$ or $>$.*

**FIGURE 3.12**

**3.** $a, \quad b$    **4.** $d, \quad c$

*For each pair of numbers, replace the comma with $<$ or $>$.*

**5.** $+3, \quad -4$    **6.** $+3, \quad +4$    **7.** $-3, \quad -4$

*Write the indicated absolute value.*

**8.** $|-3.7|$    **9.** $|+4.2|$

*Specify the number in each pair that has the lesser absolute value.*

**10.** −3;  +2      **11.** −3;  −4      **12.** +4.1;  −5.1

## Section 3.2

*Rewrite each expression as a sum, using only necessary (+) signs and parentheses, and then compute mentally.*

**13.** (−6) + (2)      **14.** (4) + (−2)

**15.** (−9) + (2)      **16.** (−7) + (−4)

*Rewrite, without using parentheses, and compute mentally.*

**17.** (12) + (−2) − (−6)      **18.** (16) − (−4) + (−1) − (+2)

*Compute, using a calculator.*

**19.** 12.9 − 23.4                          **20.** 12.9 − (−23.4)

**21.** −12.9 − 23.4                         **22.** −12.9 − (−23.4)

**23.** −5.8 + (−23.7) + (−4.63)      **24.** −15.36 + 7.96 − 10.85 − (−43.81)

## Section 3.3

*Compute mentally.*

**25.** 5 · (−6)      **26.** (−3) · (−5)      **27.** $\dfrac{12}{-3}$      **28.** $\dfrac{-10}{-2}$

*Compute. Write answers in floating decimal point.*

**29.** $\dfrac{(-6.16)(3.12)}{-7.4}$                 **30.** $\dfrac{-4.5(2.86)(-13.9)}{-10.7}$

**31.** $(-3)^4$                          **32.** $(-7)^3$

**33.** $\dfrac{7.3 - 2(-4.5)}{13}$

**34.** $\dfrac{6.7 + (3.1)(-4.1)}{1.7}$

**35.** $(9.71)^2 - 3.2(-8.2)(-2.1)$    **36.** $(-4.3)^2 + (5.1)^2(-6.34)$

## Section 3.4

*Write each expression using a positive exponent.*

**37.** $5^{-3}$    **38.** $\dfrac{1}{4^{-2}}$

*Write each expression using a negative exponent.*

**39.** $\dfrac{1}{3^4}$    **40.** $6^3$

*Compute. Round off to four significant digits.*

**41.** $(2.1)^{-3}$    **42.** $(0.76)^{-2}$    **43.** $(-1.2)^{-4}$    **44.** $(-5.3)^{-3}$

*Write each expression as a single power; evaluate the result. Round off to the nearest thousandth.*

**45.** $3^5 \cdot 3^{-7}$    **46.** $(0.87)^7 \times (0.87)^{-4}$    **47.** $\dfrac{(0.15)^{-1}}{(0.15)^2}$    **48.** $\dfrac{3^4 \cdot 3^{-3} \cdot 3^{-6}}{3^{-3} \cdot 3^{-1} \cdot 3^4}$

## Section 3.5

*Compute each of the following mentally.*

**49.** $60 + (-30)$    **50.** $(-50) + (-20)$    **51.** $50 - (-40)$    **52.** $(-10) - (-30)$

**53.** $(-20) \cdot 5$    **54.** $(-30) \cdot (-2)$    **55.** $\dfrac{(-40) + (-20)}{-10}$    **56.** $\dfrac{(-10) - (-10)}{5}$

**57.** $10^5 \times 10^{-4}$    **58.** $10^{-6} \times 10^4$    **59.** $\dfrac{10^{-2} \times 10^4 \times 10^{-3}}{10^{-2}}$    **60.** $\dfrac{10^4 \times 10^{-3} \times 10^{-2}}{10^{-3}}$

# Problem-Solving Techniques

## 4.1 Solving Equations

To solve an equation involving a variable (an *unknown*), such as

$$n + 4 = 7 \quad \text{and} \quad 3n - 19 = 218$$

means finding a replacement value (number) for the unknown so that the equation will be a true statement. Such a number is called a **solution** of the equation. The equation $n + 4 = 7$ can be readily solved by "inspection." From our knowledge of basic addition facts we see that if $n$ is replaced by 3, we obtain the true statement $3 + 4 = 7$, and so 3 is the solution of the equation. However, we cannot immediately solve the equation $3n - 9 = 218$ by inspection. Hence, we need to develop methods to solve such equations.

### ■ The Addition-Subtraction Rule

The following rule can be helpful if the solution of a given equation is not evident by inspection.

---

**Addition-Subtraction Rule**

If a number is added to, or subtracted from, each side of an equation, the resulting equation has the same solutions as the original equation.

---

A goal to keep in mind when solving equations is to obtain an equation in which the unknown appears by itself on only one side of the equation. A solution to an equation can be *checked* by substituting the solution for the unknown in the equation and verifying that the result is a true statement.

**EXAMPLE 1**   Solve each equation.
**(a)** $n - 19 = 218$      **(b)** $n + 24.9 = 81.3$

**Solutions**
**(a)**                                          $n - 19 = 218$

   Add 19 to each side.

$$n \underbrace{-19 + 19}_{} = 218 + 19$$
$$n + \quad 0 \quad = \quad 237$$
$$n = 237$$

*Check:*   Substitute 237 for $n$ in the original equation.
          Does   $(237) - 19 = 218$?      Does   $218 = 218$?   Yes.

**(b)**                                          $n + 24.9 = 81.3$

   Subtract 24.9 from each side.

$$n + \underbrace{24.9 - 24.9}_{} = \underbrace{81.3 - 24.9}_{}$$
$$n + \quad 0 \quad = \quad 56.4$$
$$n = 56.4$$

*Check:*   Substitute 56.4 for $n$ in the original equation.
          Does   $(56.4) + 24.9 = 81.3$?      Does   $81.3 = 81.3$?   Yes.

   We can sometimes simplify an equation by finding sums and differences before applying the addition-subtraction rule.

**EXAMPLE 2**
**(a)**                               $r + 25.8 = \underbrace{19.7 - 5.08}_{}$
$$r + 25.8 = \quad 14.62$$

   Subtract 25.8 from each side.

$$r + \underbrace{25.8 - 25.8}_{} = 14.62 - 25.8$$
$$r + \quad 0 \quad = \quad -11.18$$
$$r = -11.18$$

*Check:*   Substitute $-11.18$ for $r$ in the original equation.
          Does   $(-11.18) + 25.8 = 19.7 - 5.08$?      Does   $14.62 = 14.62$?   Yes.

**(b)**                               $174.5 = 68.2 + s - 49.3$
$$174.5 = s + \underbrace{68.2 - 49.3}_{}$$
$$174.5 = s + \quad 18.9$$

   Subtract 18.9 from each side

$$\underbrace{174.5 - 18.9}_{} = s + \underbrace{18.9 - 18.9}_{}$$
$$155.6 \quad = s + \quad 0$$
$$155.6 = s$$

*Check*:   Substitute 155.6 for $s$ in the original equation.
          Does   $174.5 = 68.2 + (155.6) - 49.3$?      Does   $174.5 = 174.5$?   Yes.

## ■ The Multiplication-Division Rule

It is customary practice to omit multiplication symbols between a number and a variable and between two variables; in the remainder of this text we shall frequently follow this practice. For example, we may write $1.2t$ to mean $1.2 \times t$, and $rt$ to mean $r \cdot t$.

We now consider a second rule that can be used to change an equation to a simpler equation that has the same solutions.

### Multiplication-Division Rule

If each side of an equation is multiplied by, or divided by, the same nonzero number, the resulting equation has the same solutions as the original equation.

**EXAMPLE 3** (a) $1.2r = 5.64$   (b) $\dfrac{3y}{7} = 14.4$

**Solutions**

**(a)**
$$1.2r = 5.64$$

Divide each side by 1.2.

$$\frac{1.2r}{1.2} = \frac{5.64}{1.2}$$

Because $\dfrac{\cancel{1.2}\,r}{\cancel{1.2}}$ equals $r$, then

$$r = \frac{5.64}{1.2} = 4.7$$

*Check:* Substitute 4.7 for $r$ in the original equation.
Does $1.2 \times (4.7) = 5.64$?   Does $5.64 = 5.64$?   Yes.

**(b)**
$$\frac{3y}{7} = 14.4$$

Multiply each side by 7.

$$7 \cdot \frac{3y}{7} = 7 \times 14.4$$

Because $7 \cdot \dfrac{3y}{7}$ equals $3y$,

$$3y = 7 \times 14.4$$

The unknown does not appear by itself; hence, divide each side by 3.

$$\frac{3y}{3} = \frac{7 \times 14.4}{3}$$

Because $\dfrac{\cancel{3}y}{\cancel{3}}$ equals $y$,

$$y = 33.6$$

*Check:* Substitute 33.6 for $y$ in the original equation.

Does $\dfrac{3 \times (33.6)}{7} = 14.4$?   Does $14.4 = 14.4$?   Yes.

We can solve the equation of Example 3b with fewer steps if we first note that the reciprocal of $\frac{3}{7}$ is $\frac{7}{3}$ and then proceed as follows:

$$\frac{3y}{7} = 14.4$$

$$\frac{7}{3} \cdot \frac{3y}{7} = \frac{7}{3} \times 14.4$$

Because $\frac{7}{3} \cdot \frac{3y}{7}$ equals $y$, we have $\quad y = \frac{7 \times 14.4}{3} = 33.6$.

**EXAMPLE 4**  $\dfrac{2.64t}{0.88} = 52.65$

Multiply each side by $\dfrac{0.88}{2.64}$, the reciprocal of $\dfrac{2.64}{0.88}$.

$$\frac{0.88}{2.64} \times \frac{2.64t}{0.88} = \frac{0.88}{2.64} \times 52.65$$

Because $\dfrac{0.88}{2.64} \times \dfrac{2.64\,t}{0.88}$ equals $t$,

$$t = \frac{0.88}{2.64} \times 52.65$$

First do the multiplication, and then divide by 2.64.

The solution is 17.55.

*Check:* Substitute 17.55 for $t$ in the original equation.

$$\text{Does} \quad \frac{2.64 \times (17.55)}{0.88} = 52.65? \quad \text{Does} \quad 52.65 = 52.65? \quad \text{Yes.}$$

The solution to an equation may be a nonterminating decimal, in which case the solution may be rounded off. When checking such a solution, remember to use the solution obtained *before* rounding it off.

**EXAMPLE 5**  Solve to the nearest tenth: $18x = 569$.

**Solution**                    $18x = 569$

Divide each side by 18.

$$\frac{18x}{18} = \frac{569}{18}$$

Because $\dfrac{18\,x}{18} = x$,

$$x = 31.611111$$

To the nearest tenth, the solution is 31.6.

*Check:* Substitute 31.611111 for $x$ in the original equation.
Does $\quad 18 \times (31.611111) = 569?$ Does $\quad 568.99999 = 569?$
Yes. (568.99999, is "close enough" to 569 to satisfy the check.)

    Solving some equations may require the use of both the addition-subtraction rule and the multiplication-division rule. Earlier in this section we stated that a goal in solving equations is to obtain an equation in which the unknown appears by itself on only one side. Sometimes, in order to reach this goal, we first need to obtain an equation in which a *term* that includes the unknown appears by itself on only one side of the equation.

**EXAMPLE 6**  $42.24 = 1.2t + 6$

Subtract 6 from each side so that the term $1.2t$ appears by itself on only one side.

$$42.24 - 6 = 1.2t + \underbrace{6 - 6}$$
$$36.24 = 1.2t + \quad 0$$
$$36.24 = 1.2t$$

Divide each side by 1.2.

$$\frac{36.24}{1.2} = \frac{1.2t}{1.2}$$

Because $\dfrac{\cancel{1.2}\,t}{\cancel{1.2}}$ equals $t$, then

$$t = \frac{36.24}{1.2} = 30.2$$

*Check:*  Substitute 30.2 for $t$ in the original equation.

Does  $42.24 = 1.2 \times (30.2) + 6$?
Does  $42.24 = 42.24$?  Yes.

**EXAMPLE 7**  $3.8 = \dfrac{18.6 + t}{25.5}$

Multiply each side by 25.5.

$$25.5 \times 3.8 = \frac{18.6 + t}{25.5} \times 25.5$$

Because $\dfrac{18.6 + t}{\cancel{25.5}} \times \cancel{25.5}$ equals $18.6 + t$,

$$25.5 \times 3.8 = 18.6 + t$$
$$96.9 = 18.6 + t$$

Subtract 18.6 from each side.

$$96.9 - 18.6 = 18.6 + t - 18.6$$
$$t = 78.3$$

*Check:*  Substitute 78.3 for $t$ in the original equation.

Does  $3.8 = \dfrac{18.6 \times 78.3}{25.5}$?    Does  $3.8 = 3.8$?  Yes.

## Exercises for Section 4.1

*Solve each equation. See Examples 1 and 2 for Exercises 1 to 12.*

**1.** $s - 4 = 19$        **2.** $s - 9 = 5$          **3.** $n + 8 = 9$

**4.** $n + 2 = 10$        **5.** $n + 6.9 = 5.7$      **6.** $n + 0.53 = 7.90$

**7.** $s - 5.21 = 6.32 + 1.4$        **8.** $s - 2.1 = 3.2 - 1.4$

**9.** $5.8 + d - 5.8 = 12$        **10.** $3.97 + d - 3.97 = 1.23$

**11.** $0.45 + 0.98 = t - 1.7$        **12.** $12.3 - 7.4 = t - 2.8$

*See Examples 3, 4, and 5 for Exercises 13 to 24. (a) Show results in floating decimal point. (b) Round off results to the nearest hundredth.*

**13.** $3.3n = 0.1683$        **14.** $56n = 30.8$        **15.** $16.7d = 65.1$

**16.** $25.6d = 82.4$        **17.** $2.7r = 0.97$        **18.** $4.5r = 0.76$

**19.** $\dfrac{3p}{11} = 1.8$        **20.** $\dfrac{7p}{16} = 5.6$        **21.** $\dfrac{0.5s}{0.81} = 7.2$

**22.** $\dfrac{0.9s}{0.061} = 10.7$   **23.** $\dfrac{4.2t}{6.25} = 1.63$   **24.** $\dfrac{7.2t}{8.29} = 2.14$

*See Example 6 for Exercises 25 to 40. (a) Show results in floating decimal point.*
*(b) Round off results to two significant digits.*

**25.** $1.66 = 0.56n + 0.61$   **26.** $2.53 = 3.1n + 1.25$

**27.** $0.81 = 8.3d - 7.1$   **28.** $0.16 = 6.5d - 6.1$

**29.** $9.1r + 7.8 = 2.6$   **30.** $4.6r + 8.8 = 3.5$

**31.** $0.51s - 8.2 = 0.15$   **32.** $0.77s - 4.5 = 0.21$

**33.** $\dfrac{1.6t}{3.8} + 7.5 = 6.9$   **34.** $\dfrac{9.2t}{1.4} + 4.9 = 4.4$

**35.** $\dfrac{0.53n}{0.90} - 0.70 = 9.6$   **36.** $\dfrac{0.63n}{0.99} - 0.75 = 6.0$

**37.** $2.7 = \dfrac{9.5d}{8.3} + 3.0$     **38.** $8.6 = \dfrac{3.8d}{5.2} + 9.8$

**39.** $0.99 = \dfrac{0.73r}{0.55} - 5.3$     **40.** $0.84 = \dfrac{0.85r}{0.52} - 9.0$

*See Example 7 for Exercises 41 to 46. Round off results to three significant digits.*

**41.** $6.4 = \dfrac{4.6 + r}{9.7}$     **42.** $4.3 = \dfrac{2.1 + s}{1.8}$

**43.** $0.51 = \dfrac{p - 6.7}{0.46}$     **44.** $0.82 = \dfrac{n - 5.5}{0.52}$

**45.** $\dfrac{7.6 + E}{1.2} = -3.5$     **46.** $\dfrac{4.3 + I}{8.3} = -4.8$

Temperatures in Celsius ($C$) are related to temperatures in Fahrenheit ($F$) by the equation

$$\frac{5F - 160}{9} = C$$

*In each of the following equations, C is given: solve for F.*

**47.** $\dfrac{5F - 160}{9} = 100$     **48.** $\dfrac{5F - 160}{9} = 68$

**49.** $\dfrac{5F - 160}{9} = -15$ **50.** $\dfrac{5F - 160}{9} = -50$

In machine work, if a pin is tapered as shown in Figure 4.1, the taper ($T$), in inches per foot, is given by the formula

$$T = \dfrac{12D - 12d}{L}$$

where $D$, $d$, and $L$ are measured in inches. Several variations of this formula are also used.

**FIGURE 4.1**

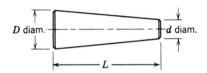

*In each of the following equations, three of the dimensions are known. Solve each equation to the nearest thousandth.*

**51.** $0.625 = \dfrac{12D - 1.5}{12}$ **52.** $0.750 = \dfrac{12D - 1.875}{12}$

**53.** $0.375 = \dfrac{13.5 - 12d}{6}$ **54.** $0.375 = \dfrac{10.5 - 12d}{7}$

**55.** $\dfrac{0.375L - 13.5}{12} = -0.875$ **56.** $\dfrac{0.375L + 6}{12} = 0.8125$

## 4.2  More Equation-Solving Methods

In Section 4.1 we solved equations by adding a number to or subtracting a number from each side of an equation or by multiplying or dividing each side of an equation by a number. Sometimes it is necessary to use similar methods with variables (unknowns).

If the unknown in an equation is preceded by a minus sign, it is often convenient to first add the unknown to each side of the equation in order to find the solution.

**EXAMPLE 1**   Solve   $11.37 - N = 4.02$.

**Solution**   Because the unknown is preceded by a minus sign, first add the unknown $N$ to each side of the equation.

$$11.37 - N + N = 4.02 + N$$

Because $11.37 - N + N$ equals 11.37, the equation becomes

$$11.37 = 4.02 + N$$

Subtract 4.02 from each side.

$$11.37 - 4.02 = 4.02 + N - 4.02$$
$$7.35 = N$$

The solution is 7.35.

*Check:*   Substitute 7.35 for $N$ in the original equation.

Does   $11.37 - 7.35 = 4.02$?   Yes.

If the unknown in an equation appears as a denominator, it is often convenient to first multiply each side of the equation by the unknown in order to find the solution.

**EXAMPLE 2**   Solve   $6.3 = \dfrac{428.6}{r}$   to the nearest tenth.

**Solution**   Because the unknown appears as a denominator, multiply each side by $r$.

$$6.3 \times r = \frac{428.6}{r} \times r$$

Because $\dfrac{428.6}{\cancel{r}} \times \cancel{r}$ equals 428.6, the equation becomes

$$6.3 \times r = 428.6$$

Divide each side by 6.3.

$$\frac{\cancel{6.3} \times r}{\cancel{6.3}} = \frac{428.6}{6.3}$$
$$r = 68.031746$$

The solution is 68.0, to the nearest tenth.

*Check:*   Substitute 68.031746 (the result obtained *before* rounding off) for $r$ in the original equation.

Does   $6.3 = \dfrac{428.6}{68.031746}$?   Yes.

In Example 1, we used only the addition-subtraction rule to solve the equation; in Example 2, we used only the multiplication-division rule. As in Section 4.1, we sometimes need to use both rules to obtain an equation in which the unknown appears by itself on only one side of the equation.

**EXAMPLE 3**  Solve  $6.8 - 3D = 5.7$  to the nearest tenth.

**Solution**  Because the term involving the unknown is preceded by a minus sign, add $3D$ to each side of the equation.

$$6.8 - 3D + 3D = 5.7 + 3D$$

Because $6.8 - 3D + 3D$ equals 6.8, the equation becomes

$$6.8 = 5.7 + 3D$$

Subtract 5.7 from each side.

$$6.8 - 5.7 = 5.7 + 3D - 5.7$$
$$1.1 = 3D$$

Divide each side by 3.

$$\frac{1.1}{3} = \frac{\cancel{3}D}{\cancel{3}}$$
$$D = 0.36666667$$

The solution is 0.4, to the nearest tenth.

*Check:*  Substitute 0.36666667 for $D$ in the original equation.

Does  $6.8 - 3(0.36666667) = 5.7$?
Yes, "close enough" to satisfy the check.

When checking solutions of equations, as shown in Examples 2 and 3, you should remember to substitute the result obtained *before* rounding off, and you should be able to recognize when the result of a computation is "close enough" to satisfy the check.

**EXAMPLE 4**  Solve  $\dfrac{2.13}{R} - 5.48 = 7.36$  to the nearest hundredth.

**Solution**  Add 5.48 to each side.

$$\frac{2.13}{R} - 5.48 + 5.48 = 7.36 + 5.48$$

$$\frac{2.13}{R} = 12.84$$

Because $R$ appears in the denominator, multiply each side by $R$.

$$\frac{2.13}{\cancel{R}} \times \cancel{R} = 12.84 \times R$$

$$2.13 = 12.84 \times R$$

Divide each side by 12.84.

$$\frac{2.13}{12.84} = \frac{\cancel{12.84} \times R}{\cancel{12.84}}$$

$$R = 0.16588785$$

The solution is 0.17, to the nearest hundredth.

*Check:*  Does  $\dfrac{2.13}{(0.16588785)} - 5.48 = 7.36?$   Yes.

## ■ *Solving Equations by Using Square Root*

In Section 1.4 we noted that 5 is a square root of 25 because $5^2$ equals 25. Because $(-5)^2$ equals 25, it follows that $-5$ is also a square root of 25. Hence, there are *two* square roots of 25. In fact, every positive number has two square roots that are equal in absolute value, but one is positive and the other is negative. The radical symbol $\sqrt{\phantom{x}}$ is used to represent the *positive* square root of the number; the symbol $-\sqrt{\phantom{x}}$ is used to represent the *negative* square root. For example,

$$\sqrt{25} = 5, \qquad -\sqrt{25} = -5$$

Now, the problem of finding all solutions of an equation such as

$$x^2 = 25$$

can be solved by answering the question, "What numbers squared are equal to 25?" The answer is, "Either the positive square root of 25 or the negative square root of 25." In symbols,

$$x = \sqrt{25} = 5 \quad \text{or} \quad x = -\sqrt{25} = -5$$

More generally, if $K$ is a positive number, then the equation

$$x^2 = K$$

has two solutions:

$$x = \sqrt{K} \quad \text{and} \quad x = -\sqrt{K}$$

For our purposes in this section, we shall need only the *positive* solution. We refer to this procedure for solving equations such as   $x^2 = K$   as the *square root method*.

**EXAMPLE 5**  Use the square root method and your calculator to solve each equation.
**(a)** $x^2 = 784$     **(b)** $x^2 = 0.784$

**Solutions**
**(a)**
$$x^2 = 784$$
$$x = \sqrt{784} = 28$$

The solution is 28.

*Check:*  Substitute 28 for $x$ in the original equation.

Does   $(28)^2 = 784?$   Yes.

**(b)**
$$x^2 = 0.784$$
$$x = \sqrt{0.784} = 0.88543774$$

The solution is 0.88543774.

*Check:*  Substitute 0.88543774 for $x$ in the original equation.

Does   $(0.88543774)^2 = 0.784?$   Yes.

In Section 1.4 we introduced the squaring key $\boxed{x^2}$. This key squares only the number in the display and does not affect any previous entries or computations.

Hence, it is particularly useful for calculations involving sums and differences of squared numbers.

**EXAMPLE 6**   Compute.
**(a)** $(11.23)^2 + (5.9)^2$      **(b)** $(11.23)^2 - (5.9)^2$

**Solutions**
**(a)** $(11.23)^2 + (5.9)^2 = 160.9229;$   this is computed by

$$11.23 \boxed{x^2} \boxed{+} 5.9 \boxed{x^2} \boxed{=} \longrightarrow 160.9229$$

**(b)** $(11.23)^2 - (5.9)^2 = 91.3029;$   this is computed by

$$11.23 \boxed{x^2} \boxed{-} 5.9 \boxed{x^2} \boxed{=} \longrightarrow 91.3029$$

**EXAMPLE 7**   Solve   $x^2 = 3.1^2 + 2.9^2$   to the nearest tenth.

**Solution**                           $x^2 = 18.02$

By the square root method,

$$x = \sqrt{18.02} = 4.244997$$

The solution is 4.2, to the nearest tenth.

*Check:*   Does   $(4.244997)^2 = 3.1^2 + 2.9^2$?   Yes.

**EXAMPLE 8**   Solve   $4.02^2 - x^2 = 1.55^2$   to the nearest hundredth.

**Solution**   Because the term including the unknown is preceded by a minus sign, add $x^2$ to both sides of the equation.

$$4.02^2 - x^2 + x^2 = 1.55^2 + x^2$$
$$4.02^2 = 1.55^2 + x^2$$

Subtract $1.55^2$ from each side.

$$4.02^2 - 1.55^2 = 1.55^2 + x^2 - 1.55^2$$
$$13.7579 = x^2$$

By the square root method,

$$x = \sqrt{13.7579} = 3.7091643$$

The solution is 3.71, to the nearest hundredth.

*Check:*   Does   $4.02^2 - (3.7091643)^2 = 1.55^2$?   Yes.

**EXAMPLE 9**   Solve   $\dfrac{3x^2}{5} + 4.2 = 83.5$   to the nearest tenth.

**Solution**   First, subtract 4.2 from each side of the equation so that the term that includes the unknown $(x)$ remains by itself.

$$\frac{3x^2}{5} + 4.2 - 4.2 = 83.5 - 4.2$$

$$\frac{3x^2}{5} = 79.3$$

Next, multiply each side by 5.

$$\not{5} \times \frac{3x^2}{\not{5}} = 79.3 \times 5$$

$$3x^2 = 396.5$$

Now, divide each side by 3.

$$\frac{\not{3}x^2}{\not{3}} = \frac{396.5}{3}$$

$$x^2 = \frac{396.5}{3}$$

By the square root method,

$$x = \sqrt{\frac{396.5}{3}} = 11.496376$$

The solution is 11.5, to the nearest tenth.

*Check:*   Does $\dfrac{3(11.496376)^2}{5} + 4.2 = 83.5$?   Yes.

## Exercises for Section 4.2

*Solve. See Example 1.*

**1.** $8.8 - N = 7.9$      **2.** $49.6 - N = 6.2$      **3.** $6.33 - P = 2.56$

**4.** $5.66 - P = 4.81$      **5.** $6.64 - R = 4.51$      **6.** $4.15 - R = 2.48$

**7.** $8.52 = 6.4 - S$      **8.** $6.9 = 5.7 - S$      **9.** $4.85 = 9.17 - t$

**10.** $1.35 = 3.93 - t$      **11.** $7.9 - x = -2.5$      **12.** $5.1 - x = -9.8$

*Solve. Round off to the nearest tenth. See Example 2.*

**13.** $7.2 = \dfrac{93.4}{r}$     **14.** $9.4 = \dfrac{6.8}{r}$     **15.** $\dfrac{6.73}{x} = 2.35$

**16.** $\dfrac{6.11}{4} = 1.46$     **17.** $-2.07 = \dfrac{6.47}{n}$     **18.** $-7.22 = \dfrac{6.89}{n}$

*Solve. Round off to the nearest hundredth. See Example 3.*

**19.** $5.24 - 7D = 2.5$     **20.** $32.2 - 9D = 9.3$

**21.** $2.17 - 3T = 3.95$     **22.** $2.69 - 6T = 15.26$

**23.** $-3.4 = 4.2 - 11y$     **24.** $-6.3 = 6.6 - 13y$

*Solve. Round off to three significant digits. See Example 4.*

**25.** $\dfrac{5.42}{R} - 2.36 = 5.38$     **26.** $\dfrac{7.02}{R} - 2.63 = 8.63$

**27.** $\dfrac{11.6}{F} - 44.1 = 13.7$     **28.** $\dfrac{98.1}{F} - 19.1 = 54.4$

**29.** $4.65 + \dfrac{1.85}{d} = 0.54$     **30.** $37.7 + \dfrac{26.8}{d} = 28.4$

*Solve. Round off to three significant digits. See Example 5.*

**31.** $x^2 = 381$     **32.** $x^2 = 762$     **33.** $y^2 = 85$

**34.** $y^2 = 78$     **35.** $R^2 = 0.0447$     **36.** $R^2 = 0.0186$

*Compute. Round off to the nearest tenth. See Example 6.*

**37.** $(6.4)^2 + (8.52)^2$     **38.** $(9.6)^2 + (5.7)^2$     **39.** $(48.5)^2 + (9.17)^2$

**40.** $(1.35)^2 + (39.3)^2$     **41.** $(79)^2 + (251)^2$     **42.** $(51)^2 + (198)^2$

*Solve. Round off to the nearest hundredth. See Examples 7, 8, and 9.*

**43.** $x^2 = (9.5)^2 + (5.9)^2$     **44.** $x^2 = (9.8)^2 + (0.77)^2$

**45.** $t^2 = (8.77)^2 + (30.6)^2$     **46.** $t^2 = (39.6)^2 + (30.8)^2$

**47.** $a^2 - (0.641)^2 = (0.983)^2$     **48.** $a^2 - (0.489)^2 = (0.719)^2$

**49.** $8.8324 - x^2 = 3.7094$     **50.** $11.0876 - x^2 = 6.8571$

**51.** $48.97 = 50.43 - s^2$     **52.** $58.23 = 72.57 - s^2$

**53.** $\dfrac{0.11b^2}{8.1} + 0.254 = 0.985$     **54.** $\dfrac{0.7b^2}{6.2} + 0.151 = 0.994$

**55.** $4.75 - \dfrac{0.23I^2}{6.4} = 1.09$     **56.** $9.46 - \dfrac{0.35I^2}{4.8} = 1.41$

## 4.3  Using Formulas

If an airplane flies at a constant rate of 485 miles per hour for 4 hours, then, by the unit-product rule, the airplane flies a distance of $485 \times 4$, or 1940 miles. If we use $d$ to represent the distance, $r$ the rate, and $t$ the time, the equation

$$d = rt$$

describes the above calculation. Such general rules are often called **formulas**. To use a formula to obtain a numerical value for an unknown value, we replace all letters except the unknown letter by appropriate numbers and solve the resulting equation.

**EXAMPLE 1**   Given the formula   $d = rt$.
**(a)** If   $r = 385$ miles per hour   and   $t = 2.75$ hours,   find $d$.
**(b)** If   $d = 758.25$ miles   and   $t = 4.5$ hours,   find $r$.

**Solutions**
**(a)** Replace $r$ by 385, $t$ by 2.75, and solve for $d$.

$$d = 385 \times 2.75 = 1058.75$$

The distance is 1058.75 miles.

**(b)** Replace $d$ by 758.25, $t$ by 4.5, and solve for $r$.

$$758.25 = r \times 4.5$$

Divide each side by 4.5.

$$\frac{758.25}{4.5} = \frac{r \times \cancel{4.5}}{\cancel{4.5}}$$

$$168.5 = r$$

*Check:*  Does   $758.25 = (168.5) \times 4.5$?     Does   $758.25 = 758.25$?   Yes.
The rate is 168.5 miles per hour.

It is not always essential to know the meaning of a formula to be able to use it. The emphasis in this section is on developing equation-solving skills needed to work with formulas.

**EXAMPLE 2**   Given the formula   $P + C = S - E$.
**(a)** If   $P = 165$,   $C = 524.52$,   and   $E = 620.74$,   find $S$.
**(b)** If   $S = 8423$,   $C = 6051$,   and   $E = 2560$,   find $P$.

**Solutions**
**(a)** Replace $P$ by 165, $C$ by 524.52, $E$ by 620.74, and solve for $S$.

$$165 + 524.52 = S - 620.74$$
$$689.52 = S - 620.74$$

Add 620.74 to each side.

$$689.52 + 620.74 = S - 620.74 + 620.74$$
$$1310.26 = S$$

*Check:*  Does   $165 + 524.52 = (1310.26) - 620.74$?
       Does        $689.52 = 689.52$?   Yes.

**(b)** Replace $S$ by 8423, $C$ by 6051, $E$ by 2560, and solve for $P$.

$$P + 6051 = 8423 - 2560$$
$$P + 6051 = 5863$$

Subtract 6051 from each side.

$$P + 6051 - 6051 = 5863 - 6051$$
$$P = -188$$

*Check:*  Does   $(-188) + 6051 = 8423 - 2560$?
       Does            $5863 = 5863$?   Yes.

**EXAMPLE 3**  Given the formula $T = \dfrac{D - d}{L}$. If $T = 0.0625$, $d = 0.658$, and $L = 7.75$, find $D$.

**Solution**  Replace $T$ by 0.0625, $d$ by 0.658, $L$ by 7.75, and solve for $D$.

$$0.0625 = \frac{D - 0.658}{7.75}$$

Multiply each side by 7.75.

$$0.0625 \times 7.75 = \frac{D - 0.658}{\cancel{7.75}} \times \cancel{7.75}$$

$$0.484375 = D - 0.658$$

Add 0.658 to each side.

$$0.484375 + 0.658 = D - 0.658 + 0.658$$
$$1.142375 = D$$

*Check:*  Does $0.0625 = \dfrac{(1.142375) - 0.658}{7.75}$?  Does $0.0625 = 0.0625$?  Yes.

**EXAMPLE 4**  Given the formula $C = \dfrac{5}{9}(F - 32)$. If $C = 33.4$, find $F$.

**Solution**  Replace $C$ by 33.4 and solve for $F$.

$$33.4 = \frac{5}{9}(F - 32)$$

Multiply each side by $\dfrac{9}{5}$, the reciprocal of $\dfrac{5}{9}$.

$$\frac{9}{5} \times 33.4 = \frac{\cancel{9}}{\cancel{5}} \times \frac{\cancel{5}}{\cancel{9}}(F - 32)$$
$$60.12 = F - 32$$

Add 32 to each side.

$$60.12 + 32 = F - 32 + 32$$
$$92.12 = F$$

*Check:*  Does $33.4 = \dfrac{5}{9}(92.12 - 32)$?  Does $33.4 = 33.4$?  Yes.

**EXAMPLE 5**  Given the formula $C = \dfrac{5}{9}(F - 32)$. If $C = -33.4$, find $F$.

**Solution**  Replace $C$ by $-33.4$ and solve for $F$.

$$-33.4 = \frac{5}{9}(F - 32)$$

Multiply each side by $\frac{9}{5}$, the reciprocal of $\frac{5}{9}$.

$$\frac{9}{5}(-33.4) = \frac{\cancel{9}}{\cancel{5}} \times \frac{\cancel{5}}{\cancel{9}}(F - 32)$$

$$-\left(\frac{9}{5} \times 33.4\right) = F - 32$$

$$-60.12 = F - 32$$

Add 32 to each side.

$$-60.12 + 32 = F - 32 + 32$$

$$-28.12 = F$$

*Check:* Does $-33.4 = \frac{5}{9}(-28.12 - 32)$? Does $-33.4 = -33.4$? Yes.

## Exercises for Section 4.3

*In this set of exercises, unless specified otherwise, give each answer to the nearest tenth unless the answer is a whole number.*

*In Exercises 1 to 6, use the formula $d = rt$. See Example 1.*

**1.** $r = 55.6$ miles per hour, $t = 8.8$ hours. Find $d$.

**2.** $r = 89.9$ kilometers per hour, $t = 6.5$ hours. Find $d$.

**3.** $d = 784.7$ miles, $t = 8.75$ hours. Find $r$.

**4.** $d = 1467.4$ kilometers, $t = 9.25$ hours. Find $r$.

**5.** $d = 2412.5$ meters, $r = 502.3$ meters per minute. Find $t$.

**6.** $d = 1843.1$ kilometers, $r = 97.6$ kilometers per hour. Find $t$.

*In Exercises 7 to 10, use the formula*  $P + C = S - E$.  *See Example 2.*

**7.** $S = 9420.37$,  $C = 5308.54$,  $E = 3067.06$.  Find $P$.

**8.** $P = 764.12$,  $S = 8234.43$,  $C = 4637.91$.  Find $E$.

**9.** $P = 34{,}037$,  $S = 163{,}362$,  $E = 52{,}144$.  Find $C$.

**10.** $S = 237{,}460$,  $C = 111{,}871$,  $E = 63{,}650$.  Find $P$.

*In Exercises 11 to 16, use the formula*  $T = \dfrac{D - d}{L}$.  *See Example 3.*

**11.** $T = 0.125$,  $d = 0.393$,  $L = 7.44$.  Find $D$.

**12.** $T = 0.375$,  $d = 0.422$,  $L = 16.1$.  Find $D$.

**13.** $T = 0.214$,  $D = 2.15$,  $L = 5.56$.  Find $d$.

**14.** $T = 0.446$,  $D = 9.13$,  $L = 22.7$.  Find $d$.

**15.** $T = 0.500$,  $D = 6.53$,  $d = 2.75$.  Find $L$.

**16.** $T = 0.115,$  $D = 46.5,$  $d = 30.9.$  Find $L$.

*In Exercises 17 to 24 use*  $C = \dfrac{5}{9}(F - 32)$  *to find F. See Examples 4 and 5.*

**17.** $C = 65$        **18.** $C = 92$        **19.** $C = 50.2$        **20.** $C = 85.3$

**21.** $C = -37$        **22.** $C = -50$        **23.** $C = -46.5$        **24.** $C = -30.9$

*In Exercises 25 to 28, use the formula*  $W = fs.$

**25.** $f = 895.6,$  $s = 95.75.$  Find $W$.        **26.** $f = 4831,$  $s = 285.5.$  Find $W$.

**27.** $W = 350,436,$  $s = 456.$  Find $f$.        **28.** $W = 92,538,$  $f = 388.$  Find $s$.

*In Exercises 29 to 32, use the formula*  $P = 0.0013333AV.$

**29.** $A = 22.5,$  $V = 110.$  Find $P$.        **30.** $A = 38.5,$  $V = 220.$  Find $P$.

**31.** $P = 2.5$,  $A = 16.3$.  Find $V$.    **32.** $P = 9.5$,  $V = 230$.  Find $A$.

*In Exercises 33 to 46, use the formula*  $C = 0.001\ Whr$.

**33.** $W = 100$,  $h = 95$,  $r = 0.214$.    **34.** $W = 550$,  $h = 8$,  $r = 0.032$.
Find $C$.    Find $C$.

**35.** $h = 13$,  $r = 0.35$,  $C = 1.32$.    **36.** $W = 420$,  $h = 15$,  $C = 1.19$.
Find $W$.    Find $r$.

*Using the formula*  $A = 3.142r^2$,  *find r for the given value of A. Round off to the nearest thousandth.*

**37.** $A = 1.892$    **38.** $A = 4.254$    **39.**  $A = 97.813$

**40.** $A = 16.422$    **41.** $A = 0.953$    **42.** $A = 0.885$

*Using the formula*  $P = I^2R$,  *find I for the given values of P and R. Round off to the nearest tenth.*

**43.** $P = 125$,  $R = 121$    **44.** $P = 216$,  $R = 87$    **45.**  $P = 5830$,  $R = 12$

**46.** $P = 980$,  $R = 8$    **47.** $P = 74$,  $R = 7.5$    **48.**  $P = 140$,  $R = 10.6$

## 4.4 Rewriting Formulas

In Section 4.3 we saw that for a given formula we can replace all the letters except the unknown letter by appropriate numbers and then solve the resulting equation. Sometimes it is more useful to first rewrite a formula so that *a specified unknown appears by itself on one side of the equation and does not appear on the other side*. The equation-solving methods introduced in Sections 4.1 and 4.2 can be used to do this.

**EXAMPLE 1**   Solve   $P = 18.64 - C$   for $C$.

**Solution**   Because the unknown $C$ is preceded by a minus sign, first add $C$ to each side of the equation.

$$P + C = 18.64 - C + C$$
$$P + C = 18.64$$

Now, subtract $P$ from each side.

$$P + C - P = 18.64 - P$$
$$C = 18.64 - P$$

**EXAMPLE 2**   Solve   $t = \dfrac{863}{r}$   for $r$.

**Solution**   Because the unknown $r$ appears in the denominator, first multiply each side of the equation by $r$.

$$t \cdot r = \frac{863}{\cancel{r}} \cdot \cancel{r}$$
$$t \cdot r = 863$$

Now, divide each side by $t$.

$$\frac{\cancel{t} \cdot r}{\cancel{t}} = \frac{863}{t}$$
$$r = \frac{863}{t}$$

**EXAMPLE 3**   Solve   $E = IR$   for $I$.

**Solution**   Divide each side of the equation by $R$.

$$\frac{E}{R} = \frac{I\cancel{R}}{\cancel{R}}$$
$$\frac{E}{R} = I$$
$$I = \frac{E}{R}$$

When rewriting formulas, it is customary to write the required unknown on the left side, as shown in the last equation of Example 3.

*EXAMPLE 4*  Solve  $I = Prt$  for $r$.

*Solution*  In order to obtain the factor $r$ by itself, divide each side of the equation by the product $Pt$.

$$\frac{I}{Pt} = \frac{\cancel{P}r\cancel{t}}{\cancel{P}\cancel{t}}$$

$$r = \frac{I}{Pt}$$

*EXAMPLE 5*  Solve  $v = 27.4 + gt$  for $t$.

*Solution*  In order to first obtain the term $(gt)$ that includes the unknown $t$ by itself, subtract 27.4 from each side of the equation.

$$v - 27.4 = 27.4 + gt - 27.4$$
$$v - 27.4 = gt$$

Now, divide each side by $g$.

$$\frac{v - 27.4}{g} = \frac{\cancel{g}t}{\cancel{g}}$$

$$t = \frac{v - 27.4}{g}$$

The square root method, as introduced in Section 4.2, can also be used for rewriting formulas.

*EXAMPLE 6*  Solve  $A = \pi r^2$  for $r$.

*Solution*  Divide each side of the equation by $\pi$.

$$\frac{A}{\pi} = \frac{\cancel{\pi}r^2}{\cancel{\pi}}$$

$$\frac{A}{\pi} = r^2$$

By the square root method,

$$r = \sqrt{\frac{A}{\pi}}$$

*EXAMPLE 7*  Solve  $a^2 = b^2 + c^2$  for $b$.

*Solution*  Subtract $c^2$ from each side of the equation.

$$a^2 - c^2 = b^2 + c^2 - c^2$$
$$a^2 - c^2 = b^2$$

By the square root method,

$$b = \sqrt{a^2 - c^2}$$

## *Exercises for Section 4.4*

*Solve each equation for the indicated unknown.*

*See Example 1 for Exercises 1 to 6.*

**1.** $d = 16 - l$,  for $l$.       **2.** $d = 48 - l$,  for $l$.

**3.** $g + 5 = f$,  for $g$.       **4.** $g + 8 = f$,  for $g$.

**5.** $x + y = 7$,  for $x$.       **6.** $x + y = 10$,  for $y$.

*See Examples 2 to 4 for Exercises 7 to 30.*

**7.** $r = \dfrac{500}{t}$,  for $t$.              **8.** $r = \dfrac{650}{t}$,  for $t$.

**9.** $P = \dfrac{37.5}{V}$,  for $V$.            **10.** $P = \dfrac{49.3}{V}$,  for $V$.

**11.** $\dfrac{E^2}{R} = P$,  for $R$.            **12.** $\dfrac{E}{I} = Z$,  for $I$.

**13.** $\dfrac{P}{R} = I^2$,  for $R$.            **14.** $\dfrac{10L}{R} = d^2$,  for $R$.

**15.** $E = IR$,  for $R$.                **16.** $C = kd$,  for $d$.

**17.** $10L = AR$, for $A$.    **18.** $200I = CE$, for $E$.

**19.** $IE = 1374H$, for $I$.    **20.** $Dr = 600d$, for $D$.

**21.** $R_2 R_x = R_1 R_3$, for $R_x$.    **22.** $V_R R_X = V_X R$, for $V_X$.

**23.** $I = Prt$, for $P$.    **24.** $I = Prt$, for $t$.

**25.** $V = lwh$, for $w$.    **26.** $V = lwh$, for $h$.

**27.** $IEP_f = P$, for $P_f$.    **28.** $EIP_f = P$, for $E$.

**29.** $IEE_p = 746H$, for $I$.    **30.** $IEE_p = 746H$, for $E_p$.

*See Example 5 for Exercises 31 to 36.*

**31.** $l = 16 + nd$, for $d$.    **32.** $l = 25 + nd$, for $n$.

**33.** $E = 220 + IR$, for $I$.    **34.** $E = 220 + IR$, for $R$.

**35.** $7.8 - RI = E$,  for $I$.     **36.** $9.3 - RI = E$,  for $R$.

*See Examples 6 and 7 for Exercises 37 to 48.*

**37.** $P = I^2R$,  *for I.*          **38.** $V = kd^2$,  for $d$.

**39.** $2.5H = D^2N$,  for $D$.       **40.** $gp = V^2d$,  for $V$.

**41.** $LI = 4\pi AN^2$,  for $N$.    **42.** $KE = 2\pi^2f^2dR^2$,  for $R$.

**43.** $c^2 = a^2 + b^2$,  for $a$.   **44.** $c^2 = a^2 + b^2$,  for $b$.

**45.** $x^2 - y^2 = z^2$,  for $y$.   **46.** $Z^2 - X_L^2 = R^2$,  for $X_L$.

**47.** $4x^2 + 9y^2 = 36$,  for $x$.  **48.** $x^2 - 26y^2 = 100$,  for $y$.

## 4.5 Problem Solving

Equations can be used as mathematical "models" for word sentences. Thus the ability to write and solve equations is an invaluable aid for solving problems. Although there is no rule for solving all problems, the following steps can be used as a guide for solving many of them.

---

1. Read the problem and note particularly what is being asked for. *Write* a short phrase to describe this quantity. Choose any letter to represent the number (unknown) you want to find.
2. Write an equation that represents the condition on the unknown.
3. Solve the equation; check the solution.
4. State the answer to the problem.

---

Some problems are so simple that you can probably determine their solutions by inspection. In such cases, the steps described above may seem unnecessary. However, it is important that you master a *problem-solving routine* that will be helpful in solving more complicated problems. Hence, you should make every effort to follow the suggested routine, as shown in the following examples, even for problems in which the solution may seem obvious.

**EXAMPLE 1**  A stamping machine is programmed to stamp out 11,580 parts before stopping. How many parts remain to be stamped out after 9674 parts have been made?

### Solution

1. Express in a phrase what you want to find; represent the unknown by a letter.

$$\text{Number of parts remaining to be stamped out: } p$$

2. Write an equation that represents the condition on the unknown: "9674 parts added to $p$ parts must equal 11,580 parts."

$$9674 + p = 11{,}580 \qquad (1)$$

3. Solve the equation. Subtract 9674 from each side.

$$9674 + p - 9674 = 11{,}580 - 9674$$
$$p = 1906$$

   *Check:*  Does   $9674 + 1906 = 11{,}580$?    Does   $11{,}580 = 11{,}580$?   Yes.
4. State the answer to the problem.

$$\text{1906 parts remain to be stamped out.}$$

It is often possible to use different equations for solving the same problem. For instance, in Example 1, you might think of using the equation

$$11{,}580 - 9674 = p$$

instead of Equation 1.

**EXAMPLE 2**  The sum of four voltages in a circuit is to be 27.8 volts. If three of the voltages are 9.6 volts, 4.8 volts, and 11.0 volts, find the fourth voltage.

### Solution

1. Express in a phrase what you want to find; represent the unknown by a letter.

$$\text{The fourth voltage: } v$$

2. Write an equation that represents the condition on the unknown: "The sum of 9.6, 4.8, 11.0, and $v$ equals 27.8."

$$9.6 + 4.8 + 11.0 + v = 27.8$$

3. Solve the equation.

$$25.4 + v = 27.8$$

Subtract 25.4 from each side.

$$25.4 + v - 25.4 = 27.8 - 25.4$$
$$v = 2.4$$

*Check:*   Does   $9.6 + 4.8 + 11.0 + (2.4) = 27.8$?
           Does   $27.8 = 27.8$?   Yes.

4. State the answer to the problem.

The fourth voltage is 2.4 volts.

**EXAMPLE 3**   Figure 4.2 shows a part of a hydraulic valve lifter system. (The symbol ″ means *inches*.) Find the missing dimension.

**Solution**

1. Express in a phrase what you want to find; represent the unknown by a letter.

The missing dimension: $d$

2. Write an equation that represents the condition on the unknown:

"The sum of 0.18750 and $d$ and 0.15625 is 0.96875."

$$0.18750 + d + 0.15625 = 0.96875$$

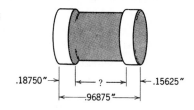

**FIGURE 4.2**

3. Solve the equation.

$$d + 0.34375 = 0.96875$$

Subtract 0.34375 from each side.

$$d + 0.34375 - 0.34375 = 0.96875 - 0.34375$$
$$d = 0.62500$$

*Check:*   Does   $0.18750 + 0.62500 + 1.5625 = 0.96875$?
           Does   $0.96875$                    $= 0.96875$?   Yes.

4. State the answer to the problem.

The missing dimension is 0.62500 inch.

Note that in Step 2 of each of the preceding three examples, we replaced

"must equal,"      "equals,"      and      "is"

by the = symbol. Although this type of replacement is not a mathematical rule, it can help us to "translate" sentences into equations.

## Exercises for Section 4.5

*Solve each exercise using the methods of Examples 1, 2, and 3.*

1. A piece 21 inches long is cut from a length of tubing 36 inches long. Find the length of the piece remaining, assuming no waste in cutting.

2. A machine is set up to turn 6.25 inches of thread on a rod. It is stopped after turning 2.75 inches. How much thread remains to be turned?

3. The total weight of four castings is 318.4 kilograms (kg). If three of the castings weigh 120.3, 41.2, and 36.4 kilograms, respectively, find the weight of the fourth casting.

4. A contractor receives $560 for doing some work. The contractor pays two employees $139.27 and $94.18, respectively, and pays $101.15 for material. How much money remains for the contractor?

*Each of the problems in the following table refers to the sketch of a gear shaft shown in Figure 4.3. Find the missing dimensions. All dimensions are in centimeters.*

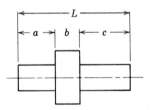

**FIGURE 4.3**

| | L | a | b | c |
|---|---|---|---|---|
| **5.** | 10.18 | 3.04 | —— | 5.62 |
| **6.** | 18.25 | —— | 2.13 | 12.41 |
| **7.** | 16.50 | 9.15 | 5.87 | —— |
| **8.** | 20.45 | 15.62 | —— | 3.05 |

*Each of the problems in the following table refers to the sketch shown in Figure 4.4. Find the missing dimensions. All dimensions are in inches.*

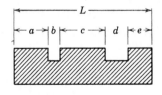

**FIGURE 4.4**

| | L | a | b | c | d | e |
|---|---|---|---|---|---|---|
| **9.** | 16.15 | 4.07 | 0.38 | 2.98 | —— | 7.97 |
| **10.** | 14.36 | 2.14 | 0.50 | —— | 0.88 | 6.15 |
| **11.** | 8.216 | —— | 0.438 | 2.125 | 1.062 | 2.125 |
| **12.** | 9.243 | 2.331 | —— | 2.331 | 1.125 | 2.331 |

13. Three pieces with lengths 1.06, 2.12, and 1.88 inches, respectively, are cut from a rod that was 10.00 inches long. If 0.06 inch is wasted on each cut, how much is left of the original rod?

**14.** A tank contains 500 liters of solvent at the start of the week. During the week, solvent is pumped from the tank in the amounts of 24.6 liters, 87.3 liters, and 107.5 liters. How much remains in the tank at the end of the week?

**15.** A dealer in scrap iron had 250 tons of iron in stock at the start of the week. During the week, he made three purchases of 115 tons, 87 tons, and 106 tons, respectively. He also made two sales of 95 tons and 180 tons. How many tons were left at the end of the week?

**16.** On the first day of June, a woman has $987.15 in a bank account. During the month, she makes deposits of $108.40, $216.32, and $50.60 and withdrawals of $89.96 and $500.17. How much remains in the account at the end of the month?

*Each set of dimensions in the following table refers to the sketch shown in Figure 4.5. Find the missing dimension as a proper fraction or mixed number. All dimensions are in inches.*

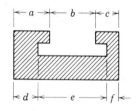

**FIGURE 4.5**

| | a | b | c | d | e | f |
|---|---|---|---|---|---|---|
| **17.** | $1\frac{1}{4}$ | $2\frac{1}{2}$ | — | $\frac{3}{4}$ | $3\frac{1}{4}$ | $\frac{3}{8}$ |
| **18.** | $\frac{7}{8}$ | $3$ | $\frac{3}{4}$ | — | $3\frac{1}{2}$ | $\frac{1}{2}$ |
| **19.** | — | $1\frac{3}{4}$ | $\frac{1}{2}$ | $\frac{1}{2}$ | $2\frac{1}{4}$ | $\frac{3}{8}$ |
| **20.** | $1\frac{1}{8}$ | $1\frac{7}{8}$ | $\frac{3}{4}$ | $\frac{11}{16}$ | — | $\frac{7}{16}$ |

One of Kirchhoff's laws states that in a closed circuit the sum of the voltages across generators and resistors must be zero.

*For each given set of voltages in Exercises 21 to 24, determine if the sum is zero. If the sum is not zero, determine a new voltage drop value for R3 so that the sum is zero.*

**21.**

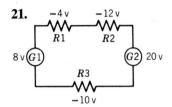

**22.**

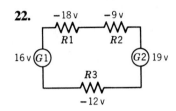

**23.**

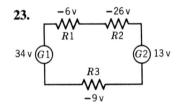

**24.**

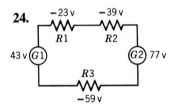

*For each of the circuits in Exercises 25 to 28, supply the missing voltage so that the sum of the voltages in the circuit is zero. (The symbol ─┤||├─ represents a battery.)*

**25.**

-24 v   ? v
+ ||| -
- ||| +   -75 v
37 v

**26.**

-15 v   12 v
+ ||| -
- ||| +   -8 v
? v

**27.**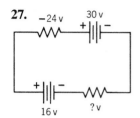

**28.**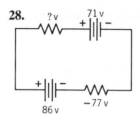

---

## 4.6  Unit-Product Rule; Unit-Quotient Rule

In this section we introduce two rules that are very useful for writing equations to solve certain types of problems.

Consider the problem, "If one bolt costs 7¢, how much do 25 bolts cost?" To solve the problem, we may correctly reason, "If one bolt costs 7¢, then 25 bolts cost 25 times as much." We then compute the product.

$$\text{(number of bolts)} \times \text{(cost per bolt)} = \text{(cost of 25 bolts)}$$
$$25 \qquad \times \qquad 7 \qquad = \qquad 175$$

Hence, 25 bolts cost 175¢, or $1.75. Note that in this problem we are *given the cost of one bolt* and are asked to *find the cost of more than one bolt.*

As another example, consider the problem, "If 1 gallon of white paint requires 3.25 pounds of titanium oxide pigment, how many pounds of pigment are required for 14 gallons of paint?" To solve this problem, we have

$$\text{(number of gallons)} \times \text{(pounds per gallon)} = \text{(total pounds of pigment)}$$
$$14 \qquad \times \qquad 3.25 \qquad = \qquad 45.5$$

Hence, 14 gallons of paint require 45.5 pounds of pigment. In this problem we are *given the amount of pigment needed for 1 gallon* and are asked to *find the amount of pigment needed for more than 1 gallon.*

### ■ The Unit-Product Rule

Although the above two problems are not too difficult, their solutions show a pattern that suggests the following rule, which can be used to solve more complicated problems.

---

**Unit-Product Rule**

If a given number is associated with one unit, the number associated with more than one such unit is equal to the product.

$$\text{(number of units)} \times \text{(number associated with one unit)}$$

---

The unit referred to in the unit-product rule can be one bolt, one gallon, one mile, one dollar, and so on.

You may be able to solve some of the following problems by methods that use fewer steps than shown. However, as in Section 4.5, it is important that you master a *problem-solving routine* rather than look for the shortest way to work a problem.

**EXAMPLE 1**  If an automobile can travel 18.4 miles per gallon (mpg) of gasoline, how far can it travel on 22.5 gallons?

**Solution**

**1.** Express in words what you want to find; represent it by a letter.

Miles a car can travel on 22.5 gallons: m

**2.** Write an equation that represents the condition on *m*. The phrase "18.4 miles per gallon" means "18.4 miles on 1 gallon." Thus, 18.4 is the number associated with *one unit* (1 gallon). Because the number *m* is associated with *more than one unit* (22.5 gallons), by the unit-product rule

$$\text{(number of miles)} = \text{(number of gallons)} \times \text{(miles per gallon)}$$
$$m \quad = \quad 22.5 \quad \times \quad 18.4$$

**3.** Solve the equation by a direct computation.

$$m = 22.5 \times 18.4 = 414$$

**4.** The car can travel 414 miles on 22.5 gallons of gasoline.

## ■ The Unit-Quotient Rule

The unit-product rule is used to solve problems in which we are given a number associated with one unit and are asked to find the number associated with more than one unit. We now consider problems in which we are *given a number associated with more than one unit* and are asked to *find the number associated with one unit.*

Consider the problem, "If 6 quarts of oil cost $5.76, find the cost of 1 quart." To solve this problem, we may correctly reason, "If $5.76 is the cost of 6 quarts of oil, then we must separate 5.76 into six equal parts to find the cost of 1 quart." Thus

$$\frac{5.76}{6} = 0.96$$

and $0.96 is the cost of 1 quart of oil. Note that in this problem we are *given the cost of more than 1 quart* and are asked to *find the cost of 1 quart.* Note also that the answer was obtained by computing the quotient

$$\frac{\text{cost of 6 quarts}}{\text{number of quarts}}$$

As another example, consider the problem, "If a machine can produce 2880 soft drink bottles in an 8-hour shift, how many bottles can be produced in 1 hour?" To solve this problem, we may reason, "If 2880 bottles are produced in 8 hours, then we must separate 2880 into eight equal parts to find how many bottles can be produced in 1 hour." Then,

$$\frac{2880}{8} = 360$$

and the machine can produce 360 bottles per hour. In this problem we are *given the number produced in more than 1 hour* and are asked to *find the number produced in 1 hour.* Note that the answer was obtained by computing the quotient

$$\frac{\text{number of bottles in 8 hours}}{\text{number of hours}}$$

These examples suggest the Unit-Quotient Rule on p. 132.

---

### Unit-Quotient Rule

If a given number is associated with more than one unit, the number associated with one such unit is equal to the quotient

$$\frac{\text{number associated with more than one unit}}{\text{number of units}}$$

---

In problems requiring division, take care to select and divide by the correct divisor. Many word problems that can be solved by the above rule contain phrases that use the word "per." As a guide, you may find it helpful to remember that when using the unit-quotient rule, the word that *follows* "per" is the divisor. For example,

"miles per hour"   indicates   "miles divided by hours"
"miles per gallon"   indicates   "miles divided by gallons"
"dollars per pound"   indicates   "dollars divided by pounds"

**EXAMPLE 2**   If an airplane flies 2470 miles in 5 hours, find the speed of the airplane in miles per hour.

### Solution

**1.** Express in words what you want to find; represent it by a letter.

Speed of the airplane in miles per hour: $s$

**2.** Write an equation that represents the condition on $s$. The number 2470 is associated with *more than one unit* (5 hours). Because we want to find the number $s$ associated with *one unit* (1 hour), we use the unit-quotient rule. The phrase "miles per hour" indicates "miles divided by hours."

$$\text{miles per hour} = \frac{\text{miles in 5 hours}}{\text{number of hours}}$$

$$s = \frac{2470}{5}$$

**3.** Solve the equation by direct computation.

$$s = \frac{2470}{5} = 494$$

**4.** The airplane flies 494 miles per hour (mph).

### Exercises for Section 4.6

*For Exercises 1 to 16, (a) write an equation, and (b) solve the equation.*

*See Example 1 for Exercises 1 to 4.*

**1.** If an automobile can travel 19 miles per gallon of gasoline, how far can it travel on 23 gallons?

**2.** If a motorcycle can travel 67 miles per gallon of gasoline, how far can it travel on 4 gallons?

**3.** If a lathe can turn 58 parts per hour, how many parts can be turned in 45 hours?

**4.** If a machine can stamp 385 parts per hour, how many parts can it stamp in 8 hours?

*See Example 2 for Exercises 5 to 8.*

**5.** At the end of a year, a company gave each employee the same amount of money as a bonus. If the company gave $4625 to 37 employees, how much did each employee receive?

**6.** A store owner purchased 36 drills for $1260. If each drill costs the same, what was the cost of each?

**7.** On a trip of 768 miles, a car used 48 gallons of gasoline. How many miles did the car travel per gallon?

**8.** An automatic machine produces 624 items in 24 hours. How many items does it produce per hour?

*Exercises 9 to 16 involve either the unit-product rule or the unit-quotient rule.*

**9.** If a high-speed printer for a computer can type 125 lines per minute, how many lines can it type in 35 minutes?

**10.** A fast-food restaurant is constructed to be able to serve 135 customers per hour. How many customers can be served in 18 hours?

**11.** A high-speed printer for a computer can print 3300 lines in 25 minutes. How many lines can it print per minute?

**12.** One of the fastest printers made can print 30,000 words in 60 seconds. How many words per second is that?

**13.** If a man pays $175 per month for a car, how much does he pay over a period of 12 months?

**14.** If a woman spends $18 per week for transportation to and from her job, how much does she spend in 26 weeks?

**15.** If 704 yards of wire weigh 16 pounds, how many yards weigh 1 pound?

**16.** One of the fastest rates of selling gasoline was 3116 gallons sold from a single pump in 24 hours. To the nearest tenth, how many gallons per hour was this?

*A company manufactures four models of bicycles. The production cost of each model and the daily production figures are shown below. Find the weekly production cost for each model. Assume a 5-day week. (Hint: Find the daily production cost for each model and multiply by 5.)*

**17.**

| Model | M-1 | M-2 | M-3 | M-4 |
|---|---|---|---|---|
| Cost per unit | $37.22 | $26.84 | $41.49 | $58.13 |
| Daily production | 98 | 102 | 87 | 125 |
| Weekly cost | ___ | ___ | ___ | ___ |

**18.**

| Model | S-1 | S-2 | S-3 | S-4 |
|---|---|---|---|---|
| Cost per unit | $22.47 | $34.98 | $47.14 | $63.85 |
| Daily production | 203 | 157 | 85 | 109 |
| Weekly cost | ___ | ___ | ___ | ___ |

*Complete each sales record.*

**19.**

| Product Number | Unit Price | Number Sold | Total Sales |
|---|---|---|---|
| A134 | $13.12 | 73 | ___ |
| A270 | $47.22 | 57 | ___ |
| B985 | $31.47 | 144 | ___ |
| | | Grand Total | ___ |

**20.**

| Product Number | Unit Price | Number Sold | Total Sales |
|---|---|---|---|
| D34A | $17.63 | 122 | ___ |
| E17A | $56.09 | 84 | ___ |
| E43B | $29.87 | 49 | ___ |
| | | Grand Total | ___ |

*Complete each table.*

**21.** For each distance and time, compute the speed in miles per hour to the nearest tenth.

|                   | a    | b    | c    | d    |
|-------------------|------|------|------|------|
| Distance (miles)  | 3090 | 3339 | 2448 | 2940 |
| Time (hours)      | 7    | 8    | 3    | 6    |
| Speed             | ——   | ——   | ——   | ——   |

**22.** For each distance and time, compute the speed in feet per minute to the nearest tenth.

|                   | a    | b    | c    | d    |
|-------------------|------|------|------|------|
| Distance (feet)   | 6677 | 6612 | 5274 | 6383 |
| Time (minutes)    | 12   | 11   | 8    | 12   |
| Speed             | ——   | ——   | ——   | ——   |

**23.** The results of a typing test for four typists are listed below. To the nearest whole number, find the typing speed of each typist in words per minute and find the average of the four typing speeds.

|                   | a    | b    | c    | d    |
|-------------------|------|------|------|------|
| Words typed       | 335  | 472  | 476  | 432  |
| Time (minutes)    | 6    | 7    | 9    | 7    |
| Typing speed      | ——   | ——   | ——   | ——   |

**24.** The results of "mileage economy" tests for four automobiles are listed below. To the nearest hundredth, find the mileage for each car in miles per gallon and find the average of the four mileages.

|                     | a      | b    | c      | d      |
|---------------------|--------|------|--------|--------|
| Distance (miles)    | 11,484 | 9503 | 13,266 | 16,733 |
| Gasoline (gallons)  | 621    | 787  | 598    | 563    |
| Mileage             | ——     | ——   | ——     | ——     |

**25.** If a car can be rented for $14.00 per day plus 16¢ per mile, what is the cost of renting a car for 7 days if it is driven 687 miles?

**26.** If a station wagon can be rented for $18.00 per day plus 18¢ per mile, what is the cost of renting it for 12 days if it is driven 963 miles?

**27.** The results of mileage tests on four different cars are listed below. To the nearest hundredth, find the results in miles per gallon for each car and find the average of the four mileages.

|                     | A     | B      | C       | D      |
|---------------------|-------|--------|---------|--------|
| Distance (miles)    | 9316  | 11,806 | 14,275  | 18,431 |
| Gasoline (gallons)  | 490.5 | 524.7  | 1089.6  | 675.2  |
| Miles per gallon    | ―     | ―      | ―       | ―      |

**28.** The following chart gives the number of months for which various automobile batteries are guaranteed to operate and the cost of each battery. To the nearest cent, find the cost per month of each battery and find the average cost per month of the four batteries.

|                | A       | B       | C       | D       |
|----------------|---------|---------|---------|---------|
| Months         | 24      | 36      | 48      | 60      |
| Cost           | $27.88  | $29.88  | $33.88  | $41.88  |
| Cost per month | ―       | ―       | ―       | ―       |

*For Exercises 29 and 30, assume that air temperature changes with altitude at the approximate rate of −0.00198° C (Celsius) for each foot increase in altitude and +0.00198° C for each foot decrease.*

**29.** If the air temperature is −24.62° C at an altitude of 20,000 feet, find the temperature to the nearest tenth of a degree (*a*) 5000 feet higher, and (*b*) 5000 feet lower.

**30.** If the air temperature is $-40.47°$ C at 28,000 feet, find the temperature to the nearest tenth of a degree (*a*) at 34,000 feet, and (*b*) at 22,000 feet.

**31.** The length of a certain bar of aluminum alloy changes with temperature at the rate of $+0.0014$ inch for a $1°$ F increase in temperature and $-0.0014$ inch for a $1°$ F decrease. If the bar is 100 inches long at $90°$ F, find its length to the nearest hundredth of an inch (*a*) at $70°$ F, and (*b*) at $120°$ F.

**32.** The length of a certain bar of stainless steel changes with temperature at the rate of $+0.0019$ inch for a $1°$ F increase in temperature and $-0.0019$ inch for a $1°$ F decrease. If the bar is 200 inches long at $200°$ F, find its length to the nearest hundredth of an inch (*a*) at $170°$ F, and (*b*) at $240°$ F.

## *4.7  Ratio and Proportion*

In this section we consider a certain type of equation that is very useful for solving many different kinds of problems.

### ■ *Ratios*

The **ratio** of one number to a second number is the quotient obtained when the first is divided by the second. Thus, a ratio is a method for comparing two numbers by division. In addition to the standard symbols for a quotient, a special notation is sometimes used for ratios. Thus, the "ratio of 3 to 5" may be written as

$$\frac{3}{5} \quad \text{or} \quad 3 \div 5 \quad \text{or} \quad 3 : 5$$

read "three to five" in each case. The symbol $3:5$ is called *colon* notation; the numbers 3 and 5 are the **terms** of the ratio.

### *EXAMPLE 1*

(a) The ratio of 7 to 9 is $\dfrac{7}{9}$. The terms are 7 and 9.

(b) The ratio of 2.25 to 3.75 is $\dfrac{2.25}{3.75}$. The terms are 2.25 and 3.75.

## ■ Proportions

A statement of equality between two ratios is called a **proportion**. For example, the equation

$$\frac{1}{2} = \frac{9}{18}$$

is a proportion, read "one is to two as nine is to eighteen." In the proportion

$$\frac{a}{b} = \frac{c}{d}$$

the terms $a$ and $d$ are called the **extremes**; the terms $b$ and $c$ are called the **means**. In Section 2.3 we stated that

$$\frac{a}{b} = \frac{c}{d} \quad \text{if} \quad ad = bc$$

In the language of proportions, we have the following rule.

---
### *Cross-Multiplication Rule*
In a proportion, the product of the extremes equals the product of the means.
---

**EXAMPLE 2**  In the proportion $\frac{2.25}{3.75} = \frac{9}{15}$, the terms 2.25 and 15 are the extremes and the terms 3.75 and 9 are the means. By the cross-multiplication rule,

$$2.25 \times 15 = 3.75 \times 9$$

or   $33.75 = 33.75$.

## ■ Solving Proportions

The cross-multiplication rule can be used to solve for an unknown term in a proportion. We refer to the process as *solving a proportion*.

**EXAMPLE 3**  Solve:  $\frac{n}{14.875} = \frac{10.5}{62.5}$

**Solution**  Apply the cross-multiplication rule.

$$62.5n = 10.5 \times 14.875$$

Divide each side by 62.5.

$$\frac{\cancel{62.5}\,n}{\cancel{62.5}} = \frac{10.5 \times 14.875}{62.5}$$

$$n = \frac{10.5 \times 14.875}{62.5} = 2.499$$

*Check:*  Does $\frac{(2.499)}{14.875} = \frac{10.5}{62.5}$?    Does  $0.168 = 0.168$?  Yes.

## ■ *Using Proportions*

Consider the problem, "If 12 gears cost $42.60, how much do 8 gears cost?" One way to solve this problem involves a two-step procedure. First, we can apply the unit-quotient rule to find the cost, $g$, of 1 gear.

$$g = \frac{42.60}{12} = 3.55$$

Hence, 1 gear costs $3.55. Next, we can apply the unit-product rule and find the cost, $C$, of 8 gears.

$$C = 8 \times 3.55 = 28.4$$

Thus, 8 gears cost $28.40.

Another method for solving problems similar to the above example combines both steps into one procedure involving a proportion. To set up such a proportion, we first set up a table with a Dollars column and a Gears column, as in part (*a*) of the accompanying table.

| Dollars  Gears | Dollars  Gears | Dollars  Gears |
|:---:|:---:|:---:|
| $\frac{\rule{1cm}{0.4pt}}{\rule{1cm}{0.4pt}} = \frac{\rule{1cm}{0.4pt}}{\rule{1cm}{0.4pt}}$ | $\dfrac{C}{\rule{1cm}{0.4pt}} = \dfrac{8}{\rule{1cm}{0.4pt}}$ | $\dfrac{C}{42.60} = \dfrac{8}{12}$ |
| (*a*) | (*b*) | (*c*) |

Then, because $C$ dollars is paired with 8 gears, we enter $C$ as the numerator in the Dollars column and 8 as the numerator in the Gears column, as in part (*b*) of the table. Next, because $42.60 is paired with 12 gears, we enter 42.60 as the denominator in the Dollars column and 12 as the denominator in the Gears column, as in part (*c*). To complete the problem, we solve the proportion obtained in (*c*).

$$\frac{C}{42.60} = \frac{8}{12}$$

Apply the cross-multiplication rule.

$$12C = 42.60 \times 8$$

Divide each side by 12.

$$\frac{\cancel{12}\,C}{\cancel{12}} = \frac{42.60 \times 8}{12}$$

$$C = 28.4$$

Thus, 8 gears cost $28.40.

Different choices can be made when setting up a proportion for solving a problem. For example, each of the proportions

$$\frac{42.60}{C} = \frac{12}{8} \qquad \text{or} \qquad \frac{8}{12} = \frac{C}{42.60} \qquad \text{or} \qquad \frac{12}{8} = \frac{42.60}{C}$$

can be used to solve the above problem correctly. The important thing is to *place like quantities in the same columns.*

**EXAMPLE 4**  If a car travels 171.2 miles on 8 gallons of gasoline, how far can it travel on 23.6 gallons?

**Solution**  Let $d$ represent the distance the car can travel on 23.6 gallons, and prepare a Miles column and a Gallons column. Because $d$ is paired with 23.6, enter $d$ as the numerator in the Miles column and 23.6 as the numerator in the Gallons column.

| Miles | Gallons | Miles | Gallons | Miles | Gallons |
|---|---|---|---|---|---|
| $\dfrac{\quad}{\quad}$ | $=\dfrac{\quad}{\quad}$ | $\dfrac{d}{\quad}$ | $=\dfrac{23.6}{\quad}$ | $\dfrac{d}{171.2}$ | $=\dfrac{23.6}{8}$ |

Next, because 171.2 is paired with 8, enter 171.2 as the denominator in the Miles column and 8 as the denominator in the Gallons column, and then solve the resulting proportion. Apply the cross-multiplication rule.

$$8d = 171.2 \times 23.6$$

Divide each side by 8.

$$\frac{\cancel{8}d}{\cancel{8}} = \frac{171.2 \times 23.6}{8}$$

$$d = 505.04$$

*Check:*  Does $\dfrac{(505.04)}{171.2} = \dfrac{23.6}{8}$?   Does $2.95 = 2.95$?   Yes.

To the nearest tenth, the car can travel 505.0 miles on 23.6 gallons of gasoline.

In the preceding examples we showed three separate stages in the setting up of a proportion. In the remainder of this book we shall show only the labels on the columns and the proportion itself.

**EXAMPLE 5**  If an electronics assembler earns \$29.25 for 7.5 hours of work, how much can she earn in 37.5 hours?

**Solution**  Let $D$ represent the amount earned in 37.5 hours, and set up a Dollars column and an Hours column. Because $D$ is paired with 37.5, enter $D$ as the numerator in the Dollars column and 37.5 as the numerator in the Hours column. Because 29.25 is paired with 7.5, enter 29.25 as the denominator in the Dollars column, 7.5 as the denominator in the Hours column, and solve the resulting proportion.

| Dollars | Hours |
|---|---|
| $\dfrac{D}{29.25}$ | $=\dfrac{37.5}{7.5}$ |

Apply the cross-multiplication rule.

$$7.5D = 29.25 \times 37.5$$

Divide each side by 7.5.

$$\frac{\cancel{7.5}\,D}{\cancel{7.5}} = \frac{29.25 \times 37.5}{7.5}$$

$$D = 146.25$$

*Check:*  Does $\dfrac{(146.25)}{29.25} = \dfrac{37.5}{7.5}$?   Does $5 = 5$?   Yes.

The assembler can earn \$146.25 for 37.5 hours of work.

## Exercises for Section 4.7

*Write each ratio as a fraction, and state the terms of the ratio. See Example 1.*

**1.** 5 to 6          **2.** 7 to 8      **3.** 1.2 to 4.3      **4.** 2.5 to 6.8

**5.** $13.2 : 2.12$      **6.** $9.2 : 6.5$      **7.** $9.7 : 3.5$      **8.** $36.3 : 6.2$

*For each proportion, (a) write the extremes, (b) write the means, and (c) verify by using the cross-multiplication rule. See Example 2.*

**9.** $\dfrac{6}{7} = \dfrac{3.75}{4.375}$      **10.** $\dfrac{8}{9} = \dfrac{2.5}{2.8125}$      **11.** $\dfrac{3.5}{7.5} = \dfrac{2.1875}{4.6875}$

**12.** $\dfrac{6.1}{8.3} = \dfrac{1.525}{2.075}$      **13.** $\dfrac{42.7}{13.2} = \dfrac{10.675}{3.3}$      **14.** $\dfrac{27.8}{18.2} = \dfrac{6.95}{4.575}$

*Solve each proportion. Round off each answer to the nearest hundredth. See Example 3.*

**15.** $\dfrac{n}{7.8} = \dfrac{8.6}{4.5}$      **16.** $\dfrac{n}{8.1} = \dfrac{1.7}{16.8}$      **17.** $\dfrac{3.7}{47.8} = \dfrac{d}{50.9}$

**18.** $\dfrac{64.8}{1.83} = \dfrac{d}{2.78}$      **19.** $\dfrac{0.655}{t} = \dfrac{0.146}{6.75}$      **20.** $\dfrac{47.0}{t} = \dfrac{9.03}{35.3}$

**21.** $\dfrac{0.394}{0.634} = \dfrac{0.420}{s}$      **22.** $\dfrac{0.910}{0.119} = \dfrac{97.2}{s}$      **23.** $\dfrac{r}{5.49} = \dfrac{5.75}{4.83}$

**24.** $\dfrac{r}{98.5} = \dfrac{31.2}{33.2}$  **25.** $\dfrac{1.24}{7.79} = \dfrac{a}{8.92}$  **26.** $\dfrac{0.710}{0.113} = \dfrac{a}{9.06}$

*For Exercises 27 to 40, see Examples 4 and 5.*

**27.** An automobile dealer sells tires at two for $86.90. Find the cost of five tires.

**28.** If 4 calculator batteries sell for $2.99, find the cost of 10 batteries.

**29.** Three packages of camera flash bulbs sell for $6.29. Find the cost of 10 packages.

**30.** A piece of aluminum building panel 6 feet long is priced at $7.95. Find the price of 26 feet of paneling.

**31.** A 54-inch by 36-inch photograph is to be reduced in size so that the 36-inch side becomes 12 inches long. How long will the 54-inch side become?

**32.** A linotype operator can set type for 11 pages of a book in 7 hours. How many hours are needed to set type for a book with 484 pages?

**33.** If a worker earns $202.80 in 24 hours, how much will he earn in 40 hours?

**34.** If 34 pins can be cut from a bar 8.5 meters (m) long, how many pins can be cut from a bar 12 meters long (assume no waste in cutting)?

**35.** If a car uses 12 gallons of gasoline
to travel 216 miles, how many
gallons are needed to travel 2970
miles?

**36.** If the car of Exercise 35 has a
gasoline tank that holds 23 gallons,
how far can the car travel on a
tankful of gasoline?

**37.** A 6-inch by 8-inch photograph is to
be enlarged so that the 6-inch side
becomes 24 inches long. How long
will the 8-inch side become?

**38.** A faulty automobile speedometer reads 36 miles per hour when the true
speed is 32 miles per hour. Solve for the missing data.

|             | a  | b  | c  | d  |
|-------------|----|----|----|----|
| Speedometer | 54 | 63 | —— | —— |
| True speed  | —— | —— | 64 | 40 |

**39.** A length of 1 inch on a road map corresponds to 32 actual highway miles.
Determine how many highway miles correspond to each map length
indicated in the table.

|              | a | b | c |        d        |       e        |
|--------------|---|---|---|-----------------|----------------|
| Map (inches) | 4 | 5 | 8 | $7\frac{3}{16}$ | $11\frac{3}{4}$ |
| Highway miles | —— | —— | —— | —— | —— |

**40.** On a floor plan of a house and lot, 1 inch on the drawing corresponds to 4
actual feet. Find the house and lot dimensions for each of the lengths on the
floor plan in Figure 4.6.

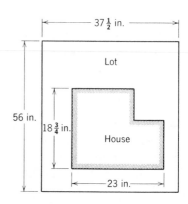

**FIGURE 4.6**

**EXAMPLE** The number of teeth on two gears has a ratio of 2:3. If the larger gear has 72 teeth, how many teeth are on the smaller gear?

**Solution** Let $n$ represent the number of teeth on the smaller gear. Then, since they have a gear ratio of 2:3, we have

$$\frac{n}{72} = \frac{2}{3}$$
$$3n = 144$$
$$n = 48$$

The smaller gear has 48 teeth.

**41.** Two gears have a gear ratio of 3:5. If the larger gear has 125 teeth, how many teeth does the smaller gear have?

**42.** Two gears have 42 and 90 teeth, respectively. If the larger gear is changed to one with 105 teeth, how many teeth must the smaller gear have to keep the same ratio?

**43.** If a gasoline engine has a horsepower-to-weight ratio of 5:6 and weighs 360 pounds, how many horsepower can it produce?

**44.** If an engine has a weight-to-horsepower ratio of 4:7 and produces 357 horsepower, what is its weight (in pounds)?

## 4.8  Mental Calculations

### Exercises for Section 4.8

*Each exercise in this section is to be done mentally.*
*Solve each equation.*

**1.** $x - 10 = 10$   **2.** $x - 25 = 10$   **3.** $t + 30 = 10$

**4.** $t + 55 = 45$   **5.** $30 - 10 = r - 10$   **6.** $25 - 15 = r - 15$

**7.** $40 - 20 = s - 30$   **8.** $60 - 30 = s - 20$   **9.** $20p = 80$

**10.** $30p = 90$   **11.** $15n = 45$   **12.** $25n = 100$

**13.** $\dfrac{10x}{20} = 6$   **14.** $\dfrac{20x}{60} = 10$   **15.** $15 = 3t + 3$

**16.** $25 = 7t + 4$   **17.** $\dfrac{5t}{10} + 6 = 8$   **18.** $\dfrac{3t}{9} + 1 = 4$

**19.** A contractor starts a job with a supply of 4000 feet of wire. After using 500 feet, how many feet of wire are left?

**20.** If a press stamps out 16,000 pieces in 8 hours, how many pieces can it stamp out in 1 hour?

**21.** A store sells motor oil at 12 quarts for $12.00. Find the cost of 16 quarts.

**22.** A machine turns out 24,000 pieces in 8 hours. How many pieces can it turn out in 6 hours?

Use the formula $V = \dfrac{wh}{3}$ in Exercises 23 and 24.

**23.** Find $V$ if $w = 30$ and $h = 40$.

**24.** Find $h$ if $V = 100$ and $w = 75$.

## ■ Review Exercises

### Section 4.1

*Solve each equation. For Exercises 5 to 8, write your answer (a) using floating decimal notation, (b) rounded off to the nearest thousandth, (c) rounded off to three significant digits.*

**1.** $r - 3.2 = 7.09$

**2.** $s + 9.3 = 6.4$

**3.** $0.21 - 0.97 = t - 6.32$

**4.** $4.1 + d - 4.1 = 3.21$

**5.** $6.52n = 42.8$

**6.** $\dfrac{2.70s}{9.28} = 1.41$

**7.** $5.66 = 0.65n + 1.52$

**8.** $\dfrac{0.29d}{4.10} - 0.57 = 0.06$

### Section 4.2

*Solve each equation. Round off to the nearest tenth.*

**9.** $94.6 - N = 5.62$

**10.** $4.9 = \dfrac{8.6}{r}$

**11.** $6.29 - 7T = 26.51$

**12.** $\dfrac{2.07}{R} - 6.32 = 3.68$

**13.** $x^2 = (8.9)^2 + (8.1)^2$      **14.** $\dfrac{5b^2}{9} + 7.61 = 38.24$

## Section 4.3

*For Exercises 15 to 18, round off each answer to the nearest thousandth.*

*Use the formula* $\quad H = \dfrac{62.4fd}{550}\quad$ *in Exercises 15 and 16.*

**15.** $f = 67.2,\quad d = 3.45.\quad$ Find $H$.

**16.** $H = 21.341,\quad d = 6.412.\quad$ Find $f$.

*Use the formula* $\quad B = \dfrac{6.2832nwR}{33,000}\quad$ *in Exercises 17 and 18.*

**17.** $n = 78.4,\quad w = 8.75,\quad R = 146.7.\quad$ Find $B$.

**18.** $B = 32,\quad n = 81.2,\quad w = 9.16.\quad$ Find $R$.

## Section 4.4

*Solve for the indicated unknown.*

**19.** $l = 12 + d,\quad$ for $d$.      **20.** $E = mc^2,\quad$ for $m$.

**21.** $V = lwh,\quad$ for $l$.      **22.** $E = mc^2,\quad$ for $c$.

## Section 4.5

**23.** A contractor starts a building with a supply of 2980 feet of baseboard. After using 1153 feet, how many feet are left in the supply?

**24.** Four thickness gauges with thicknesses of 0.1000 inch, 0.2500 inch, 0.5000 inch, and 0.0025 inch are placed on top of each other. What is their combined thickness?

*Find the missing dimension (in inches) in each drawing.*

**25.**
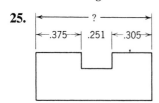

**26.**
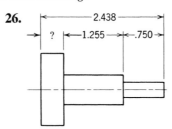

## Section 4.6

*For Exercises 27 to 30, (a) write an equation, and (b) solve the equation.*

**27.** If a person watches television 20 hours each week, how many hours will the person watch in 52 weeks?

**28.** If one fitting for a machine requires 141.3 grams (g) of aluminum during manufacturing, how many grams are required to make 12 fittings?

**29.** If a punch press stamps out 27,800 pieces in 8 hours, how many are stamped in 1 hour?

**30.** A carpenter divides a board into six pieces of equal length. If the board is 84 inches long, how long is each piece? Make no allowance for saw cuts.

## Section 4.7

**31.** Write the ratio $5:7$ as a fraction and state the terms of the ratio.

**32.** Write (*a*) the extremes and (*b*) the means of the proportion:

$$\frac{30.6}{16.2} = \frac{5.1}{2.7},$$

and (*c*) verify the equality by using the cross-multiplication rule.

*Solve each proportion in Exercises 33 and 34. Round off each answer to three significant digits.*

**33.** $\dfrac{n}{1.19} = \dfrac{3.09}{0.834}$    **34.** $\dfrac{8.61}{74.8} = \dfrac{5.33}{r}$

**35.** A service station operator sells motor oil at 12 quarts for $5.95. Find the cost of 8 quarts, to the nearest cent.

**36.** An automatic machine can turn out 40,816 pieces in 8 hours. How many pieces can it make in 3 hours?

## Section 4.8

*Each of the following exercises is to be done mentally.*
*Solve each equation.*

**37.** $x - 5 = 7$    **38.** $t + 8 = 20$

**39.** $6p = 24$     **40.** $\dfrac{5x}{3} = 10$

**41.** A machine turns out 16,000 pieces in 4 hours. How many pieces can it turn out in 6 hours?

**42.** If  $V = 10lw$,  find $V$ if  $l = 8$  and  $w = 3$.

# Percent

## 5.1 Percent Equivalents

In this section we review certain relations between percents, decimals, and fractions.

### ■ Decimal-Percent Equivalents

The percent symbol, %, is used to mean *per hundred*. For example,

$$78 \text{ per } 100 \longrightarrow \frac{78}{100} = 0.78 = 78\%$$

and

$$6 \text{ per } 100 \longrightarrow \frac{6}{100} = 0.06 = 6\%$$

The following two rules can be used to change from a decimal to a percent and from a percent to a decimal.

---

**To Change a Decimal to a Percent**

Move the decimal point two places to the right. Add or drop zeros as needed and add the percent symbol to follow the last digit.

---

**EXAMPLE 1**
**(a)** $0.29 = 29\%$    **(b)** $0.067 = 6.7\%$
**(c)** $0.125 = 12.5\%$    **(d)** $1 = 1.00 = 100\%$

---

**To Change a Percent to a Decimal**

Move the decimal point two places to the left. Add or drop zeros as needed and drop the percent symbol.

---

### EXAMPLE 2
(a) $54\% = 0.54$          (b) $0.02\% = 0.0002$
(c) $112\% = 1.12$          (d) $200\% = 2.00 = 2$

Percents that involve fractional parts can be changed to decimal forms by first changing the fractional part to its decimal equivalent.

### EXAMPLE 3
(a) $6\frac{3}{4}\% = 6.75\%$          (b) $33\frac{1}{3}\% = 33.3333333\%*$
$= 0.0675$                   $= 0.333333333$
(c) $\frac{3}{4}\% = 0.75\%$          (d) $\frac{1}{3}\% = 0.3333333\%*$
$= 0.0075$                   $= 0.003333333$

## ■ Fraction-Percent Equivalents

Fractions can be changed to percents by first changing the fraction to its decimal equivalent.

### EXAMPLE 4
(a) $\dfrac{3}{5} = 0.6 = 60\%$          (b) $\dfrac{13}{5} = 2.6 = 260\%$

(c) $\dfrac{7}{8} = 0.875 = 87.5\%$          (d) $\dfrac{1}{200} = 0.005 = 0.5\%$

(e) $\dfrac{8}{11} = 0.72727273* = 72.73\%$, to the nearest hundredth of a percent.

Percents can be changed to fractions by writing an appropriate fraction with the denominator 100 and reducing to lowest terms, if necessary.

### EXAMPLE 5
(a) $80\% = \dfrac{80}{100} = \dfrac{4}{5}$          (b) $112\% = \dfrac{112}{100} = \dfrac{28}{25}$

(c) $0.7\% = \dfrac{0.7}{100} = \dfrac{7}{1000}$          (d) $0.07\% = \dfrac{0.07}{100} = \dfrac{7}{10,000}$

Percents with fractional parts can also be changed to fractions. Example 6a shows a method that can be used when the fractional part results in a terminating decimal; Example 6b shows a method for use when the fractional part results in a nonterminating decimal.

### EXAMPLE 6
(a) $12\frac{1}{2}\% = 12.5\% = \dfrac{12.5}{100}$          (b) $16\frac{2}{3}\% = \dfrac{16\frac{2}{3}}{100} = 16\frac{2}{3} \div 100$

$= \dfrac{125}{1000} = \dfrac{1}{8}$                   $= \dfrac{50}{3} \cdot \dfrac{1}{100} = \dfrac{1}{6}$

---

*Very close approximation.

## *Exercises for Section 5.1*

*Write each decimal as a percent. See Example 1.*

**1.** 0.33    **2.** 0.54    **3.** 0.70    **4.** 0.06

**5.** 0.08    **6.** 0.212    **7.** 0.504    **8.** 0.686

**9.** 0.787    **10.** 0.074    **11.** 0.021    **12.** 0.005

**13.** 0.008    **14.** 0.0008    **15.** 5.5    **16.** 6

*Write each percent as a decimal. See Example 2.*

**17.** 15%    **18.** 25%    **19.** 0.27%    **20.** 0.62%

**21.** 6.8%    **22.** 1.1%    **23.** 2.7%    **24.** 5.8%

**25.** 119%    **26.** 652%    **27.** 259%    **28.** 528%

*Write each percent as a decimal. See Example 3.*

**29.** $3\frac{1}{4}\%$    **30.** $6\frac{1}{8}\%$    **31.** $7\frac{1}{2}\%$    **32.** $1\frac{5}{8}\%$

**33.** $2\frac{3}{8}\%$    **34.** $8\frac{3}{4}\%$    **35.** $8\frac{2}{5}\%$    **36.** $9\frac{3}{5}\%$

**37.** $16\frac{1}{3}\%$    **38.** $14\frac{1}{6}\%$    **39.** $15\frac{2}{3}\%$    **40.** $18\frac{5}{6}\%$

**41.** $\frac{1}{2}\%$    **42.** $\frac{1}{4}\%$    **43.** $\frac{1}{6}\%$    **44.** $\frac{2}{3}\%$

*Write each fraction as a percent. See Example 4.*

**45.** $\frac{3}{4}$    **46.** $\frac{1}{5}$    **47.** $\frac{3}{8}$    **48.** $\frac{5}{5}$

**49.** $\dfrac{3}{5}$    **50.** $\dfrac{3}{20}$    **51.** $\dfrac{9}{4}$    **52.** $\dfrac{8}{5}$

**53.** $\dfrac{12}{5}$    **54.** $\dfrac{11}{4}$    **55.** $\dfrac{1}{250}$    **56.** $\dfrac{1}{500}$

*Write each fraction as a percent rounded off:* (*a*) *to the nearest tenth of a percent, and* (*b*) *to the nearest hundredth of a percent. See Example 4e.*

**57.** $\dfrac{2}{7}$    **58.** $\dfrac{1}{3}$    **59.** $\dfrac{5}{12}$    **60.** $\dfrac{7}{12}$

**61.** $\dfrac{11}{24}$    **62.** $\dfrac{1}{6}$    **63.** $\dfrac{5}{6}$    **64.** $\dfrac{13}{24}$

**65.** $\dfrac{13}{6}$    **66.** $\dfrac{17}{12}$    **67.** $\dfrac{1}{24}$    **68.** $\dfrac{1}{360}$

*Write each percent as a fraction. See Examples 5 and 6.*

**69.** 25%    **70.** 10%    **71.** 75%    **72.** 87%

**73.** 125%    **74.** 150%    **75.** 0.9%    **76.** 0.11%

**77.** $37\frac{1}{2}\%$    **78.** $87\frac{1}{2}\%$    **79.** $15\frac{1}{4}\%$    **80.** $12\frac{3}{4}\%$

**81.** $33\frac{1}{3}\%$    **82.** $66\frac{2}{3}\%$    **83.** $83\frac{1}{3}\%$    **84.** $15\frac{1}{7}\%$

## 5.2   Finding a Percent of a Number

Of the many problems that involve the use of percent, three types are basic. In this section we consider the problem of "finding a percent of a number" (often referred to as Case I).

### ■ Case I Computations with Decimal Forms

Recall from Section 2.6 that the word "of" is often used to indicate multiplication and from Section 4.5 that the word "is" often translates into the = symbol. Hence, a sentence such as

$$83\% \text{ of } 731.6 \text{ is what number?}$$

translates into the equation

$$0.83 \times 731.6 = N$$

where 83% has been changed to decimal form and $N$ represents the required number. Because this type of problem is so frequently encountered, it is practical to use the following rule:

---

### To Find a Percent of a Number
**1.** Change the percent to a decimal.
**2.** Multiply the number by the decimal.

---

### EXAMPLE 1
**(a)** 83% of 731.6 = 0.83 × 731.6
$$= 607.228$$

**(b)** $112\frac{1}{2}\%$ of 731.6 = 112.5% of 731.6
$$= 1.125 \times 731.6$$
$$= 823.05$$

**(c)** 0.08% of 731.6 = 0.0008 × 731.6
$$= 0.58528$$

**(d)** $\frac{1}{5}\%$ of 731.6 = 0.2% of 731.6
$$= 0.002 \times 731.6$$
$$= 1.4632$$

### ■ Case I Computations with Fractions

For certain percents, if the fraction form is known, it may be simpler to use the fraction form rather than the decimal form.

In Example 6*b* of Section 5.1 we changed $16\frac{2}{3}\%$ to the fraction $\frac{1}{6}$. To avoid such computations it is helpful to be familiar with the following useful conversions when making Case I computations involving these percents:

$$33\tfrac{1}{3}\% = \frac{1}{3}, \qquad 66\tfrac{2}{3}\% = \frac{2}{3},$$

$$16\tfrac{2}{3}\% = \frac{1}{6}, \quad \text{and} \quad 83\tfrac{1}{3}\% = \frac{5}{6}$$

**EXAMPLE 2**  Using the fact that  $16\tfrac{2}{3}\% = \frac{1}{6}$,

$$16\tfrac{2}{3}\% \text{ of } 843.02 = \frac{1}{6} \times 843.02$$

$$= \frac{843.02}{6} = 140.50333$$

### ■ Case I Applications

Throughout this chapter we consider technical applications in which the three basic types of percent problems are used. However, it is not essential for you to be completely familiar with the technical details for each of these applications.

**EXAMPLE 3**  The "effective value" of the current in an AC electrical circuit, as shown on an ammeter, is approximately equal to 70.7% of the maximum (or peak) value of the current. To the nearest tenth, find the effective value of the current in a circuit with a maximum current of 35 amperes (amp).

**Solution**  Let $I$ represent the effective value of the current. Then

$$I = 70.7\% \text{ of } 35$$
$$= 0.707 \times 35 = 24.745$$

The effective value of the current is 24.7 amperes, to the nearest tenth.

**EXAMPLE 4**  A machine produces 18,560 parts per day. At the end of each day, if more than 1.2% of the parts are defective, the machine must be readjusted. What is the smallest number of defective parts that will require the machine to be readjusted?

**Solution**  Let $P$ represent the smallest number of defective parts. Then

$$P = 1.2\% \text{ of } 18,560$$
$$= 0.012 \times 18,560 = 222.72$$

The machine will require adjustment if at least 223 parts (to the nearest whole number) are defective.

## Exercises for Section 5.2

*Find each result to the nearest hundredth. See Example 1.*

**1.** 40% of 60.15

**2.** 70% of 198.86

**3.** 29% of 169.12

**4.** 72% of 816.7

**5.** 115% of 53.8

**6.** 232% of 1.18

**7.** 6.5% of 59.65

**8.** 8.07% of 283.41

**9.** 7.75% of 3025.63

**10.** 9.04% of 12,476.09

**11.** $37\frac{1}{2}\%$ of 407.134

**12.** $87\frac{1}{2}\%$ of 74.33

**13.** 0.002% of 3648.5

**14.** 0.0015% of 2349.15

**15.** $\frac{1}{4}\%$ of 16.64

**16.** $\frac{3}{4}\%$ of 214

**17.** $\frac{2}{5}\%$ of 79.63

**18.** $\frac{1}{2}\%$ of 4563.22

**19.** $116\frac{1}{2}\%$ of 981.06     **20.** $266\frac{1}{4}\%$ of 2304

*Find each result to the nearest hundredth. See Example 2.*

**21.** $33\frac{1}{3}\%$ of 111.69     **22.** $33\frac{1}{3}\%$ of 472.5

**23.** $66\frac{2}{3}\%$ of 70.26     **24.** $66\frac{2}{3}\%$ of 81.09

**25.** $16\frac{2}{3}\%$ of 18.32     **26.** $16\frac{2}{3}\%$ of 50.96

**27.** $33\frac{1}{3}\%$ of 56.95     **28.** $33\frac{1}{3}\%$ of 33.45

**29.** $66\frac{2}{3}\%$ of 46.26     **30.** $16\frac{2}{3}\%$ of 2236.32

*To the nearest tenth, find the effective value of a current having the given maximum current. See Example 3.*

**31.** 27 amperes     **32.** 15 amperes

**33.** 37.4 amperes     **34.** 46.0 amperes

*The maximum value of the voltage of an electric current is approximately 141% of the effective value of the voltage, as shown on a voltmeter. To the nearest tenth, find the maximum value of the voltage of a current having the given effective value.*

**35.** 165 volts     **36.** 220 volts

**37.** 200 volts     **38.** 117 volts

*A metal alloy, sometimes called "brazing metal," is composed of 85% copper and 15% zinc. To the nearest hundredth, find the number of kilograms of copper in the given weight of brazing metal.*

**39.** 7.12 kilograms     **40.** 31.0 kilograms

**41.** 11.19 kilograms     **42.** 9.89 kilograms

**43.** 51.1 kilograms        **44.** 26.1 kilograms

**45.** In building a house, a contractor allows 2% of the total cost for excavation. How much is allowed for excavating if the total cost of the house is $94,760?

**46.** On a commercial building, a contractor allows 9% of the total cost for carpenter work. How much is allowed for carpentry if the total cost is $76,807?

**47.** A belt-driven lathe is set to turn at 2500 revolutions per minute (see Figure 5.1). If the belt has a 12% slippage, how fast is the lathe actually turning?

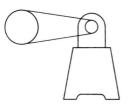

**FIGURE 5.1**

**48.** A casting is poured using an alloy that shrinks $1\frac{1}{2}$% as it cools. If the length of a casting is 95 centimeters, what is the cooled length of the casting to the nearest tenth of a centimeter?

**49.** When machining a part from a forging, as shown in Figure 5.2, 18% of the material is removed. If the forging weighed 2.800 kilograms originally, find the weight of the finished part.

**FIGURE 5.2**

**50.** A brick mason estimates that, as a result of breakage, 8% of the bricks delivered to a job site are not usable. If 3750 bricks are delivered, how many bricks can be used?

## 5.3 Percent: Case II and Case III

In this section we consider the second and third types of basic percent problems. The second basic percent problem (Case II) can be described as:

"To find what percent one number is of another"

### ∎ Case II Computations

Two different phrases can be used to express this type of problem. For example, both

"132 is what percent of 150"     and     "What percent of 150 is 132"

express the same condition on the percent. Thus, if $P$ represents the required percent, we translate "of" to indicate multiplication and "is" to indicate the $=$ symbol to obtain the respective equations:

$$132 = P \times 150 \quad \text{and} \quad P \times 150 = 132$$

From either equation (we shall use the latter)

$$\frac{P \times 150}{150} = \frac{132}{150}$$

$$P = 0.88 = 88\%$$

To *check*, we apply Case I and verify that

$$88\% \text{ of } 150 = 0.88 \times 150 = 132$$

### EXAMPLE 1
**(a)** 115 is what percent of 250?     **(b)** What percent of 440 is 27.5?

**Solutions**  Let $P$ represent the required percent.

**(a)** $115 = P \times 250$

$$\frac{115}{250} = \frac{P \times 250}{250}$$

$$0.46 = P$$

$$P = 46\%$$

Check:
$$46\% \text{ of } 250 = 0.46 \times 250$$
$$= 115$$

**(b)** $P \times 440 = 27.5$

$$\frac{P \times 440}{440} = \frac{27.5}{440}$$

$$P = 0.0625$$

$$P = 6.25\%$$

Check:
$$6.25\% \text{ of } 440 = 0.0625 \times 440$$
$$= 27.5$$

In each of the above examples, a quotient is computed to obtain the required percent, and the computation is checked by finding a product. If the quotient is a nonterminating decimal, recall that we check the computation by using the quotient *before* it is rounded off.

**EXAMPLE 2**  37 is what percent of 65? Express the result to the nearest tenth of a percent.

**Solution**  Let $P$ represent the required percent. Then

$$37 = P \times 65$$

$$\frac{37}{65} = \frac{P \times 65}{65}$$

$$0.56923077 = P$$

$$P = 56.923077\%$$

*Check*:  $56.923077\%$  of  $65 = 0.56923077 \times 65 = 36.999996.$
To the nearest tenth,  $P = 56.9\%$.

## ■ *Case II Applications*

The following examples indicate how Case II computations can be used to solve problems.

**EXAMPLE 3**   The "efficiency" of a motor is the percent the power output is of the power input of the motor. To the nearest tenth of a percent, find the efficiency of a motor that produces a power output of 3.6 kilowatts (kw) with a power input of 5.25 kilowatts.

**Solution**   Consider the question, "The output is what percent of the input?"

$$3.6 \text{ is what percent of } 5.25?$$

and let $E$ represent the required percent. Then

$$3.6 = E \times 5.25$$
$$\frac{3.6}{5.25} = \frac{E \times 5.25}{5.25}$$
$$0.68571429 = E$$
$$E = 68.571429\%$$

*Check*:   $0.68571429 \times 5.25 = 3.6$
The efficiency of the motor is 68.6%, to the nearest tenth.

When a change has occurred, we are often interested in the *percent increase* or the *percent decrease* of the change. To make this computation, we first find the amount of change that has occurred and then compute what percent the amount of change is of the *original* number.

**EXAMPLE 4**   The mileage of a certain car is listed as 12.7 miles per gallon (mpg). After replacing certain steel parts with aluminum and plastic, the car weighed less and the mileage increased to 15.6 mpg. To the nearest tenth of a percent, find the percent increase of the mileage.

**Solution**   The change in mileage is   $15.6 - 12.7 = 2.9$ miles per gallon.   Because 12.7 is the original number, we consider the question

$$2.9 \text{ is what percent of } 12.7?$$

and let $P$ represent the required percent. Then

$$2.9 = P \times 12.7$$
$$\frac{2.9}{12.7} = \frac{P \times 12.7}{12.7}$$
$$0.22834646 = P$$
$$P = 22.834646\%$$

*Check*:   $0.22834646 \times 12.7 = 2.9$
To the nearest tenth, the mileage increased by 22.8%.

The steel pin in Figure 5.3 is to be machined to a length of 4.825 inches ±0.025 inch (the symbol ± is read "plus or minus"). The dimension 4.825 inches is the *basic size* of the pin. To allow for errors in manufacturing, the length of the pin can vary from an *oversize* of 4.825 inches +0.025 inch, or 4.850 inches, to an *undersize* of 4.825 inches minus 0.025 inch, or 4.800 inches. These two dimensions are called the *dimension limits*. The difference of the dimension limits is called the

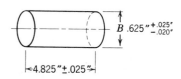

**FIGURE 5.3**

*tolerance*. Thus the tolerance in the length of the pin is 4.850 − 4.800, or 0.050 inch. The level of precision of the machining of a part is specified by the *percent of relative error*, which is computed by

$$\frac{1}{2} \text{ of the tolerance as a percent of the basic size}$$

Low values for percents of relative error indicate high levels of precision.

**EXAMPLE 5**   Find the percent of relative error of dimension *B* of Figure 5.3.

**Solution**   The basic size is 0.625 inch.

$$\begin{aligned} \text{Oversize:} \quad & 0.625 + 0.025 = 0.650 \text{ inch} \\ \text{Undersize:} \quad & 0.625 - 0.020 = 0.605 \text{ inch} \\ \text{Tolerance:} \quad & 0.650 - 0.605 = 0.045 \text{ inch} \end{aligned}$$

Now consider the question

$$\frac{1}{2} \text{ of } 0.045 \text{ is what percent of } 0.625?$$

Let *P* represent the required percent. Then

$$\frac{1}{2}(0.045) = P \times 0.625$$

$$0.0225 = P \times 0.625$$

Divide each side by 0.625.

$$\frac{0.0225}{0.625} = \frac{P \times 0.625}{0.625}$$

$$0.036 = P$$

Hence, the percent of relative error is 3.6%.

## ∎ *Case III Computations*

The third type of basic percent problem can be described as

"Given a percent of a number, find the number."

Two different phrases can be used to express this type of problem. For example, both

"144 is 90% of what number?"      and      "90% of what number is 144?"

express the same condition on the number. Thus, if *n* represents the required number and 90% is changed to 0.9, we translate "of" to indicate multiplication and "is" to indicate the = symbol to obtain the respective equations

$$144 = 0.9 \times n \quad \text{or} \quad 0.9 \times n = 144$$

From either equation (we shall use the latter)

$$\frac{0.9 \times n}{0.9} = \frac{144}{0.9}$$

$$n = 160$$

To *check*, we apply Case I and verify that

$$90\% \text{ of } 160 = 0.9 \times 160 = 144$$

### EXAMPLE 6
**(a)** 46.8 is 60% of what number?      **(b)** $4\frac{1}{2}$% of what number is 292.5?

### Solutions
**(a)** Change 60% to 0.6.

$$46.8 = 0.6 \times n$$

Divide each side by 0.6.

$$\frac{46.8}{0.6} = \frac{0.6 \times n}{0.6}$$

$$78 = n$$

*Check*:
60% of 78 = 0.6 × 78 = 46.8

**(b)** Change $4\frac{1}{2}$% to 0.045

$$0.045 \times n = 292.5$$

Divide each side by 0.045.

$$\frac{0.045 \times n}{0.045} = \frac{292.5}{0.045}$$

$$n = 6500$$

*Check*:
$4\frac{1}{2}$% of 6500 = 0.045 × 6500 = 292.5

**EXAMPLE 7**   17% of what number is 58? Express the answer to the nearest tenth.

**Solution**   Let $n$ represent the required number, and change 17% to 0.17. Then

$$0.17 \times n = 58$$

$$\frac{0.17 \times n}{0.17} = \frac{58}{0.17}$$

$$n = 341.17647$$

*Check*:   17% of 341.17647 = 0.17 × 341.17647 = 57.999999;   which result is very "close" to 58.

To the nearest tenth, the required number is 341.2.

**EXAMPLE 8**   832.3 is $66\frac{2}{3}$% of what number?

**Solution**   Let $n$ represent the required number. Change $66\frac{2}{3}$% to $\frac{2}{3}$.

$$832.3 = \frac{2}{3} \times n$$

Multiply each side by $\frac{3}{2}$, the reciprocal of $\frac{2}{3}$.

$$\frac{3}{2} \times 832.3 = \frac{3}{2} \times \frac{2}{3} \times n$$

$$1248.45 = n$$

*Check*:   $66\frac{2}{3}$% of 1248.45 = $\frac{2}{3}$ × 1248.45 = 832.3

### ■ Case III Applications

The following example indicates how Case III computations can be used to solve problems.

**EXAMPLE 9**   The effective value of the voltage in an AC electric circuit, as shown on a voltmeter, is approximately 70.7% of the maximum (or peak) value of the voltage. To the nearest tenth, find the maximum voltage in a circuit with an effective voltage of 86.5 volts.

**Solution**   Consider the equation

$$86.5 \text{ is } 70.7\% \text{ of what number?}$$

and let *V* represent the required number. Then

$$86.5 = 70.7\% \text{ of } V$$
$$86.5 = 0.707 \times V$$
$$\frac{86.5}{0.707} = \frac{0.707 \times V}{0.707}$$
$$122.34795 = V$$

*Check*:   $0.707 \times 122.34795 = 86.500001$
To the nearest tenth, the maximum voltage is 122.3 volts.

## Exercises for Section 5.3

*For Exercises 1 to 8, (a) write an equation, and (b) solve the equation to find the percent. Express each answer to the nearest tenth of a percent. See Examples 1 and 2.*

**1.** 8 is what percent of 25?      **2.** 7 is what percent of 8?

**3.** What percent of 64 is $16\frac{1}{2}$?      **4.** What percent of 80 is $22\frac{1}{2}$?

**5.** 102 is what percent of 42?      **6.** 245 is what percent of 80?

**7.** What percent of 63 is 1.56?      **8.** What percent of 53 is 4.09?

*To the nearest tenth of a percent, find the efficiency of an electric motor having the given power input and output in kilowatts. See Example 3.*

**9.** 5.2 input,   3.7 output      **10.** 8.0 input,   5.1 output

**11.** 7.9 input,   4.5 output      **12.** 9.7 input,   7.2 output

*For Exercises 13 to 18, express answers to the nearest tenth of a percent.*

**13.** A gear weighing 127 pounds is machined from a casting weighing 160 pounds. What percent of the original weight of the casting is removed in finishing?

**14.** A metal casting is 210.3 centimeters long when it is cast and 209.7 centimeters long after it has cooled. What percent is the cooled length of the cast length?

*For Exercises 15 to 18, see Example 4.*

**15.** After some adjustments were made, the power output of a motor increased from 24.6 horsepower to 27.4 horsepower. Find the percent of increase.

**16.** By simplifying procedures, the number of bushings produced in a shop was increased from 3640 a day to 4025 a day. Find the percent of increase.

**17.** By redesigning the mold, the metal removed when machining a casting was reduced from 3.6 kilograms to 3.1 kilograms. Find the percent of decrease.

**18.** On a new lathe, the time required to produce a part was reduced from 17.3 minutes to 14.6 minutes. Find the percent of decrease.

*In Exercises 19 to 22, find the percent of relative error (to the nearest tenth of a percent) if the dimension has the given tolerance. See Example 5 and Figure 5.4.*

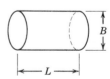

**FIGURE 5.4**

**19.** $L = 2.829 \begin{smallmatrix} +0.010 \\ -0.015 \end{smallmatrix}$ inch

**20.** $L = 3.210 \begin{smallmatrix} +0.045 \\ -0.007 \end{smallmatrix}$ inch

**21.** $B = 0.427 \begin{smallmatrix} +0.005 \\ -0.030 \end{smallmatrix}$ inch

**22.** $B = 0.625 \begin{smallmatrix} +0.065 \\ -0.012 \end{smallmatrix}$ inch

*For Exercises 23 to 32, (a) write an equation, and (b) solve the equation to find the number. Express answers to the nearest tenth. See Examples 6, 7, and 8.*

**23.** 68 is 27% of what number?

**24.** 80 is 30% of what number?

**25.** 60% of what number is 80?

**26.** 75% of what number is 16?

**27.** 212 is $7\frac{3}{4}$% of what number?

**28.** 136 is $8\frac{1}{2}$% of what number?

**29.** 6.05% of what number is 443?

**30.** 8.06% of what number is 693?

**31.** 106% of what number is 555.069?

**32.** 121% of what number is 30.855?

*For Exercises 33 to 36, see Example 9. Round off answers to three significant digits.*

**33.** The effective value of an alternating current is approximately 63.7% of the maximum value. Find the maximum value of an alternating current if the effective value is 36.3 amperes.

**34.** The effective value of an alternating current is 43.4 amperes. Find the maximum value. See Exercise 33.

**35.** In a manufacturing plant, $2\frac{1}{2}$% of the screwdrivers produced are defective. In one lot, 117 were defective. How many screwdrivers were in the entire lot?

**36.** A certain alloy contains 11% tin. If one casting contains 12 kilograms of tin, what is the weight of the casting?

*Exercises 37 to 46 include Case I, Case II, and Case III types of percent problems. Round off percent answers to the nearest tenth of a percent and other answers to the nearest hundredth.*

**37.** 63 is what percent of 53?       **38.** 21.2% of what number is 73.5?

**39.** What is 6.08% of 93.6?       **40.** What percent of 3.7 is 5.2?

**41.** 65.1 is 46% of what number?       **42.** 9.4% of 27.3 is what number?

**43.** A certain automobile engine burns only about 88% of the gasoline it uses. If it uses 95 liters of gasoline during a performance test, how many liters did it actually burn?

**44.** A tool salesman is paid 15% of his sales. If the amount of sales for 1 week is $1674, how much was he paid?

**45.** If a shipment of 800 gallons of fuel contains 250 gallons of type Y fuel, what percent of the fuel is type Y?

**46.** A certain satellite is in orbit around the earth such that its minimum distance from the earth is 1230 miles and its maximum distance is 2010 miles. The minimum distance is what percent of the maximum distance?

## 5.4 Further Applications of Percent

In this section we consider some applications of percent that are related to business.

### ■ Discount

The specified price of an article is called the **list price** (or **selling price**). When a dealer buys an article from a manufacturer, the dealer usually pays the **net price** (or **cost price**). The net price of an article can be determined by subtracting a certain percent of the list price from the list price. The amount to be subtracted is called the **discount**. The percent used to compute the discount is called the **discount rate**.

**EXAMPLE 1**  The list price of a set of four shock absorbers is $49.95. If a discount rate of 35% is given to the dealer, find the net price of the shock absorbers.

**Solution**  Let $D$ represent the discount. Then

$$D = 35\% \text{ of } 49.95$$
$$D = 0.35 \times 49.95 = 17.4825$$

To the nearest cent, the discount is $17.48. Now let $N$ represent the net price. Then

$$N = 49.95 - 17.48 = 32.47$$

The net price is $32.47.

In some cases, more than one discount is offered on an article.

**EXAMPLE 2**  On the $49.95 shock absorbers of Example 1, a dealer offers a regular 35% discount and an additional discount of 5% of the first net price if the buyer pays cash. Find the net price for a cash sale.

**Solution**  From Example 1, the shock absorbers are priced at $32.47 after the first (35%) discount has been subtracted. Let $D$ represent the second (5%) discount. Then

$$D = 5\% \text{ of } 32.47$$
$$D = 0.05 \times 32.47 = 1.6235$$

The second discount is $1.62, to the nearest cent. Let $N$ represent the net price. Then,

$$N = 32.47 - 1.62 = 30.85$$

To the nearest cent, the net price after two discounts is $30.85.

### ■ Markup

In most retail stores, the selling price of an article is determined by "marking up" the cost price. Thus, a "markup of 30%" means that 30% of the cost price is to be *added* to the cost price to determine the selling price.

**EXAMPLE 3**  To the nearest cent, find the selling price of a set of metric socket wrenches if the cost price is $17.37 and the markup is 42%.

**Solution**  Let $S$ represent the selling price. Then

$$S = (42\% \text{ of } 17.37) + 17.37$$
$$S = 0.42 \times 17.37 + 17.37 = 24.6654$$

The selling price is $24.67, to the nearest cent.

## ■ Depreciation

*Depreciation* is a term used to indicate that the value of an article decreases.

**EXAMPLE 4**   The original value of a turret lathe is $28,500. If the value of the lathe depreciates by 5% each year, what is its value (to the nearest dollar) at the end of 2 years?

**Solution**   Let $F$ represent the value at the end of the first year. Thus, $F$ equals the original value less 5% of the original value:

$$F = 28{,}500 - 5\% \text{ of } 28{,}500$$
$$F = 28{,}500 - 0.05 \times 28{,}500 = 27{,}075$$

The value at the end of the first year is $27,075. Now let $V$ represent the value at the end of the second year. Then

$$V = 27{,}075 - 5\% \text{ of } 27{,}075$$
$$V = 27{,}075 - 0.05 \times 27{,}075 = 25{,}721.25$$

The value at the end of 2 years is $25,721, to the nearest dollar.

## ■ Simple Interest

The charge paid for the use of borrowed money is called the **interest**; the amount of money borrowed is called the **principal**. The **interest rate** is usually given as a *percent per year*.

   If the principal does not change over the entire period of time of a loan, the interest $I$ is called **simple interest** and can be computed by the formula

$$I = Prt$$

where $P$ is the principal, $r$ is the yearly interest rate expressed as a decimal, and $t$ is the time expressed *in years* (or as a fraction of a year).

**EXAMPLE 5**   To the nearest cent, find the simple interest on a principal of $8564 at the yearly interest rate of 9.4% for a time of 3 years.

**Solution**   $P = 8564;$   $r = 0.094;$   $t = 3.$   Thus,

$$I = Prt$$
$$I = (8564)(0.094)(3) = 2415.048$$

The interest is $2415.05, to the nearest cent.

   If the time is a number of months, then $t$ is expressed as a fraction with denominator 12. If the time is a number of days, then we express $t$ as a fraction with denominator 365.

**EXAMPLE 6**   To the nearest cent, find the simple interest on a principal of $6079.85 at the yearly interest rate of 8.07% for (*a*) 13 months, and (*b*) 305 days.

**Solutions**   In each case,   $P = 6079.85$   and   $r = 0.0807.$

(a) The time $t$ is $\dfrac{13}{12}$. Thus,

$$I = Prt$$
$$I = 6079.85 \times 0.0807 \times \frac{13}{12} = \frac{6079.85 \times 0.0807 \times 13}{12}$$
$$I = 531.53089$$

The interest is $531.53, to the nearest cent.

**(b)** The time $t$ is $\dfrac{305}{365}$. Thus,

$$I = Prt$$

$$I = 6079.85 \times 0.0807 \times \frac{305}{365} = \frac{6079.85 \times 0.0807 \times 305}{365}$$

$$I = 409.99011$$

The interest is $409.99, to the nearest cent.

## Exercises for Section 5.4

*For each given list price (l.p.) and discount rate (d.r.), find the net price to the nearest cent. See Example 1.*

**1.** l.p. $73.57
   d.r. 8.07%

**2.** l.p. $53.20
   d.r. 31.8%

**3.** l.p. $141.54
   d.r. 13.1%

**4.** l.p. $188.44
   d.r. 9.1%

**5.** Find the net price of each item.

| Name | List Price | Discount Rate | Net Price |
|------|-----------|---------------|-----------|
| Timing chain | $30.49 | 15% | —— |
| Camshaft | $84.99 | 12% | —— |
| Water pump | $36.95 | 20% | —— |
| Fuel pump | $27.39 | 18% | —— |

**6.** Find the net price of each item.

| Name | List Price | Discount Rate | Net Price |
|------|-----------|---------------|-----------|
| Incinerator | $196.95 | 20% | —— |
| Electric range | $379.95 | 22% | —— |
| Sewing machine | $315.00 | 16% | —— |
| Vacuum cleaner | $225.95 | 15% | —— |

**7.** The net price of an item after a 25% discount is $93.75. Find the list price of the item. (Hint: The net price is 75% of the list price because 25% has been deducted.)

8. The net price of an item after a 32% discount is $348.70. Find the list price, to the nearest cent.

*In Exercises 9 to 16, use the method of Example 2 to find the cost of each item for the given list price and discount rates. Express answers to the nearest cent.*

9. l.p. $81.52
   d.r. 29%,  6%

10. l.p. $72.95
    d.r. 20%,  5%

11. l.p. $483.99
    d.r. 39%,  4%

12. l.p. $248.73
    d.r. 26%,  7%

13. The list price of a drill press is $595. There is a discount to the dealer of 24% and an additional discount of 2% if the drill press is paid for in cash. What is the cost price if a cash payment is made?

14. The list price of paint is $7.95 a gallon. There are discounts of 25% and 20%. What is the cost price of the paint?

15. The list price of a grinder is $670.85. One supplier offers two discounts of 25% and 10%. Another supplier offers one discount of $33\frac{1}{3}\%$. Which is the better bargain?

16. The list price of an arc welder is $165. One supplier offers discounts of 26% and 9%. Another supplier offers discounts of 21% and 14%. Which is the better bargain?

*Find the selling price of an article with the given cost and markup. See Example 3.*

**17.** $418.38;   12%      **18.** $312.22;   20%      **19.** $346.34;   16.7%

**20.** $2160.64;   26.1%      **21.** $6594.13;   13.4%      **22.** $9832.69;   19.8%

**23.**

| Name | Cost | Markup | Selling Price |
|------|------|--------|---------------|
| Axe | $7.88 | 38% | —— |
| Miter box | $74.50 | 41% | —— |
| Bench vise | $36.75 | 27% | —— |
| Hand saw | $9.62 | 36% | —— |

**24.**

| Name | Cost | Markup | Selling Price |
|------|------|--------|---------------|
| Hand drill | $45.83 | 28% | —— |
| Drill bit set | $32.70 | 37% | —— |
| Router bit set | $20.18 | 42% | —— |
| Router | $54.50 | 32% | —— |

*In Exercises 25 to 30, the original value and the yearly rate of depreciation are given. To the nearest dollar, find the value (a) at the end of 1 year, and (b) at the end of the indicated number of years. See Example 4.*

**25.** $986.91;   10%;   after 2 years

**26.** $619.95;   15%;   after 2 years

**27.** $2229.68;   9%;   after 3 years

**28.** $2353.59;   13%;   after 3 years

**29.** The original value of a machine shop was $152,000. If the value depreciates by 7% each year, what is its value (to the nearest dollar) after 3 years?

**30.** The original value of a milling machine was \$35,000. If its value depreciates by 6% each year, what is its value (to the nearest dollar) after 5 years?

*To the nearest cent, find the simple interest for the given principal and yearly interest rate for the indicated time. See Examples 5 and 6.*

**31.** \$494.87;  5%;  2 years        **32.** \$528.02;  8%;  6 years

**33.** \$294.80;  9%;  15 years       **34.** \$463.17;  6%;  8 years

**35.** \$8600;  $16\frac{1}{2}$%;  18 months    **36.** \$3358;  $14\frac{1}{4}$%;  70 months

**37.** \$2431;  8.5%;  5 months       **38.** \$7728;  7.9%;  4 months

**39.** \$590;  9%;  220 days          **40.** \$518;  8%;  164 days

**41.** \$165;  9.3%;  237 days        **42.** \$183;  6.5%;  299 days

*In Exercises 43 to 46, express the answer to the nearest cent.*

**EXAMPLE**  A shop owner borrows \$2000 to pay for equipment. There is a 10% yearly rate of interest on the unpaid balance. If the shop owner makes a payment of \$150 each month, find:
**(a)** The interest for the first month.

**(b)** The unpaid balance remaining after the first payment.
**(c)** The interest for the second month.
**(d)** The unpaid balance remaining after the second payment.

### Solution

**(a)** $P = \$2000;\quad r = 0.10;\quad t = \dfrac{1}{12}$

$$I = Prt = 2000 \times 0.10 \times \frac{1}{12} = 16.6666667$$

The interest is \$16.67 for the first month.

**(b)** Because \$16.67 is the interest paid on the loan, the amount applied to the loan itself is

$$\$150 - \$16.67 = \$133.33$$

Thus, after the first payment, the amount remaining in the unpaid balance is

$$\$2000 - \$133.33 = \$1866.67$$

**(c)** $P = \$1866.67;\quad r = 0.10;\quad t = \dfrac{1}{12}$

$$I = 1866.67 \times 0.10 \times \frac{1}{12} = 15.555583$$

The interest is \$15.56 for the second month.

**(d)** Because \$15.56 is the interest paid on the loan, the amount applied to the loan itself is
$$\$150 - \$15.56 = \$134.44$$

Thus, after the second payment, the amount remaining in the unpaid balance is

$$\$1866.67 - \$134.44 = \$1732.23$$

**43.** From the results of the above example, find:
    **(a)** The interest charged for the third month,
    **(b)** The unpaid balance after the third payment.

**44.** From the results of Exercise 43, find:
    **(a)** The interest charged for the fourth month,
    **(b)** The unpaid balance after the fourth payment.

**45.** Do the above example if the amount borrowed is \$8000, the interest rate is 11%, and the monthly payment is \$450.

**46.** Do the above example if the amount borrowed is \$5000, the interest rate is $9\frac{1}{2}\%$, and the monthly payment is \$300.

## 5.5 Mental Calculations

The following conversions occur so often that you should try to memorize them.

| Fraction | $\frac{1}{4}$ | $\frac{1}{2}$ | $\frac{3}{4}$ | $\frac{1}{3}$ | $\frac{2}{3}$ | $\frac{1}{8}$ | $\frac{3}{8}$ | $\frac{5}{8}$ | $\frac{7}{8}$ | $\frac{1}{6}$ | $\frac{5}{6}$ |
|---|---|---|---|---|---|---|---|---|---|---|---|
| Decimal | 0.25 | 0.5 | 0.75 | $0.33\frac{1}{3}$ | $0.66\frac{2}{3}$ | $0.12\frac{1}{2}$ | $0.37\frac{1}{2}$ | $0.62\frac{1}{2}$ | $0.87\frac{1}{2}$ | $0.16\frac{2}{3}$ | $0.83\frac{1}{3}$ |
| Percent | 25% | 50% | 75% | $33\frac{1}{3}\%$ | $66\frac{2}{3}\%$ | $12\frac{1}{2}\%$ | $37\frac{1}{2}\%$ | $62\frac{1}{2}\%$ | $87\frac{1}{2}\%$ | $16\frac{2}{3}\%$ | $83\frac{1}{3}\%$ |

### Exercises for Section 5.5

*Each exercise in this section is to be done without a calculator. Compute.*

**1.** 10% of 67.2

**2.** 10% of 71.8

**3.** 10% of 89.16

**4.** 10% of 33.47

**5.** 20% of 100

**6.** 30% of 1000

**7.** 200% of 100

**8.** 300% of 1000

**9.** 1% of 1000

**10.** 1% of 10,000

**11.** 50 is what percent of 100?

**12.** 5 is what percent of 20?

**13.** 20 is what percent of 30?

**14.** 20 is what percent of 60?

**15.** 30 is what percent of 40?

**16.** 60 is what percent of 200?

**17.** 80 is 50% of what number?

**18.** 25 is 50% of what number?

**19.** 20 is 25% of what number?

**20.** 12 is 25% of what number?

**21.** 60 is $33\frac{1}{3}\%$ of what number?

**22.** 40 is $33\frac{1}{3}\%$ of what number?

**23.** 25% of 20

**24.** 75% of 40

**25.** $12\frac{1}{2}$% of 16

**26.** $37\frac{1}{2}$% of 16

**27.** $66\frac{2}{3}$% of 24

**28.** $33\frac{1}{3}$% of 45

**29.** 1 is what percent of 6?

**30.** 5 is what percent of 6?

*Estimate each of the following.*

**31.** 11% of 88.1

**32.** 9% of 212.2

**33.** $33\frac{1}{3}$% of 65.4

**34.** $66\frac{2}{3}$% of 29.7

**35.** 7.1 is what percent of 21.9?

**36.** 8.9 is what percent of 44.8?

**37.** 19.9 is what percent of 200?

**38.** 0.91 is what percent of 1.84?

■ *Review Exercises*

**Section 5.1**

*Write each decimal as a percent.*

**1.** 0.35    **2.** 0.087    **3.** 5

*Write each percent as a decimal.*

**4.** 16%    **5.** 215%    **6.** 0.03%

**7.** $2\frac{1}{4}$%    **8.** $\frac{3}{4}$%    **9.** $21\frac{2}{3}$%

*Write each fraction as a percent.*

**10.** $\frac{2}{5}$    **11.** $\frac{1}{400}$

*Write each fraction as a percent rounded off: (a) to the nearest tenth of a percent, and (b) to the nearest hundredth of a percent.*

**12.** $\dfrac{4}{7}$     **13.** $\dfrac{15}{11}$

*Write each percent as a fraction.*

**14.** 75%       **15.** $87\frac{1}{2}\%$

## Section 5.2

*Find each result to the nearest hundredth.*

**16.** 47% of 89.16       **17.** 7.18% of 38.5

**18.** $7\frac{3}{4}\%$ of 321.52       **19.** $\frac{3}{5}\%$ of 4386.1

**20.** 115% of 47.67       **21.** $231\frac{1}{3}\%$ of 33.47

**22.** $33\frac{1}{3}\%$ of 6.911       **23.** $66\frac{2}{3}\%$ of 91.62

**24.** $16\frac{2}{3}\%$ of 409.64       **25.** $83\frac{1}{3}\%$ of 76.54

## Section 5.3

*For Exercises 26 and 27, (a) write an equation, and (b) solve the equation to find the percent to the nearest tenth of a percent.*

**26.** 17 is what percent of 32?

**27.** What percent of 70 is 16.3?

**28.** To the nearest tenth of a percent, find the efficiency of an electric motor having an output of 3.6 kilowatts for an input of 5.3 kilowatts.

*For Exercises 29 and 30, (a) write an equation, and (b) solve the equation to find the number. Round off to three significant digits.*

**29.** 58.2 is 17% of what number?

**30.** 7.25% of what number is 9.18?

**31.** A certain alloy contains 13% copper. If one casting contains 4.3 pounds of copper, what is the weight of the casting? Round off to the nearest tenth.

**32.** A fuel contains 29% of type Y fuel. If a shipment contains 375 gal of type Y, how many gallons of fuel are in the entire shipment? Round off to the nearest tenth.

## Section 5.4

**33.** Find the net price to the nearest cent of a pump with a list price of $81.50, if the discount rate is 12.5%.

**34.** Find the net cost to the nearest cent of a calculator with a list price of $39.89, if the discount rate is 22%.

**35.** Find the price to the nearest cent of a set of tools having a list price of $105.95, if there are two discount rates of 26% and 9%.

**36.** Find the selling price to the nearest cent of a drill press that costs $345.25, if there is a markup of 19.5%.

**37.** Find the value to the nearest dollar
at the end of 2 years of a boring
machine with an original value of
$27,000, if its value depreciates by
7% each year.

*To the nearest cent, find the simple interest for the given principal and yearly interest
rate for the indicated time.*

**38.** $6400;   8.6%;   4 years          **39.** $950;   15½%;   16 months

**40.** $8277.20;   9%;   320 days        **41.** $5385;   4⅓%;   8 years

*Exercises 42 to 47 include Cases I, II, and III. Express percent answers to the nearest
tenth of a percent and other answers to the nearest hundredth.*

**42.** 74% of 19.68                      **43.** 8.17% of ? = 58.3

**44.** ?% of 66.47 = 21.23              **45.** 16⅔% of 86.1

**46.** 9.61% of ? = 15.22               **47.** ?% of 19.26 = 5.64

## Section 5.5

*Exercises 48 to 60 are to be done mentally.*

**48.** 10% of 16.7          **49.** 20% of ? = 2        **50.** ?% of 16 = 4

**51.** 33⅓% of 12          **52.** 50% of ? = 6        **53.** ?% of 175 = 1.75

**54.** 75% of 40

**55.** 25% of 160

**56.** $12\frac{1}{2}$% of 32

**57.** $62\frac{1}{2}$% of 32

**58.** ?% of 12 = 4

**59.** ?% of 12 = 8

**60.** ?% of 64 is 48?

# Measurement

**Measurement** is the process of comparing a quantity with a standard unit that depends on the *system* of measurement being used. There are two systems of measurement in common use—the United States system and the metric system. In this chapter we consider fundamental relationships in each system, together with techniques for changing units from one system to the other.

Numbers that involve units of measure, such as 11 miles and 14 quarts, are called **denominate numbers** (or **measurement numbers**, if you like). Numbers that do not involve units of measure are **abstract numbers**.

## 6.1 The United States System

The basic unit of length in the U.S. system is the *foot*; the basic unit of weight is the *pound*. These basic units are not always the most convenient for every measurement. For example, we would generally use the *mile* as the unit of measurement for the distance between two cities, the *yard* (or foot) as the unit of measurement for the dimensions of a building, and the *inch* as the unit of measurement for the dimensions of parts of machinery. Relationships between these units, and some standard abbreviations, are given in Table A.4 (taken from Appendix A); relationships between these units and others commonly used in the U.S. system are given in the more extensive tables in Appendix A.

In Table A.4, note that the number 12 is associated with *one* foot for inch-feet conversions, the number 3 is associated with *one* yard for feet-yards conversions, and so forth. We refer to the numbers associated with one unit, such as 12 and 3, as **conversion numbers**. Although you are probably familiar with many of the common conversion numbers we use in this section, we will refer to the appropriate appendix table in each example.

**TABLE A.4**

| |
|---|
| 12 inches (in.) = 1 foot (ft) |
| 3 feet = 36 inches = 1 yard (yd) |
| 5280 feet = 1760 yards = 1 mile (mi) |

## ∎ *Changing Units*

The following rules can be used to change denominate numbers from one unit to another.

> **Unit-Conversion Rules**
>
> **1.** To change a number $S$ of smaller units to a number of larger units, divide $S$ by the conversion number.
> **2.** To change a number $L$ of larger units to a number of smaller units, multiply $L$ by the conversion number.

### *EXAMPLE 1*
**(a)** 3660 min = ? hr          **(b)** 54 yd = ? ft

### *Solutions*
**(a)** Note that the minutes-hour conversion number in Table A.2 is 60. Because the conversion is from smaller to larger units, divide the number of minutes by 60. Thus,

$$3660 \text{ min} = \frac{3660}{60} \text{ hr} = 61 \text{ hr}$$

**(b)** Note that the feet-yards conversion number in Table A.4 is 3. Because the conversion is from larger to smaller units, multiply the number of yards by 3. Thus,

$$54 \text{ yd} = (3 \times 54) \text{ ft} = 162 \text{ ft}$$

## ∎ *Denominate Numbers; More Than One Unit*

Measurements in the U.S. system are often given in terms of more than one unit. Such numbers can be changed to denominate numbers with only one unit.

### *EXAMPLE 2*   24 gal 2 qt = ? qt

**Solution**   Change 24 gallons to quarts and add 2 quarts to the result.
  From Table A.8, the conversion number (gallons-quarts) is 4.
Hence,

$$24 \text{ gal } 2 \text{ qt} = (24 \times 4 + 2) \text{ qt} = 98 \text{ qt}$$

  We sometimes need to change a denominate number with one unit to a denominate number with more than one unit. Consider the problem

$$41 \text{ ft} = ? \text{ yd } ? \text{ ft}$$

The feet-yards conversion number in Table A.4 is 3. Because we are changing from smaller to larger units, we divide the given number of feet (41) by 3:

$$\frac{41}{3} = 13 + \text{remainder } 2$$

Hence,

$$41 \text{ ft} = 13 \text{ yd } 2 \text{ ft}$$

**EXAMPLE 3**   139 min = ? hr ? min

**Solution**   From Table A.2, the conversion number is 60. Hence we divide 139 by 60:

$$\frac{139}{60} = 2 + \text{remainder } 19$$

Hence,

$$139 \text{ min} = 2 \text{ hr } 19 \text{ min}$$

## ■ Decimal Form of a Denominate Number

In the examples above we changed a denominate number with one unit to a denominate number with more than one unit. We can also change the one unit number directly to a decimal. For example,   $41 \text{ ft} = \frac{41}{3} \text{ yd}$   can be changed to 13.666667 yd. *Decimal forms* of denominate numbers are particularly useful for calculator applications.

**EXAMPLE 4**   Write each conversion as a decimal form, to the nearest hundredth.
**(a)** 11,624 ft = ? mi      **(b)** 28 min = ? hr

**Solutions**
**(a)** From Table A.4, the conversion number (feet-miles) is 5280. Hence,

$$11,624 \text{ ft} = \frac{11,624}{5280} \text{ mi} = 2.2015152 \text{ mi}$$

Thus,   11,624 ft = 2.20 mi,   to the nearest hundredth.

**(b)** From Table A.2, the conversion number (minutes-hours) is 60. Hence,

$$28 \text{ min} = \frac{28}{60} \text{ hr} = 0.46666667 \text{ hr}$$

Thus,   28 min = 0.47 hr,   to the nearest hundredth.

**EXAMPLE 5**   Write 6 ft 9 in. as a number of feet in decimal form.

**Solution**   Change 9 in. to feet (in decimal form) and add the result to 6 ft. From Table A.4, the conversion number (inches-feet) is 12. Hence,

$$6 \text{ ft } 9 \text{ in.} = \left(6 + \frac{9}{12}\right) \text{ ft} = (6 + 0.75) \text{ ft}$$
$$= 6.75 \text{ ft}$$

## Exercises for Section 6.1

*In the following exercises, refer to the appropriate table in Appendix A as necessary.*

*Convert each measurement. See Example 1.*

  **1.** 168 in. = ? ft      **2.** 3 mi = ? yd        **3.** 27 ft = ? yd

**4.** 256 oz = ? lb      **5.** 20 T = ? lb      **6.** 96 qt = ? gal

**7.** 15 qt = ? pt      **8.** 12,320 yd = ? mi      **9.** 15 hr = ? min

**10.** 40 pt = ? fl oz      **11.** 24 mi = ? ft      **12.** 31,680 ft = ? mi

*Each of the following requires more than one conversion.*

**13.** 2 mi = ? in.      **14.** 1024 fl oz = ? qt      **15.** 640 pt = ? gal

**16.** 15 gal = ? pt      **17.** 2 wk = ? min      **18.** 10,080 min = ? wk

**19.** 17 pk = ? pt      **20.** 90 qt = ? fl oz      **21.** 24 bu = ? qt

**22.** 3 wk = ? hr      **23.** 15 da = ? min      **24.** 2 da = ? sec

*Convert each measurement. See Example 2.*

**25.** 9 lb 14 oz = ? oz      **26.** 2 T 516 lb = ? lb

**27.** 17 yd 2 ft = ? ft      **28.** 5 mi 368 yd = ? yd

**29.** 8 hr 52 min = ? min    **30.** 9 wk 4 da = ? da

**31.** 5 gal 3 qt = ? qt    **32.** 13 qt 1 pt = ? pt

*Convert each measurement. See Example 3.*

**33.** 77 in. = ? ft ? in.    **34.** 200 oz = ? lb ? oz

**35.** 59 qt = ? gal ? qt    **36.** 30,000 ft = ? mi ? ft

**37.** 12,000 ft = ? mi ? ft    **38.** 113 qt = ? gal ? qt

**39.** 130 oz = ? lb ? oz    **40.** 110 in. = ? ft ? in.

*Express each result as a decimal rounded off to the nearest tenth. See Example 4.*

**41.** 67 fl oz = ? pt    **42.** 71 qt = ? gal    **43.** 127 oz = ? lb

**44.** 124 wk = ? yr    **45.** 38 mo = ? yr    **46.** 873 lb = ? T

**47.** 205 in. = ? ft    **48.** 91 in. = ? yd    **49.** 3146 yd = ? mi

**50.** 350 min = ? hr        **51.** 2700 min = ? da        **52.** 2700 sec = ? hr

*Convert each measurement to decimal form. Round off to the nearest hundredth. See Example 5.*

**53.** 4 ft 7 in. = ? ft        **54.** 12 ft 5 in. = ? ft

**55.** 5 mi 700 ft = ? mi        **56.** 3 mi 960 ft = ? mi

**57.** 6 lb 11 oz = ? lb        **58.** 4 lb 3 oz = ? lb

**59.** 5 gal 3 qt = ? gal        **60.** 2 gal 1 qt = ? gal

**61.** Ocean depths are often specified in *fathoms*, 1 fathom being equal to 6 ft. A depth of 4200 fathoms is equivalent to how many feet? To how many miles (to the nearest tenth)?

**62.** To the nearest tenth, a depth of 3179 ft is equivalent to how many fathoms? (See Exercise 61.)

**63.** In surveying, a *chain* is a distance of 22 yd. A distance of 60.5 chains is equivalent to how many yards? To how many feet?

**64.** To the nearest tenth, a distance of 115 yd is equivalent to how many chains? (See Exercise 63.)

**65.** If a ship is traveling at a speed of 1 *knot*, it travels 6082.2 ft in 1 hr. To the nearest tenth, how many miles is this?

**66.** A boat is traveling with a speed of 15 mi in 1 hr. Express this equivalently in knots. (See Exercise 65.)

**67.** In the printing trade, 1 *pica* is 0.166044 in. Express 16 picas equivalently in inches (to the nearest hundredth).

**68.** To the nearest hundredth, express 12 in. in picas. (See Exercise 67.)

## 6.2  *Arithmetic of Denominate Numbers*

In this section we consider methods for finding sums, differences, products, and quotients of denominate numbers.

### ■ *Addition and Subtraction*

Denominate numbers can be added or subtracted only if they have the same units.

**EXAMPLE 1**
**(a)** Compute   8 ft 6 in. + 11 ft 10 in. + 3 ft 9 in.
**(b)** Express the result in decimal form (using the larger unit).

**Solution**
**(a)**   8 ft   6 in.        Change 25 in. to 2 ft 1 in. Then,
  11 ft 10 in.        22 ft 25 in. = 22 ft + 2 ft 1 in.
   3 ft   9 in.                      = 24 ft 1 in.
  ‾‾‾‾‾‾‾‾‾‾‾‾
  22 ft 25 in.

**(b)** 24 ft 1 in. = $\left(24 + \dfrac{1}{12}\right)$ ft = 24.08 ft, to the nearest hundredth.

### EXAMPLE 2
**(a)** Compute   9 lb 3 oz − 5 lb 11 oz.
**(b)** Express the result in decimal form (using the larger unit).

### Solution
**(a)** Because 11 oz cannot be subtracted from 3 oz, "borrow" 1 lb from the 9 lb and change the 1 lb to 16 oz. Then,

$$
\begin{array}{l}
\phantom{-}9\text{ lb }\ 3\text{ oz} = 8\text{ lb} + 16\text{ oz} + 3\text{ oz} = \\
-5\text{ lb } 11\text{ oz}
\end{array}
\qquad
\begin{array}{r}
8\text{ lb } 19\text{ oz} \\
-5\text{ lb } 11\text{ oz} \\
\hline
3\text{ lb }\ \ 8\text{ oz}
\end{array}
$$

**(b)** $3\text{ lb } 8\text{ oz} = \left(3 + \dfrac{8}{16}\right)\text{ lb} = 3.5\text{ lb}$

## ■ Multiplication and Division

The product of an abstract number and a denominate number with more than one unit can be found by multiplying the number associated with each different unit by the abstract number.

**EXAMPLE 3**   $8 \times (5\text{ mi } 723\text{ yd})$

**Solution**   First multiply the number associated with each unit by 8.

$$
\begin{array}{r}
5\text{ mi } 723\text{ yd} \\
\times 8 \\
\hline
40\text{ mi } 5784\text{ yd}
\end{array}
$$

Because 1 mi = 1760 yd (see Table A.4)

$$
\frac{5784\text{ yd}}{1760} = 3\text{ mi} + 504\text{ yd}
$$

Hence,

$$
\begin{aligned}
40\text{ mi } 5784\text{ yd} &= 40\text{ mi} + 3\text{ mi} + 504\text{ yd} \\
&= 43\text{ mi } 504\text{ yd}
\end{aligned}
$$

In a division, denominate numbers with more than one unit are usually first changed to denominate numbers with only one unit.

### EXAMPLE 4
**(a)** (13 ft 6 in.) ÷ 9        **(b)** (13 ft 6 in.) ÷ 9 in.

**Solutions**   In each case,

$$
13\text{ ft } 6\text{ in.} = (12 \times 13 + 6)\text{ in.} = 162\text{ in.}
$$

**(a)** $\dfrac{13\text{ ft } 6\text{ in.}}{9} = \dfrac{162\text{ in.}}{9}$        **(b)** $\dfrac{13\text{ ft } 6\text{ in.}}{9\text{ in.}} = \dfrac{162\text{ in.}}{9\text{ in.}}$

$\phantom{(a)}\quad = 18\text{ in.}$                $\phantom{(b)}\quad = 18$

In Example 4a, note that the quotient involves units (inches), while the quotient in (b) is an abstract number (no units). Example 4a could provide the answer to the problem: "If a 13 ft 6 in. aluminum rod is cut into 9 pieces (with no waste), how long is each piece?" The answer is that each piece is 18 in. long. Example 4b could provide the answer to the problem: "If a 13 ft 6 in. aluminum rod is cut into

9 in. pieces, how many such pieces can be obtained?'' Assuming no waste, the answer is that 18 such pieces can be obtained.

## Exercises for Section 6.2

*For each of the following, (a) compute, and (b) express the result in decimal form rounded off to the nearest hundredth (using the larger unit). See Examples 1 and 2.*

**1.** 10 ft + 4 ft + 8 ft 5 in.

**2.** 15 qt + 3 qt + 6 qt 1 pt

**3.** 22 lb + 6 lb + 8 lb 7 oz

**4.** 4 yd + 65 yd + 73 yd 2 ft

**5.** 39 ft 9 in. + 9 ft 5 in.

**6.** 27 lb 9 oz + 8 lb 12 oz

**7.** 6 hr 47 min + 5 hr 33 min + 7 hr

**8.** 2 bu 3 pk + 9 bu 3 pk + 8 bu 2 pk

**9.** 34 yd 1 ft + 9 yd 1 ft + 6 yd 2 ft

**10.** 8 da 12 hr + 4 da 8 hr + 8 da 11 hr

**11.** 948 mi − 391 mi

**12.** 89 gal − 63 gal

**13.** 16 da 5 hr − 8 da 2 hr

**14.** 65 ft 5 in. − 36 ft 4 in.

**15.** 92 ft 2 in. − 37 ft 9 in.

**16.** 75 wk 1 da − 27 wk 4 da

**17.** 32 lb 9 oz − 13 lb 12 oz          **18.** 42 ft 9 in. − 15 ft 11 in.

**19.** 275 gal 1 qt − 95 gal 2 qt          **20.** 4 da 6 hr − 3 da 17 hr

*Find each product or quotient. See Examples 3 and 4.*

**21.** 37 × (31 ft)                    **22.** 92 × (44 qt)

**23.** 8 × (98 ft 2 in.)               **24.** 5 × (74 lb 6 oz)

**25.** 8 × (14 gal 2 qt)               **26.** 18 × (19 gal 3 qt)

**27.** (24 da) ÷ 6                     **28.** (36 lb) ÷ 12

**29.** (171 hr 36 min) ÷ 9             **30.** (242 gal) ÷ 11

**31.** (133 lb 8 oz) ÷ 12              **32.** (86 ft 4 in.) ÷ 7

**33.** (8 ft 9 in.) ÷ 15               **34.** (14 yd 2 ft) ÷ 11

**35.** (16 pt 2 fl oz) ÷ 6 fl oz  **36.** (50 gal 1 qt) ÷ 3 qt

**37.** (70 ft 6 in.) ÷ 9 in.  **38.** (90 lb 8 oz) ÷ 8 oz

**39.** How many 4-oz applications can be obtained from 15 lb of wheel-bearing grease?

**40.** How many 6-oz applications can be obtained from 27 lb 12 oz of polishing compound?

*In Exercises 41 to 44, assume that there is no waste in cutting.*

**41.** A board 3 ft 4 in. long is cut from a board that is 8 ft 3 in. long. How long is the remaining piece?

**42.** A length of pipe measures 16 ft 8 in. Two pieces, one 2 ft 6 in. long and the other 5 ft 9 in. long, are cut from the original pipe. How long is the remaining piece?

**43.** If a piece of lumber 19 ft 6 in. long is cut into six pieces of equal length, how long is each of the resulting pieces?

**44.** If a metal ingot weighing 49 lb 12 oz is cut into four equal parts, what is the weight of each part?

**45.** Certain concrete blocks are 16 in.
long. How many of these blocks,
laid end to end, will it take to cover
a distance of 53 ft 4 in.?

**46.** How many 9-in.-long pieces of
asphalt tile will it take to make a
row of tile 17 ft. 3 in. long?

*In Exercises 47 to 50, find the indicated dimensions in the foundation plans shown
in Figures 6.1 and 6.2.*

**47.** Dimension *A*.

**48.** Dimension *B*.

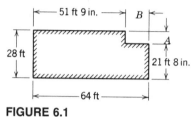

**FIGURE 6.1**

**49.** Dimension *C*.

**50.** Dimension *D*.

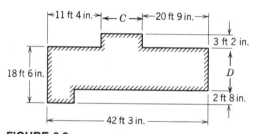

**FIGURE 6.2**

## 6.3  The Metric System;
## U.S.–Metric Conversions

The metric system of measurement is currently used by most of the countries in
the world. The basic unit of length is the **meter** (or **metre**), which is about 3 in.
longer than a yard (see Figure 6.3). The basic unit of weight is the **gram**, which is
slightly less than $\frac{1}{28}$ of an ounce (see Figure 6.4). The basic unit of capacity is the
**liter** (or **litre**), which is slightly more than a quart (see Figure 6.5).

**FIGURE 6.3**

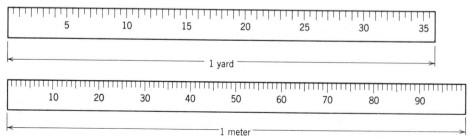

**FIGURE 6.4**

1 gram        1 ounce

**FIGURE 6.5**

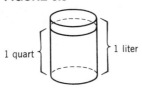

1 quart ⎰ ⎱ 1 liter

## ■ Metric Units and Tables

The names of the basic metric units (meter, gram, and liter) are combined with certain prefixes, as listed in Table 6.1, to provide names for other metric units. Each prefix indicates a particular relation between the unit named and the basic unit.

### EXAMPLE 1

**(a)** "Kilometer" means "one thousand meters."
**(b)** "Kilogram" means "one thousand grams."
**(c)** "Milliliter" means "one thousandth of a liter."
**(d)** "Millimeter" means "one thousandth of a meter."

In our work we will limit the metric unit prefixes to the three most common units listed in a useful form in Table 6.2.*

**TABLE 6.1** Metric Unit Prefixes

| PREFIX | DECIMAL MEANING |
|--------|-----------------|
| Milli  | 0.001 |
| centi  | 0.01 |
| deci   | 0.1 |
| deka   | 10 |
| hecto  | 100 |
| kilo   | 1000 |

**TABLE 6.2**

$$
1 \text{ centi} \begin{cases} \text{meter} \\ \text{gram} \\ \text{liter} \end{cases} = 10 \text{ milli} \begin{cases} \text{meters} \\ \text{grams} \\ \text{liters} \end{cases}
$$

$$
1 \begin{cases} \text{meter} \\ \text{gram} \\ \text{liter} \end{cases} = 100 \text{ centi} \begin{cases} \text{meters} \\ \text{grams} \\ \text{liters} \end{cases}
$$

$$
1 \text{ kilo} \begin{cases} \text{meter} \\ \text{gram} \\ \text{liter} \end{cases} = 1000 \begin{cases} \text{meters} \\ \text{grams} \\ \text{liters} \end{cases}
$$

A major advantage of the metric system is that one unit can be converted to a smaller unit by multiplying by a power of 10; to a larger unit by dividing by a power of 10.

The tables introduced in Section 1.5, which are reproduced here, simplify the process of multiplying and dividing by a power of 10 when we want to convert a metric measurement from one unit to another.

---

*More complete metric conversions for length, weight, and liquid capacity are given in Tables 9, 10, and 11 in Appendix A.

| TO MULTIPLY BY | MOVE THE DECIMAL POINT | TO DIVIDE BY | MOVE THE DECIMAL POINT |
|---|---|---|---|
| 10 | one place to the right | 10 | one place to the left |
| 100 | two places to the right | 100 | two places to the left |
| 1000 | three places to the right | 1000 | three places to the left |
| etc. | etc. | etc. | etc. |
|  | Add or drop zeros as needed |  | Add or drop zeros as needed. |

**EXAMPLE 2**   (a) 3.09 cm = ? mm        (b) 4200 m = ? km

*Solutions*   See Table 6.2.
(a) Because   1 cm = 10 mm,   multiply 3.09 by 10.

$$3.09 \text{ cm} = 30.9 \text{ m}$$

(b) Because   1 km = 1000 m,   divide 4200 by 1000.

$$4200 \text{ m} = 4.200 \text{ km} = 4.2 \text{ km}$$

**EXAMPLE 3**
(a) 609 cg = ? g        (b) 14.39 ℓ = ? ml

*Solutions*   See Table 6.2.
(a) Because   1 g = 100 cg,   divide 609 by 100.

$$609 \text{ cg} = 6.09 \text{ g}$$

(b) Because   1 ℓ = 1000 ml,   multiply 14.39 by 1000.

$$14.39 \text{ ℓ} = 14390 \text{ ml}$$

## ■ U.S.–Metric Conversions

Tables B.1 through B.4 in Appendix B list relationships between commonly used U.S. and metric units. Because the entries in these tables are mainly approximations, in the following examples the word "equals" in phrases such as "0.454 kilogram equals 1 pound" is to be understood to mean "is approximately equal to."

Proportions, together with an appropriate table from Appendix B, can be used to compute U.S.–metric conversions.

**EXAMPLE 4**   Convert 6.8 lb to kilograms.

*Solution*                          Number of kilograms: $K$

From Table B.2, note that 1 lb equals 0.454 kg. Set up and solve a proportion.

**kilograms    pounds**

$$\frac{K}{0.454} = \frac{6.8}{1}$$

By the cross-multiplication rule,

$$K \times 1 = 0.454 \times 6.8 = 3.0872$$

To the nearest tenth, 6.8 lb equals 3.1 kg.

**EXAMPLE 5**  Convert 500 m to yards (to the nearest tenth).

**Solution**                    Number of yards: $Y$

From Table B.1, note that 1 yd equals 0.914 m. Set up and solve a proportion.

$$\frac{\overset{\textbf{yards}}{Y}}{1} = \frac{\overset{\textbf{meters}}{500}}{0.914}$$

By direct division,

$$Y = \frac{500}{0.914} = 547.0 \text{ yd}$$

to the nearest tenth.

We sometimes need to convert measurements that involve different types of units.

**EXAMPLE 6**  21.4 mpg = ? kilometers per liter (to the nearest tenth).

**Solution**  First change 21.4 mi to kilometers and then 1 gal to liters.

**1.**                    Number of kilometers: $K$

From Table B.1, note that 1.609 km equals 1 mi. Set up and solve a proportion.

$$\frac{\overset{\textbf{kilometers}}{K}}{1.609} = \frac{\overset{\textbf{miles}}{21.4}}{1}$$

By the cross-multiplication rule,

$$K \times 1 = 1.609 \times 21.4 = 34.4326$$

Thus, 21.4 mi equals 34.4326 km.

**2.** From Table B.4, note that 1 gal equals 3.785ℓ. Now, to obtain the number of kilometers per liter, apply the unit-quotient rule and compute the quotient

$$\frac{34.4326}{3.785} = 9.09$$

to the nearest hundredth. Hence, 21.4 mpg equals 9.1 km per liter (to the nearest tenth).

## Exercises for Section 6.3

*In the following exercises, use the appropriate table in Appendix A or Appendix B as necessary. Write the meaning of each of the following in terms of meter, gram, or liter. See Example 1.*

**1.** millimeter      **2.** centiliter      **3.** centigram

**4.** kiloliter      **5.** kilometer      **6.** milligram

*Convert each measurement as indicated. See Examples 2 and 3.*

**7.** 75 cm = ? mm          **8.** 118 cm = ? m          **9.** 0.38 g = ? mg

**10.** 4210 g = ? kg          **11.** 148 ml = ? ℓ          **12.** 6.1 g = ? cg

**13.** 1.2 kg = ? g          **14.** 0.6 ℓ = ? mℓ

*In the following exercises, use the appropriate table in the Appendix, as necessary.*
*For Exercises 15–38, see Examples 4 and 5. Round off answers to the nearest tenth.*

**15.** 8.7 lb = ? kg          **16.** 17.9 lb = ? kg          **17.** 5.4 lb = ? g

**18.** 2.5 lb = ? g          **19.** 9.1 oz = ? g          **20.** 5.2 oz = ? g

**21.** 4.7 in. = ? cm          **22.** 7.9 in. = ? cm          **23.** 14.5 ft = ? m

**24.** 12.4 ft = ? m          **25.** 47.6 yd = ? m          **26.** 16.9 yd = ? m

**27.** 8.7 cm = ? in.          **28.** 57.2 cm = ? in.          **29.** 9.3 m = ? ft

**30.** 7.3 m = ? ft          **31.** 98.9 m = ? yd          **32.** 48.1 m = ? yd

**33.** 559 km = ? mi          **34.** 4.6 km = ? mi          **35.** 13.8 g = ? oz

**36.** 8.5 g = ? oz    **37.** 6.5 kg = ? lb    **38.** 89 kg = ? lb

**39.** 100 m = ? ft    **40.** 10 lb = ? kg    **41.** 1.4 ℓ = ? gal

**42.** 3.4 gal = ? ℓ

**43.** A machine-shop operator has a tank containing 97 ℓ of coolant. To the nearest tenth, how many gallons is this?

**44.** The length of the hub on a gear is 5.68 cm. To the nearest hundredth, what is this in inches?

*Each of the following exercises gives the rating of various automobiles in miles per gallon. Convert each to kilometers per liter to the nearest tenth of a kilometer. See Example 6.*

**45.** 18.3 mpg                 **46.** 15.1 mpg

**47.** 14.6 mpg                 **48.** 23.5 mpg

*Convert each of the following to miles per gallon to the nearest tenth of a mile.*

**49.** 8.7 km per liter         **50.** 9.3 km per liter

**51.** 7.9 km per liter         **52.** 10.1 km per liter

**53.** Many cameras use film that is 35 mm wide. Convert this width to inches, to the nearest tenth.

**54.** Motion picture film is made in various widths, two of them being 8 mm and 16 mm. Convert these widths to inches, to the nearest hundredth.

*The American Wire Gage is a numbering system used to indicate the diameter of wire. Convert each dimension to millimeters, to the nearest tenth.*

| | **GAGE NUMBER** | **DIAMETER, IN INCHES** |
|---|---|---|
| **55.** | 000 | 0.410 |
| **56.** | 0 | 0.325 |
| **57.** | 6 | 0.162 |
| **58.** | 16 | 0.051 |

## 6.4 Formulas Involving Industrial Measurements

Many technical formulas are used in industry. In this section we consider some formulas that involve the concepts of *work* and *power*, with emphasis on the measure of each of these quantities.

### ■ Work

If an object is lifted or moved, we say that **work** has been done (see Figure 6.6). The *measure* of work ($w$) done is defined as follows:

$$w = f \cdot d \qquad (1)$$

where $f$ is the force being applied and $d$ is the distance the object is moved. A commonly used unit of measure of work is the *foot-pound* (ft-lb), which is the amount of work done when a force of one pound is applied to an object over a distance of one foot.

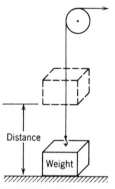

**FIGURE 6.6**

**EXAMPLE 1** A force of 527 lb is applied to an object over a distance of 625 ft. Find the work done.

**Solution** Substitute 527 for $f$ and 625 for $d$ in Formula 1.

$$w = f \cdot d = 527 \cdot 625 = 329{,}375$$

The work done is 329,375 ft-lb.

**EXAMPLE 2**   Find the work done in lifting a 3100-lb car to a height of 23 ft.

**Solution**   Substitute 3100 for $f$ and 23 for $d$ in Formula 1.

$$w = f \cdot d = 3100 \cdot 23 = 71{,}300$$

The work done is 71,300 ft-lb.

## ■ Power

**Power** is the measure, or the amount, of work done per unit of time. Thus,

$$P = \frac{\text{work}}{\text{time}} = \frac{w}{t} = \frac{f \cdot d}{t} \qquad (2)$$

where the units for the power $P$ depend on the units used for $f$, $d$, and $t$.

**EXAMPLE 3**   To the nearest tenth, find the power needed by a gasoline engine to raise 83.8 lb a distance of 63.5 ft in 12 sec.

**Solution**   Substitute 83.8 for $f$, 63.5 for $d$, and 12 for $t$ in Formula 2.

$$P = \frac{f \cdot d}{t} = \frac{83.8 \times 63.5}{12} = 443.44167$$

To the nearest tenth, the power is 443.4 ft-lb/sec.*

**EXAMPLE 4**   If an engine can deliver 85 ft-lb/min of power, find the amount of work done by the engine in 123.5 min.

**Solution**   Substitute 85 for $P$ and 123.5 for $t$ in Formula 2 in the form   $P = \dfrac{w}{t}$.

$$85 = \frac{w}{123.5}$$

Multiply each side by 123.5.

$$85 \times 123.5 = \frac{w}{123.5} \times 123.5$$
$$w = 10{,}497.5$$

The amount of work done is 10,497.5 ft-lb.

## ■ Horsepower

In the U.S. system, **horsepower** is a commonly used unit of power, equal to 33,000 ft-lb/min. That is, one horsepower is the amount of power needed to raise 33,000 pounds to a height of one foot (or 1 pound to a height of 33,000 feet) in one minute. If the force ($f$) is measured in *pounds*, the distance ($d$) in *feet*, and the time ($t$) in *minutes*, we can compute horsepower by dividing units of power by 33,000.

$$\text{hp} = \frac{P}{33{,}000} \qquad (3)$$

---

*The term "ft-lb/sec" is read "foot pounds per second."

From Formula 2 we have $P = \dfrac{f \cdot d}{t}$. Hence, we can obtain an alternate form for Formula 3.

$$hp = \frac{\dfrac{f \cdot d}{t}}{33{,}000} = \frac{f \cdot d}{33{,}000 \cdot t} \qquad \textbf{(3a)}$$

**EXAMPLE 5** The total weight of a loaded elevator is 4625 lb. To the nearest tenth, find the horsepower required to lift the elevator to a height of 362 ft in 1.8 min.

**Solution** Substitute 4625 for $f$, 362 for $d$, and 1.8 for $t$ in Formula 3a.

$$hp = \frac{f \cdot d}{33{,}000 \cdot t} = \frac{4625 \times 362}{33{,}000 \times 1.8} = 28.186027$$

To the nearest tenth, 28.2 hp are required.

### ■ *Measuring Horsepower of Gasoline Engines*

The horsepower rating of a gasoline engine can be calculated in various ways. One way is the N.A.C.C.* horsepower rating, given by

$$hp = \frac{D^2 \cdot N}{2.5} \qquad \textbf{(4)}$$

where $D$ is the *bore* of each cylinder in inches (see Figure 6.7) and $N$ is the number of cylinders.

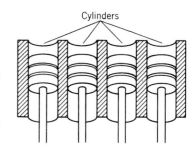
Cylinders

**FIGURE 6.7**

**EXAMPLE 6** To the nearest tenth, find the N.A.C.C. horsepower of a 6-cylinder gasoline engine if the bore of each cylinder is 3.25 in.

**Solution** Substitute 3.25 for $D$ and 6 for $N$ in Formula 4.

$$hp = \frac{D^2 \cdot N}{2.5} = \frac{3.25^2 \times 6}{2.5} = 25.35$$

The N.A.C.C. rating is 25.4 hp, to the nearest tenth.

A second way of measuring the horsepower of a gasoline engine depends on the force that the engine can exert, as measured by a special device. Figure 6.8 shows a Prony brake (also called a dynamometer), which is used for measuring *brake horsepower*—the horsepower output of an engine. The brake horsepower (bhp) is given by

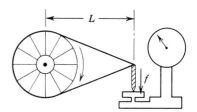

**FIGURE 6.8**

$$bhp = \frac{2\pi L R f}{33{,}000} \qquad \textbf{(5)}$$

where $L$ is the length of the arm in *feet*; $R$ is the engine speed in *revolutions per minute* (rpm); $f$ is the force on the scale, in *pounds*; and $\pi$ is a symbol for a number approximately equal to 3.14159.

**EXAMPLE 7** To the nearest tenth, find the brake horsepower of an engine if $L$ is 5 ft, the engine is running at 475 rpm, and the force on the scale measures 360 lb.

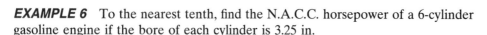

*National Automobile Chamber of Commerce. This rating is also known as the SAE horsepower (Society of Automotive Engineers).

***Solution*** Substitute 5 for $L$, 475 for $R$, and 360 for $f$ in Formula 5.

$$\text{bhp} = \frac{2\pi LRf}{33,000} = \frac{2\pi \times 5 \times 475 \times 360}{33,000} = 162.79162$$

The brake horsepower is 162.8 hp, to the nearest tenth.

## ■ Water Power

The power generated by water flowing over a dam can also be measured in horse-power. Water weighs (approximately) 62.4 lb per cubic foot (cu ft). Now, suppose that we have a dam $s$ feet in height. If $N$ represents the number of cubic feet of water falling, then 62.4$N$ represents the force (weight) of $N$ cubic feet of water, and $(62.4N) \cdot s$ represents the work done by the water falling $s$ feet. Hence the power generated (work per minute) is given by

$$P = \frac{62.4N \cdot s}{t}$$

To obtain results in horsepower, we divide by 33,000 to obtain the formula

$$\text{hp} = \frac{62.4N \cdot s}{33,000t} \tag{6}$$

**EXAMPLE 8**  To the nearest whole number, find the horsepower generated by a dam 33 ft high if the water flows at the rate of 48,000 cu ft/min.

***Solution***  Substitute 48,000 for $N$, 33 for $s$, and 1 for $t$ in Formula 6.

$$\text{hp} = \frac{62.4 \times 48,000 \times 33}{33,000 \times 1} = 2995.2$$

The dam generates 2995 hp, to the nearest whole number.

## Exercises for Section 6.4

*Find the work done if the indicated force is applied in moving an object the given distance. See Example 1.*

**1.** 123 lb;  15 ft  **2.** 427 lb;  91 ft

**3.** 634 lb;  3 ft  **4.** 35 lb;  40 ft

*Find the work done in lifting the object with the given weight to the stated height. See Figure 6.9 and Example 2.*

**5.** 30 lb;  8 ft  **6.** 24 lb;  9 ft

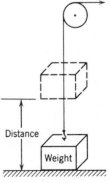

**FIGURE 6.9**

**7.** 246 lb;   86 ft     **8.** 377 lb;   26 ft

*Find the power needed to raise each weight. See Example 3. Round off to the nearest tenth.*

**9.** 23 lb;   98 ft in 6 sec          **10.** 49 lb;   42 ft in 9 sec

**11.** 194 lb;   80 ft in 7 min     **12.** 814 lb;   528 ft in 37 min

*Given the power that an engine can deliver, find the work done in the given time. See Example 4. Round off to the nearest tenth.*

**13.** 219 ft-lb/min;   83 min     **14.** 467 ft-lb/min;   42 min

**15.** 468 ft-lb/sec;   2.9 sec     **16.** 577 ft-lb/sec;   7.1 sec

*Find the horsepower required to lift each weight. See Example 5. Round off to the nearest tenth.*

**17.** 2702 lb;   67 ft in 4.6 min          **18.** 4285 lb;   92 ft in 9.6 min

**19.** 8390 lb;   207 ft in 4.9 min     **20.** 9491 lb;   74 ft in 2.3 min

*Find the N.A.C.C. horsepower of each engine. See Example 6. Round off to the nearest tenth.*

**21.** 4 cylinders;   3.8 in. bore     **22.** 4 cylinders;   4.1 in. bore

**23.** 6 cylinders;  4.25 in. bore     **24.** 6 cylinders;  3.75 in. bore

**25.** 8 cylinders;  3.6 in. bore     **26.** 8 cylinders;  4.32 in. bore

*Find the brake horsepower of each engine. See Example 7. Round off to the nearest tenth. (Use the* $\boxed{\pi}$ *key, or 3.14159.)*

| | L | R | f |
|---|---|---|---|
| **27.** | 4 ft | 845 rpm | 212 lb |
| **28.** | 6 ft | 566 rpm | 250 lb |
| **29.** | 5 ft | 880 rpm | 273 lb |
| **30.** | 8 ft | 768 rpm | 477 lb |

*Find the horsepower generated by water flowing at the indicated rate over each dam. See Example 8. Round off to the nearest tenth.*

| | Dam Height | Water Flow |
|---|---|---|
| **31.** | 65 ft | 43,000 cu ft/min |
| **32.** | 80 ft | 37,000 cu ft/min |
| **33.** | 95 ft | 38,300 cu ft/min |
| **34.** | 164 ft | 160,000 cu ft/min |

The **indicated horsepower** of a gasoline engine (or a steam engine) is determined by the average pressure, $P$, on the piston in *pounds per square inch* (psi); the length, $L$, of each power stroke in *feet*; the bore, $B$, in *inches*; and the number of power *strokes per minute*, $N$. The indicated horsepower (ihp) is given by the formula

$$\text{ihp} = \frac{PLB^2N}{42{,}017}$$

*Find the ihp of each engine, to the nearest tenth.*

| | P | L | B | N |
|---|---|---|---|---|
| **35.** | 120 psi | 0.5 ft | 6 in. | 5000 |
| **36.** | 130 psi | 0.6 ft | 5.8 in. | 6000 |
| **37.** | 115 psi | 5 in. | 4.5 in. | 6400 |
| **38.** | 125 psi | 6.3 in. | 5.2 in. | 6200 |

**39.** The ihp of a certain gasoline engine is 48 hp. If the length of stroke is 0.23 ft, the pressure on the piston is 105 psi, and there are 3000 power strokes per minute, find the bore (to the nearest tenth of an inch).

**40.** Find the ihp of a steam engine with a 24-in. bore and a 40-in. stroke if $N = 180$ and there is a pressure of 60 psi.

Figure 6.10 illustrates a *differential chain hoist.* As indicated, it consists of two pulleys (called *sheaves*) with diameters $D_1$ and $D_2$ on top, and one sheave on the bottom connected by an endless chain. The force necessary to raise a weight $w$ is given by

$$F = \frac{w}{2}\left(\frac{D_1 - D_2}{D_1}\right).$$

*Find the force necessary to lift the given weight with a hoist having sheaves with the given diameters. Round off to the nearest whole number.*

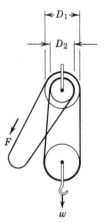

**FIGURE 6.10**

|      | $w$      | $D_1$  | $D_2$  |
|------|----------|--------|--------|
| **41.** | 1500 lb  | 10 in. | 8 in.  |
| **42.** | 2000 lb  | 12 in. | 11 in. |
| **43.** | 760 kg   | 25 cm  | 22 cm  |
| **44.** | 1000 kg  | 30 cm  | 27 cm  |

---

## 6.5 Formulas Involving Electrical Measurements

Electric currents are transmitted by conductors, usually metal wires. The rate of flow of a current is measured in *amperes*, the pressure under which the current flows is measured in *volts*, and the opposition (or resistance) of an electric circuit to the flow of current is measured in *ohms*.

### ■ Ohm's Law

Amperes, volts, and ohms are related by the fact that it takes one volt to "push" one ampere of current through a resistance of one ohm. This relationship is expressed by the formula known as *Ohm's law*,

$$E = I \cdot R \tag{1}$$

where $E$ is the number of volts, $I$ is the number of amperes, and $R$ is the number of ohms in a circuit.

**EXAMPLE 1**  If the resistance in a 110-v circuit is 48.5 ohms, find the number of amperes in the circuit (to the nearest tenth).

**Solution**  Substitute 110 for $E$ and 48.5 for $R$ in Formula 1.

$$E = I \cdot R$$
$$110 = I \times 48.5$$

Divide each side by 48.5.

$$\frac{110}{48.5} = \frac{I \times 48.5}{48.5}$$
$$I = 2.2680412$$

To the nearest tenth,  $I = 2.3$ amperes.

## ■ Resistors

Devices that provide opposition in a circuit are called *resistors*. In simple circuits, two or more resistors may be connected in either of two ways: in *series*, as shown in Figure 6.11*a*; in *parallel*, as shown in Figure 6.11*b*. In each figure, $R_1$, $R_2$, and $R_3$ represent resistors.

When resistors are connected in series, the total resistance $(R)$ of the resulting circuit is the sum of the individual resistances.

$$R = R_1 + R_2 + R_3 + \cdots \qquad \textbf{(2)}$$

**EXAMPLE 2**  Three resistors measuring 4.6 ohms, 68.3 ohms, and 45.2 ohms are connected in series. Find the total resistance of the circuit.

**Solution**  The total resistance is the sum of the resistances.

$$R = 4.6 + 68.3 + 45.2 = 118.1 \text{ ohms}$$

When resistors are connected in parallel, the total resistance $(R)$ of the circuit is related to the resistances in the circuit by the formula

$$\frac{1}{R} = \frac{1}{R_1} + \frac{1}{R_2} + \frac{1}{R_3} + \cdots \qquad \textbf{(3)}$$

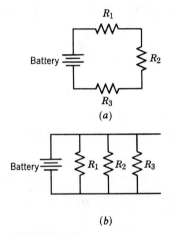

*(a)*

*(b)*

**FIGURE 6.11**

**EXAMPLE 3**  Three resistors measuring 4.6 ohms, 68.3 ohms, and 45.2 ohms are connected in parallel. Find the total resistance of the circuit.

**Solution**  Substitute 4.6 for $R_1$, 68.3 for $R_2$, and 45.2 for $R_3$ in Formula 3.

$$\frac{1}{R} = \frac{1}{4.6} + \frac{1}{68.3} + \frac{1}{45.2}$$

$$\frac{1}{R} = 0.2173913 + 0.01464129 + 0.02212389 = 0.25415648$$

Hence,

$$R = \frac{1}{0.25415648} = 3.9345839$$

To the nearest tenth, the total resistance is 3.9 ohms.

The necessary computations can be done efficiently by use of the $\boxed{1/x}$ key. Follow the sequence:

$$4.6 \ \boxed{1/x} \ \boxed{+} \ 68.3 \ \boxed{1/x} \ \boxed{+} \ 45.2 \ \boxed{1/x} \ \boxed{=} \ \boxed{1/x} \longrightarrow 3.9345839$$

## ■ Electric Power

Power used by an electric current is measured by a unit called a *watt*. The number of watts used by an electric circuit is given by the formula

$$\textbf{watts = volts} \times \textbf{amperes}$$

A more useful unit of electrical power is the *kilowatt* (kw), which is equal to 1000 watts. Thus, a formula for computing the power of an electric circuit in kilowatts is

$$\text{kw} = \frac{\text{volts} \times \text{amperes}}{1000} \qquad \qquad \textbf{(4)}$$

**EXAMPLE 4**   To the nearest tenth, find the kilowatts used by an electric motor running on a 220-v line and using 31.8 amp of current.

**Solution**   Substitute 220 for volts and 31.8 for amperes in Formula 4.

$$\text{kw} = \frac{\text{volts} \times \text{amperes}}{1000} = \frac{220 \times 31.8}{1000} = 6.996$$

The power used is 7.0 kw, to the nearest tenth.

Electrical power can also be measured in horsepower units. It can be shown that 1 horsepower is (approximately) equal to 746 watts. Hence, we have the formula

$$\text{hp} = \frac{\text{volts} \times \text{amperes}}{746} \qquad \qquad \textbf{(5)}$$

**EXAMPLE 5**   A 5-hp electric motor is connected to a 220-volt line. To the nearest whole number, how many amperes of current does it require?

**Solution**   Substitute 5 for hp and 220 for volts in Formula 5.

$$\text{hp} = \frac{\text{volts} \times \text{amperes}}{746}$$

$$5 = \frac{220 \times \text{amperes}}{746}$$

Multiply each side by 746.

$$5 \times 746 = \frac{220 \times \text{amperes}}{746} \times 746$$

$$5 \times 746 = 220 \times \text{amperes}$$

Divide each side by 220.

$$\frac{5 \times 746}{220} = \frac{220 \times \text{amperes}}{220}$$

$$\text{amperes} = \frac{5 \times 746}{220} = 16.95454$$

To the nearest whole number, the motor requires 17 amperes.

## Exercises for Section 6.5

*In each exercise, find the missing measurement. See Example 1. Round off to the nearest tenth.*

|    | E         | I        | R         |
|----|-----------|----------|-----------|
| **1.**  | 120 volts | —        | 32.2 ohms |
| **2.**  | 115 volts | —        | 20.5 ohms |
| **3.**  | —         | 42.5 amp | 57.3 ohms |
| **4.**  | —         | 56.3 amp | 42.1 ohms |
| **5.**  | 220 volts | 18.5 amp | —         |
| **6.**  | 230 volts | 9.6 amp  | —         |
| **7.**  | 110 volts | —        | 26.4 ohms |
| **8.**  | 120 volts | —        | 19.6 ohms |
| **9.**  | —         | 62.7 amp | 57.9 ohms |
| **10.** | —         | 7.5 amp  | 83.4 ohms |
| **11.** | 300 volts | 28.2 amp | —         |
| **12.** | 350 volts | 5.4 amp  | —         |

*Find the total resistance of each set of resistors if they are connected (a) in series, and (b) in parallel. See Figure 6.12 and Examples 2 and 3. Round off to the nearest tenth. All measurements are in ohms.*

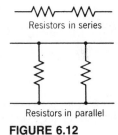
Resistors in series

Resistors in parallel

**FIGURE 6.12**

**13.** 70.7;  57.1          **14.** 64.5;  84.9

**15.** 6.6;  34.6          **16.** 4.2;  41.6

**17.** 76.6;  55.6;  62.0          **18.** 65.8;  55.7;  79.1

**19.** 2.8;  8.4;  18.0          **20.** 3.3;  6.1;  1.5

**21.** 34.9;   52.3;   78.8;   8.3      **22.** 29.0;   8.0;   92.5;   79.6

**23.** 4.6;   5.6;   50.4;   10.2      **24.** 7.0;   6.6;   3.8;   8.8

*Find the number of kilowatts used by each of the following. See Example 4. Round off to the nearest tenth.*

**25.** An electric motor running on 120 volts and using 12.5 amp.

**26.** A hair dryer running on 110 volts and using 10.9 amp.

**27.** An automobile engine starter running on 12 volts and using 300 amp.

**28.** An electric drill running on 120 volts and using 4.2 amp.

**29.** An electric hoist running on 220 volts and using 44.6 amp.

**30.** An industrial machine running on 240 volts and using 53.4 amp.

*For each specified electrical device, find the missing measurement. See Example 5. Round off to the nearest tenth.*

| | Device | Volts | Amperes | Horsepower |
|---|---|---|---|---|
| **31.** | Motor | 115 | —— | 4 |
| **32.** | Sprayer | 110 | —— | 2.6 |
| **33.** | Car battery | 12 | 75 | —— |
| **34.** | Submersible pump | 220 | 18 | —— |
| **35.** | Chain saw | —— | 11.8 | 1.9 |
| **36.** | Industrial machine | —— | 53.2 | 17.1 |

Electric energy is usually sold to the consumer in *kilowatt-hours* (kwh), given by

$$\text{kwh} = \frac{E \cdot I \cdot t}{1000}$$

where $E$ is the voltage; $I$, the amperes; and $t$, the time in hours.

**EXAMPLE**   The number of kilowatt-hours used by a circuit drawing 2.7 amps from a 115-volt line for 3.3 hr is

$$\text{kwh} = \frac{115 \times 2.7 \times 3.3}{1000} = 1.02465$$

or 1.0 kwh to the nearest tenth.

*Find the number of kilowatt-hours used by each system. Round off to the nearest tenth.*

**37.** A motor drawing 16.8 amp from a 220-volt line for $2\frac{1}{2}$ hr.

**38.** A soldering iron drawing 2.2 amp from a 120-volt line for 1.6 hr.

**39.** A heater rated at 7.5 amp in a 220-volt circuit for 4.6 hr.

**40.** An electric iron drawing 5.1 amp from a 115-volt line for 1.3 hr.

Two useful power formulas are

$$P = \frac{E^2}{R} \qquad \text{and} \qquad P = I^2 \cdot R$$

where $P$ is the electrical power in watts, $E$ is the voltage, $I$ is the amperes, and $R$ is the resistance.

**EXAMPLE**   A certain ammeter (an instrument that measures the number of amperes in a current) has an internal resistance of 0.019 ohms. When it reads 7.6 amp, the power it is using is given by

$$P = I^2 \cdot R = (7.6)^2 \times 0.019$$
$$= 1.09744$$

The ammeter uses 1.1 watts, to the nearest tenth.

**41.** Find the power used by an ammeter
having an internal resistance of
0.023 ohms when it reads 9.2 amp.
Round off to the nearest tenth.

**42.** Find the power used by an ammeter
having an internal resistance of
0.021 ohms when it reads 12.5 amp.
Round off to the nearest tenth.

**43.** The internal resistance of a certain
voltmeter (an instrument used to
measure the number of volts in a
circuit) is 260,000 ohms. Find the
power used when it reads 210 volts.
Round off to the nearest
thousandth.

**44.** Find the power used by a voltmeter
having an internal resistance of
240,000 ohms when it reads 256
volts. Round off to the nearest
thousandth.

## 6.6  Formulas Involving
## Some Simple Machines

In this section we consider formulas related to some simple, but basic, mechanical devices.

### ■ Levers

Figure 6.13 shows a bar balanced upon a pivot point called the **fulcrum**. The symbols $w_1$ and $w_2$ represent weights (or forces) pulling on the bar; $d_1$ and $d_2$ represent distances from the fulcrum to the weights. Each of these distances is called a **lever arm**. The product of the weight (or force) and its related lever arm is called a **torque**. For example, if a 20-lb weight is hung at a distance of 8 in. from the fulcrum, the resulting torque is

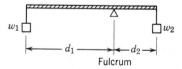

**FIGURE 6.13**

$$8 \text{ in.} \times 20 \text{ lb} = 160 \text{ in.-lb}$$

(the symbol "in.-lb" is read "inch pounds.") If the distance is measured in feet, the torque is in foot-pounds. Thus,

$$8 \text{ ft} \times 20 \text{ lb} = 160 \text{ ft-lb}$$

In general, a torque $T$ is a measure of the tendency of the bar to rotate about the fulcrum and can be computed by the formula

$$T = dw \qquad\qquad (1)$$

**EXAMPLE 1**  What force (to the nearest pound) must be applied to the end of a 12.4-in. wrench in order to develop a torque of 795 in.-lb? (See Figure 6.14.)

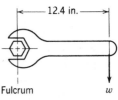

**FIGURE 6.14**

**Solution**  In Formula 1, substitute 795 for $T$, 12.4 for $d$, and solve for $w$.

$$T = dw$$
$$795 = 12.4w$$

Divide each side by 12.4 to obtain

$$w = \frac{795}{12.4} = 64.112903$$

To the nearest pound, the force needed is 64 lb.

Consider again the bar in Figure 6.13 and note that there is a torque $d_1w_1$ to the left of the fulcrum and a torque $d_2w_2$ to the right of the fulcrum. If these two torques are equal, the bar will be balanced (this condition is called "equilibrium"). In symbols, this basic principle of the lever is given by

$$d_1w_1 = d_2w_2 \qquad\qquad (2)$$

**EXAMPLE 2**  An 875-lb crate is hanging at the end of a bar, 10 in. from the fulcrum.
**(a)** To the nearest tenth of a pound, what force must be applied at a distance 11.5 ft from the fulcrum, on the opposite side of the lever, to balance the crate?
**(b)** To the nearest pound, find the least force that will raise the crate upward from the balanced position.

**Solutions**  Make a sketch and label with the given information. (See Figure 6.15.)

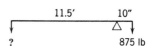

**FIGURE 6.15**

**(a)** Change 10 in. to feet: $\frac{10}{12} = 0.83333333$ ft.   In Formula 2, substitute 875 for $w_1$, 0.83333333 for $d_1$, 11.5 for $d_2$, and solve for $w_2$.

$$d_1w_1 = d_2w_2$$
$$0.83333333 \times 875 = 11.5 \times w_2$$

Divide each side by 11.5

$$w_2 = \frac{0.83333333 \times 875}{11.5} = 63.405797$$

To the nearest tenth, 63.4 lb will balance the crate.

**(b)** A force greater than 63.4 lb will raise the crate. Thus, to the nearest pound, 64 lb is sufficient.

■ *Gears*

Figure 6.16 shows gear $A$ meshed with gear $B$. If gear $A$ has 16 teeth and rotates at a speed of 1 rpm (revolution per minute), then gear $B$ with 8 teeth will rotate twice as fast—2 rpm—but in the *opposite* direction. More generally, if $t_1$ and $t_2$ are the respective numbers of *teeth* of two gears in mesh, and $r_1$ and $r_2$ are their respective

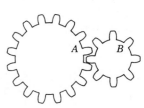

**FIGURE 6.16**

*speeds*, then the basic relationship between the gears (for any unit of speed) is given by the formula

$$t_1 r_1 = t_2 r_2 \qquad (3)$$

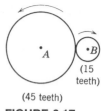

FIGURE 6.17

If a gear turns in the same direction as the hands of a clock, it is said to be rotating *clockwise.* A rotation opposite to that of the hands of a clock is said to be *counterclockwise.*

In diagrams, gears are frequently represented by figures such as that shown in Figure 6.17. The arrows indicate counterclockwise rotation (gear *A*) and clockwise rotation (gear *B*). In a two-gear system, the gear that supplies the power is called the *driver*, and the other gear is called the *driven*, gear.

**EXAMPLE 3**   Driver gear *A* rotates 282 rpm in a clockwise direction (see Figure 6.18). Compute the speed of driven gear *B* and specify the direction.

**Solution**   In Formula 3, substitute 12 for $t_1$, 282 for $r_1$, 48 for $t_2$, and solve for $r_2$.

$$t_1 r_1 = t_2 r_2$$
$$12 \cdot 282 = 48 r_2$$
$$r_2 = \frac{12 \cdot 282}{48} = 70.5$$

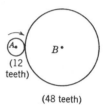

FIGURE 6.18

Thus, the speed of gear *B* is 70.5 rpm. Because its direction is opposite that of gear *A*, gear *B* rotates counterclockwise.

**EXAMPLE 4**   A driver gear with 90 teeth and a speed of 550 rpm is meshed with a larger driven gear with a speed of 198 rpm (see Figure 6.19). Find the number of teeth on the driven gear.

**Solution**   Make a sketch showing the given information. Let $t_2$ be the number of teeth on the larger gear. In Formula 3,

$$t_1 r_1 = t_2 r_2$$

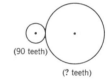

FIGURE 6.19

substitute 90 for $t_1$, 550 for $r_1$, 198 for $r_2$, and solve for $t_2$.

$$90 \times 550 = t_2 \times 198$$
$$\frac{90 \times 550}{198} = t_2$$
$$t_2 = 250$$

Thus the driven gear has 250 teeth.

If the distance between the centers of a driver gear and a driven gear is so great that the required gear diameters are too large, a gear called an *idler gear* may be inserted, as shown in Figure 6.20, where gear *A* is the driver gear, gear *B* is the idler gear, and gear *C* is the driven gear. An idler gear may also be inserted between two gears if the driver gear and the driven gear are required to rotate in the same direction. In Figure 6.20, note that if gear *A* rotates clockwise, then gear *B* rotates counterclockwise and gear *C* rotates clockwise.

Although idler gears can change the *direction* of rotation of a driven gear, it can be shown that idler gears cannot affect the *speed* of either the driver or the driven gear (see Exercise 6.6–47). Thus, given a system of meshed gears that includes idler gears, we can omit consideration of the idlers in the computations needed to find the speed of the driver gear or the driven gear.

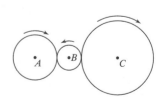

FIGURE 6.20

**EXAMPLE 5**   For the system shown in Figure 6.21, the speed of driven gear $C$ is 414 rpm. Compute the speed and direction of driver gear $A$.

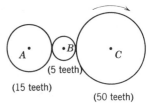

**FIGURE 6.21**

**Solution**   Consider only gears $A$ and $C$. In Formula 3, substitute 15 for $t_1$, 50 for $t_2$, 414 for $r_2$, and solve for $r_1$.

$$t_1 r_1 = t_2 r_2$$
$$15 r_1 = 50 \times 414$$
$$r_1 = \frac{50 \times 414}{15} = 1380$$

Thus, the speed of gear $A$ is 1380 rpm. Gear $C$ rotates clockwise. Hence, gear $B$ rotates counterclockwise and gear $A$ rotates clockwise.

## ■ Pulleys

Pulleys, such as those shown in Figure 6.22, operate very much like gears, except that power is transmitted from a driver pulley to a driven pulley by belts instead of by meshing teeth. We refer to the size of a pulley by its *diameter,* which is the distance measured across the pulley through its center. Figure 6.23 indicates a 12-in. pulley connected to a 6-in. pulley by a belt.

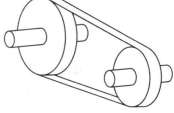

**FIGURE 6.22**

Because of the similarity of operation of gears and pulleys, Formula 3 can be supplied to systems of pulleys. We need only replace $t_1$ and $t_2$ by diameters $D_1$ and $D_2$, respectively. Thus, we have a basic relationship between pulleys:

$$D_1 r_1 = D_2 r_2 \qquad\qquad (4)$$

**EXAMPLE 6**   If the speed of pulley $A$ in Figure 6.24 is 125 rpm, find the diameter of pulley $B$, to the nearest tenth, so that the speed of pulley $B$ is 325 rpm.

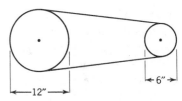

**FIGURE 6.23**

**Solution**   In Formula 4, substitute 18 for $D_1$, 125 for $r_1$, 325 for $r_2$; solve for $D_2$.

$$D_1 r_1 = D_2 r_2$$
$$18 \times 125 = D_2 \times 325$$
$$D_2 = \frac{18 \times 125}{325} = 6.9230769$$

To the nearest tenth, the diameter of the required pulley is 6.9 cm.

**EXAMPLE 7**   A driver pulley with a 28.4-cm diameter and a speed of 180 rpm is connected by a belt to a driven pulley with a diameter of 11.5 cm (see Figure 6.25). To the nearest whole number, find the speed of the driven pulley.

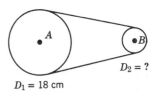

$D_1 = 18$ cm

**FIGURE 6.24**

**Solution**   Make a sketch showing the given information. Let $r_2$ be the speed of the smaller pulley. In Formula 4,

$$D_1 r_1 = D_2 r_2$$

substitute 28.4 for $D_1$, 180 for $r_1$, 11.5 for $D_2$, and solve for $r_2$.

$$28.4 \times 180 = 11.5 \times r_2$$
$$\frac{28.4 \times 180}{11.5} = r_2$$
$$r_2 = 444.52174$$

To the nearest whole number, the speed of the smaller pulley is 445 rpm.

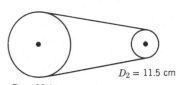

$D_1 = 28.4$ cm

**FIGURE 6.25**

## Exercises for Section 6.6

*Find the force, w (to the nearest pound), necessary to develop the given torque. See Example 1 and Figure 6.26.*

**1.** $T = 380$ in.-lb;   $d = 8$ in.        **2.** $T = 700$ in.-lb;   $d = 12$ in.

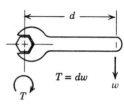

**FIGURE 6.26**

**3.** $T = 900$ in.-lb;   $d = 14$ in.        **4.** $T = 1050$ in.-lb;   $d = 16$ in.

*Find the torque, T (to the nearest inch-pound), developed.*

**5.** $d = 10.5$ in.;   $w = 50$ lb        **6.** $d = 15$ in.;   $w = 67$ lb

**7.** $d = 19\frac{3}{4}$ in;   $w = 31$ lb        **8.** $d = 10\frac{3}{16}$ in.;   $w = 48$ lb

*Find the length of a wrench, d (to the nearest tenth of an inch), necessary to develop the given torque, T, with the indicated force, w.*

**9.** $T = 1500$ in.-lb;   $w = 70$ lb        **10.** $T = 880$ in.-lb;   $w = 62$ lb

**11.** $T = 610$ in.-lb;   $w = 48$ lb        **12.** $T = 940$ in.-lb;   $w = 59$ lb

*Determine if each system of weights is in balance. See Example 2.*

|  | 6 ft | 3 ft |
|---|---|---|

**13.** 50 lb                          100 lb

|  | 10.4 ft | 5.3 ft |
|---|---|---|

**14.** 98 lb                          192 lb

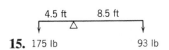

**15.** 175 lb               93 lb     **16.** 297 lb            198 lb

**17.** Find, to the nearest tenth of a pound, the force, $w$, necessary to just balance the 700-lb weight of a milling machine, using the level shown. Also find, to the nearest pound, the least force necessary to raise the machine. (See Figure 6.27.)

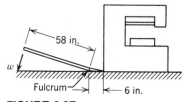

**FIGURE 6.27**

**18.** Solve Exercise 17 if the machine weight is 1200 lb, the fulcrum is 5 in. from the machine, and the force, $w$, is being applied 64 in. from the fulcrum.

**19.** Find, to the nearest pound, the least force, $w$, that must be exerted by the hydraulic jack to raise the 400-lb die shown in Figure 6.28.

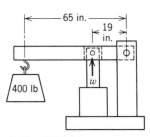

**FIGURE 6.28**

**20.** Solve Exercise 19 for a die weighing 520 lb.

*Gears A and B are meshed, as shown in Figure 6.29. Complete the following table. See Examples 3 and 4. (Under "Rotation," specify "clockwise" or "counterclockwise.")*

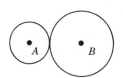

**FIGURE 6.29**

| | Gear | Rotation | Teeth | RPM |
|---|---|---|---|---|
| **21.** | A | ↺ | 12 | 240 |
| | B | — | 36 | — |
| **22.** | A | ↻ | 15 | 150 |
| | B | — | 25 | — |
| **23.** | A | — | 16 | 64 |
| | B | ↺ | 8 | — |
| **24.** | A | — | 72 | 100 |
| | B | ↻ | 20 | — |
| **25.** | A | ↻ | 10 | 120 |
| | B | — | — | 40 |
| **26.** | A | ↺ | 18 | 200 |
| | B | — | — | 72 |
| **27.** | A | — | 36 | 80 |
| | B | ↻ | — | 144 |
| **28.** | A | — | 48 | 90 |
| | B | ↺ | — | 216 |

*Gears A and B are connected by a set of idler gears. If gear A is rotating in a clockwise direction, make a sketch and determine the direction of rotation of gear B for a system having the given number of idlers. See Example 5 and Figure 6.30.*

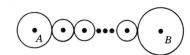

**FIGURE 6.30**

**29.** 1 idler        **30.** 2 idlers        **31.** 3 idlers

**32.** 5 idlers        **33.** 4 idlers        **34.** 6 idlers

*Gears A and B are meshed with a single idler gear, as shown in Figure 6.31. Complete the following table. See Example 5.*

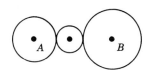

**FIGURE 6.31**

|  | Gear | Rotation | Teeth | RPM |
|---|---|---|---|---|
| **35.** | A | ↻ | 25 | 50 |
|  | B |  | 10 | — |
| **36.** | A | ↺ | 72 | 100 |
|  | B |  | 18 | — |
| **37.** | A |  | 14 | — |
|  | B | ↻ | 21 | 120 |
| **38.** | A |  | 18 | — |
|  | B | ↺ | 30 | 150 |

*Pulleys A and B are connected by a belt, as shown in Figure 6.32. Complete the following table. See Examples 6 and 7. Round off diameters to the nearest tenth and pulley speeds to the nearest whole number.*

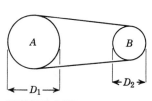

**FIGURE 6.32**

|  | Pulley | Diameter (in.) | RPM |
|---|---|---|---|
| **39.** | A | 25 | 175 |
|  | B | — | 295 |
| **40.** | A | 32 | 180 |
|  | B | — | 300 |
| **41.** | A | — | 210 |
|  | B | 12 | 400 |
| **42.** | A | — | 195 |
|  | B | 8 | 370 |
| **43.** | A | 28 | 90 |
|  | B | 18 | — |
| **44.** | A | 30 | 125 |
|  | B | 24 | — |
| **45.** | A | 16 | — |
|  | B | 9 | 100 |
| **46.** | A | 18 | — |
|  | B | 12 | 200 |

**47.** Gear *A* is rotating at 360 rpm in a counterclockwise direction (see Figure 6.33).

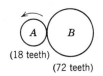

(a) Find the speed and direction of gear *B*.

(b) Insert an idler gear *C* between *A* and *B*, and find the speed and direction of gear *B*.

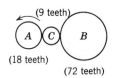

**FIGURE 6.33**

(c) Does the idler gear affect the speed of gear *B*?

(d) Does the idler gear affect the direction of gear *B*?

## 6.7 Mental Calculations

### Exercises for Section 6.7

*Each exercise in this section is to be done without a calculator.*

**1.** 120 in. = ? ft

**2.** 5 gal = ? qt

**3.** 32 oz = ? lb

**4.** 30 yd = ? ft

**5.** 2 ft 6 in. = ? in.

**6.** 100 pt = ? qt

**7.** 25 oz = ? lb ? oz

**8.** 5 ft 3 in. = ? ft (to nearest tenth)

**9.** 3 ft 2 in. + 5 ft 3 in.

**10.** 4 da 8 hr + 5 da 16 hr

**11.** 6 gal − 3 gal 1 qt

**12.** 10 lb − 5 lb 6 oz

**13.** 4 × (2 ft 2 in.)

**14.** 8 × (3 ft 3 in.)

**15.** 100 nm = ? cm

**16.** 1000 cm = ? m

**17.** 10,000 m = ? km

**18.** 100 cm = ? mm

**19.** 100 m = ? cm

**20.** 100 km = ? m

**21.** 100 mg = ? cg

**22.** 1000 cg = ? g

**23.** 10,000 g = ? kg

**24.** 100 cg = ? mg

**25.** 100 g = ? cg          **26.** 100 kg = ? g          **27.** 100 ml = ? cl

**28.** 1000 cl = ? $\ell$          **29.** 10,000 $\ell$ = ? kl          **30.** 100 cl = ? ml

**31.** 100 $\ell$ = ? cl          **32.** 100 kl = ? $\ell$

*In each of the following, decide which of the choices is the most reasonable.*

**33.** The width of a thumb is about
    **(a)** 2 cm  **(b)** 2 mm  **(c)** 2 m

**34.** The height of a door in a house is about
    **(a)** 2 cm  **(b)** 2 mm  **(c)** 2 m

**35.** A baseball bat is a little longer than
    **(a)** 1 m  **(b)** 10 m  **(c)** 100 m

**36.** The waist measurement of a beauty-contest winner might be about
    **(a)** 0.6 m  **(b)** 6 m  **(c)** 60 m

**37.** An automobile might weigh about
    **(a)** 1300 mg  **(b)** 1300 g
    **(c)** 1300 kg

**38.** The weight of a baseball is about
    **(a)** 14 g  **(b)** 140 g  **(c)** 14,000 g

**39.** A glass of water contains about
    **(a)** 25 $\ell$  **(b)** 2.5 $\ell$  **(c)** 0.25 $\ell$

**40.** An automobile gasoline tank might hold about
    **(a)** 68 kl  **(b)** 68 ml  **(c)** 68 $\ell$

■ **Review Exercises**

**Section 6.1**

*Convert each measurement.*

**1.** 324 in. = ? ft          **2.** 100 qt = ? fl oz

**3.** 6 gal 2 qt = ? qt          **4.** 221 oz = ? lb ? oz

**5.** 4196 yd = ? mi (to nearest tenth)          **6.** 8 ft 7 in. = ? ft (to nearest tenth)

## Section 6.2

*For Exercises 7 to 9, (a) compute, and (b) express the result in decimal form rounded off to the nearest hundredth.*

**7.** 14 ft + 6 ft 2 in. + 4 ft 5 in.

**8.** 2 hr 25 min + 3 hr 9 min + 7 hr 5 min

**9.** 29 da 8 hr − 8 da 12 hr

*Find each product or quotient.*

**10.** 24 × (32 ft)  **11.** (182 gal) ÷ 13  **12.** (32 pt 2 fl oz) ÷ 6 fl oz

## Section 6.3

**13.** Write the meaning of (*a*) milligram, and (*b*) kilogram in terms of "gram."

*Convert each measurement.*

**14.** 113 cm = ? mm  **15.** 750 cg = ? mg

**16.** 74.9 km = ? m  **17.** 3.65 ℓ = ? ml  **18.** 14 kg = ? mg

*Round off each answer to the nearest tenth.*

**19.** 7.4 in. = ? cm  **20.** 100 yd = ? m  **21.** 17.2 pt = ? ℓ

**22.** 3.7 m = ? ft  **23.** 5.6 kg = ? lb  **24.** 30 ℓ = ? gal

## Section 6.4

*For Exercises 25 to 30, refer to the formula used in the indicated example in Section 6.4.*

**25.** Find the work done if a force of
92 lb is applied a distance of 67 ft
(Example 1).

**26.** Find the power needed to raise
94 lb a distance of 24 ft in 7 sec
(Example 3). Round off to the
nearest tenth.

**27.** Find the work done in 38 min by an
engine that can deliver power at the
rate of 376 ft-lb/min (Example 4).
Round off to the nearest tenth.

**28.** Find the horsepower required to lift
2854 lb a distance of 29 ft in 6.9 min
(Example 5). Round off to the
nearest tenth.

**29.** Find the N.A.C.C. horsepower of a
six-cylinder engine with a 4.2-in.
bore (Example 6). Round off to the
nearest tenth.

**30.** Find the brake horsepower of an
engine if   $L = 6$ ft,   $R = 584$
rpm,   and   $f = 372$ lb   (Example
7). Round off to the nearest tenth.
(Use the $\boxed{\pi}$ key, or 3.14159.)

## Section 6.5

*For Exercises 31 to 36, refer to the formula used in the indicated example in Section
6.5. Round off each answer to the nearest tenth.*

**31.** Find $I$ if   $E = 115$ volts   and
$R = 50.2$ ohms   (Example 1).

**32.** Find $R$ if   $E = 220$ volts   and
$I = 72.6$ amp.   (Example 1).

**33.** Find the total resistance of three resistors when $R_1 = 54.6$ ohms, $R_2 = 17.5$ ohms, and $R_3 = 64.3$ ohms, if they are connected (*a*) in series, and (*b*) in parallel (Examples 2 and 3).

**34.** Find the number of kilowatts used by an electric motor running on 225 volts with a current of 37.7 amp (Example 4).

**35.** A 2.5-hp electric motor is connected to a 120-volt line. How many amperes of current does it require? (Example 5).

**36.** A 4-hp electric motor uses 13.6 amp. Find the voltage in the line (Example 5).

## Section 6.6

*For Exercises 37 to 40, refer to the formula used in the indicated example in Section 6.6.*

**37.** If $d = 18$ in., find the force necessary to develop a torque of 1224 in.-lb (Example 1).

**38.** If a weight of 2400 lb is 5 in. from the fulcrum of a lever, find the distance from the fulcrum at which a force of 125 lb would balance the weight (Example 2).

**39.** Gear $A$ is driving gear $B$. If $A$ has 36 teeth and is rotating at 100 rpm, and $B$ has 40 teeth, find the speed of $B$ (Example 3).

**40.** Pulleys *A* and *B* are connected by a
belt. If *A* has a diameter of 28 cm
and is rotating at 210 rpm, and *B*
has a diameter of 18 cm, find the
speed of *B* correct to the nearest
whole number (Example 7).

## Section 6.7

*Each of the following exercises is to be done mentally.*

**41.** 48 in. = ? ft          **42.** 160 oz = ? lb          **43.** 2 gal 2 qt + 3 gal 2 qt

**44.** 6 × (2 ft 4 in.)       **45.** 3000 m = ? km          **46.** 2 m = ? cm

**47.** 5 kg = ? g             **48.** 10 cl = ? ml

# Geometric Figures

## 7.1 Angles and Their Measure

In this section we first introduce several geometric figures, together with the symbols used to name them, and then we consider some of the basic properties of the figures.

### ■ Naming Points, Lines, Rays, and Segments

- **Points** are named with capital letters A, B, C, and so on, as shown in Figure 7.1*a* on page 226.
- **Lines** are named by using any two points of a line. Figure 7.1*b* shows line PQ; the arrowheads at each end indicate that the line does not have any endpoints.
- A **ray** is part of a line that includes only one endpoint. Figure 7.1*c* shows ray RS; the first letter (R) specifies the endpoint, and the second letter (S) names any other point on the ray.
- A **segment** is a part of a line between (and including) two endpoints. The names of its endpoints are used to name a segment. For example, Figure 7.1*b* shows segment $PQ$ (as part of line $PQ$); Figure 7.1*c* shows segment $RS$ (as part of ray $RS$). The names of the endpoints are also used to indicate the *length* of a segment. For example, we use the symbol "$PQ$" to mean "the length of segment $PQ$." If segment $PQ$ is 4.7 in. long, we write $PQ = 4.7$ in.

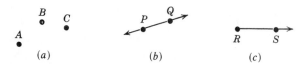

**FIGURE 7.1**

## ■ Naming Angles

An **angle** is a figure formed by two rays (the **sides**) with a common endpoint (the **vertex**). Using the symbol "∠" to mean "angle," the angle of Figure 7.2 may be named ∠B, ∠CBA, or ∠ABC. When using three letters to name an angle, the middle letter names the vertex. Numbers are also used to name angles, as shown in Figure 7.3, where ∠1, ∠2, ∠3, and ∠4 are indicated.

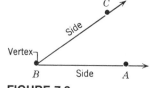

**FIGURE 7.2**

## EXAMPLE 1
**(a)** Name the vertex and sides of ∠5 in Figure 7.4.
**(b)** Name the vertex and sides of ∠6 in Figure 7.4.

**FIGURE 7.3**

### Solutions
**(a)** Point $A$ is the vertex and rays $AE$ and $AF$ are the sides of ∠5.
**(b)** Point $B$ is the vertex and rays $BC$ and $BD$ are the sides of ∠6.

## ■ Measuring Angles

Angles are commonly measured in **degree** units.* One degree is written as 1°; there are 360° in a circle (see Figure 7.5).

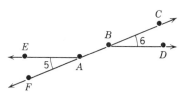

**FIGURE 7.4**

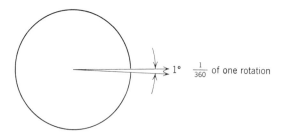

**FIGURE 7.5**

We can use a *protractor* scaled in degrees to obtain the measure of any angle between 0° and 180° (see Figure 7.6).

## EXAMPLE 2
In Figure 7.6,

$$\angle AEB = 42°$$
$$\angle AEC = 115°$$
$$\angle AED = 180°$$

Degrees can be subdivided into *minutes* (1° = 60′) and minutes can be subdivided into *seconds* (1′ = 60″). However, for calculator use, the *decimal form* of angle measure is more convenient than the degree-minute form. The methods introduced in Section 6.1 can be used to change an angle measure from degree-minute to decimal form.

---

*Another unit of measure for angles, called a *radian*, is considered in Section 8.6.

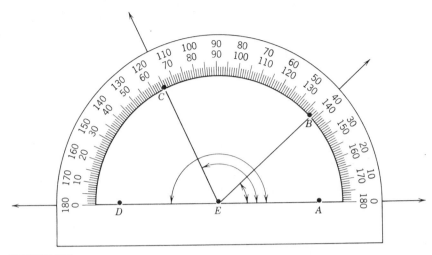

**FIGURE 7.6**

**EXAMPLE 3** Write 54°28′ as a number of degrees in decimal form.

**Solution** The conversion number (minute-degree) is 60. Hence,

$$54°28' = \left(54\frac{28}{60}\right)^° = 54.46666667°$$

Thus, 54°28′ = 54.47°, to the nearest hundredth.

We sometimes need to change the decimal form of an angle measure to degree-minute form.

**EXAMPLE 4** Write 128.74° in degree-minute form.

**Solution** First, change 0.74° to minutes.
$$0.74° = (0.74 \times 60)' = 44.4'$$
Then,
$$128.74° = 128° + 0.74° = 128° + 44.4'$$
$$= 128°44.4'$$

Thus, 128.74° = 128°44′, to the nearest minute.

## ■ Classifying Angles

Angles can be classified according to their measures, as shown in Table 7.1 on page 228.

Note the use of the symbol ∟ in Table 7.1 to indicate a *right angle* (∠C). Any angle marked in this manner is understood to be a right angle.

**EXAMPLE 5** Specify whether each angle is acute, right, obtuse, or straight.
**(a)** ∠F = 173°     **(b)** ∠G = 180°     **(c)** ∠H = 90°     **(d)** ∠J = 73°

**Solutions** In each case, refer to Table 7.1.
**(a)** ∠F is obtuse because 173° is between 90° and 180°.
**(b)** ∠G is a straight angle because ∠G equals 180°.
**(c)** ∠H is a right angle because ∠H equals 90°.
**(d)** ∠J is acute because 73° is between 0° and 90°.

**TABLE 7.1**

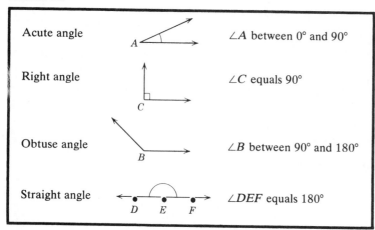

| | | |
|---|---|---|
| Acute angle | | $\angle A$ between 0° and 90° |
| Right angle | | $\angle C$ equals 90° |
| Obtuse angle | | $\angle B$ between 90° and 180° |
| Straight angle | | $\angle DEF$ equals 180° |

## ■ *Perpendicular and Parallel Lines*

If two intersecting lines form a right angle, the two lines are said to be **perpendicular**. For example, in Figure 7.7a, $\angle RPT$ is a right angle. Hence, line $RS$ is perpendicular to line $QT$. If two lines in a plane never intersect, the two lines are said to be **parallel**. In Figure 7.7b, lines $AB$ and $CD$ are parallel.

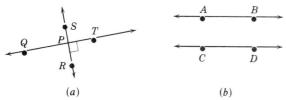

(a)                              (b)

**FIGURE 7.7**

## *Exercises for Section 7.1*

*Refer to Figure 7.8 to name the vertex and sides of each angle. See Example 1.*

**1.** $\angle 1$       **2.** $\angle 2$       **3.** $\angle 3$

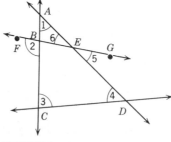

**FIGURE 7.8**

**4.** $\angle 4$       **5.** $\angle 5$       **6.** $\angle 6$

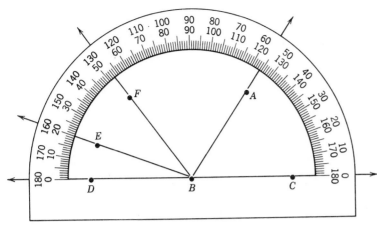

**FIGURE 7.9**

*Estimate the measure of each angle in Figure 7.9 to the nearest degree. See Example 2.*

**7.** ∠ABC **8.** ∠DBE **9.** ∠DBF **10.** ∠CBF

**11.** ∠EBF **12.** ∠ABF **13.** ∠ABE **14.** ∠ABD

*Write each angle in decimal form. See Example 3. Round off to the nearest hundredth.*

**15.** 40°23′ **16.** 20°44′ **17.** 23°4′ **18.** 6°10′

**19.** 16°38′ **20.** 34°47′ **21.** 123°21′ **22.** 87°46′

*Write each angle in degree-minute form. See Example 4. Round off to the nearest minute.*

**23.** 90.72° **24.** 64.36° **25.** 89.62° **26.** 150.12°

**27.** 115.66° **28.** 164.08° **29.** 173.11° **30.** 57.91°

*Specify whether an angle with the given measure is acute, right, straight, or obtuse. See Example 5.*

**31.** 40°        **32.** 80°        **33.** 90°        **34.** 37°

**35.** 110°        **36.** 179°        **37.** 180°        **38.** 92°

*For Exercises 39 to 46, refer to Figure 7.10.*

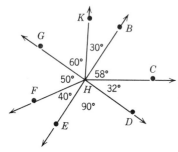

**FIGURE 7.10**

*Specify the measure of each angle.*

**39.** ∠EHC        **40.** ∠FHK        **41.** ∠KHD        **42.** ∠FHB

**43.** ∠KHE        **44.** ∠KHC        **45.** ∠FHC        **46.** ∠GHC

---

## 7.2  Triangles

A **triangle** is a closed figure formed by three segments. Each of the three segments is called a **side** of the triangle. Each of the three points at which the sides of a triangle meet is called a **vertex** (plural: **vertices**) of the triangle. Triangles are named by using the symbol "△" (read "triangle") together with the names of the three vertices. For example, Figure 7.11 shows △ABC, △DEF, and △MNP.

- ■ A **scalene triangle** is a triangle in which no two sides are equal (Figure 7.11a).
- ■ An **equilateral triangle** is a triangle with three equal sides (Figure 7.11b).

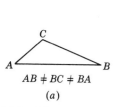
$AB \neq BC \neq BA$
(a)

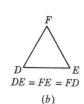
$DE = FE = FD$
(b)

$MP = NP$
(c)

**FIGURE 7.11**

■ An **isosceles triangle** is a triangle with two equal sides: The angle formed by the two equal sides is called the **vertex angle** (Figure 7.11*c*).

**EXAMPLE 1** Classify each triangle in Figure 7.12 according to the lengths of its sides. If a triangle is isosceles, name the vertex angle.

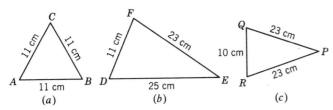

**FIGURE 7.12**

**Solutions**

**(a)** △*ABC* is equilateral because it has three equal sides.

**(b)** △*DEF* is scalene because no two sides are equal.

**(c)** △*PQR* is isosceles because it has two equal sides. The vertex angle is ∠*P*.

Triangles can also be classified according to the measures of their angles. An **acute triangle** is a triangle with each of its angles acute (Figure 7.13*a*); an **obtuse triangle** is a triangle with one obtuse angle (Figure 7.13*b*); a **right triangle** is a triangle with one right angle. The side opposite the right angle of a right triangle is called the **hypotenuse** (Figure 7.13*c*).

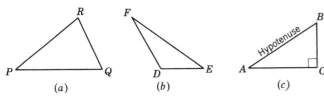

**FIGURE 7.13**

**EXAMPLE 2** Classify each triangle in Figure 7.14 according to the measures of its angles. If a triangle is a right triangle, name the hypotenuse.

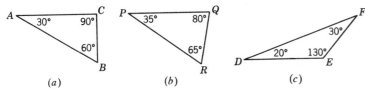

**FIGURE 7.14**

**Solutions**

**(a)** △*ABC* is a right triangle because it includes one right angle (∠*C*). The hypotenuse is *AB*.

**(b)** △*PQR* is an acute triangle because each of its angles is acute.

**(c)** △*DEF* is an obtuse triangle because it includes one obtuse angle (∠*DEF*).

## ■ Altitudes and Bases

The line segment from any vertex of a triangle perpendicular to the opposite side
(or the opposite side extended) of the triangle is called an **altitude** of the triangle;
the side to which it is perpendicular is then referred to as the **base**. Figure 7.15
shows several examples. In the right triangle (Figure 7.15*c*), note that one side (not
the hypotenuse) is already perpendicular to the second side. Hence, in a right
triangle, each side (other than the hypotenuse) can be considered as an altitude of
the triangle. The *length* of an altitude is often represented by *h*, as in the figure.
This length is also commonly referred to as the altitude.

**FIGURE 7.15**

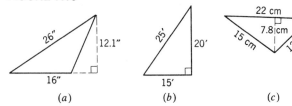

**EXAMPLE 3**   For each triangle in Figure 7.16, specify the length *h* of the altitude
and the length *b* of the related base.

**FIGURE 7.16**

 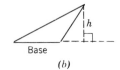 

*(a)*                          *(b)*                          *(c)*

### Solutions
**(a)** $h = 12.1$ in.   and   $b = 16$ in.
**(b)** Either   $h = 20$ ft   and   $b = 15$ ft,   or   $h = 15$ ft   and   $b = 20$ ft.
**(c)** $h = 7.8$ cm   and   $b = 22$ cm.

## ■ Properties of Triangles

Following are two useful properties of triangles. The first property refers to any
kind of triangle.

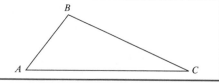

**The sum of the three angles of a triangle is 180°.**

$$\angle A + \angle B + \angle C = 180°$$

Because one of the angles of a right triangle is a 90° angle, by this property
we see that the sum of the other two angles must equal 90°. Thus, two angles (not
the right angle) of a right triangle are *acute* angles.

### EXAMPLE 4
**(a)** In $\triangle FGH$ in Figure 7.17, if   $\angle F = 49.5°$   and   $\angle G = 38.7°$,   then $\angle H$ can
be obtained by using the equation

$$49.5° + 38.7° + \angle H = 180°$$

Solving this equation yields

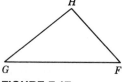

**FIGURE 7.17**

$$88.2° + \angle H = 180°$$
$$88.2° + \angle H - 88.2° = 180° - 88.2°$$
$$\angle H = 91.8°$$

**(b)** In right triangle *ABC* in Figure 7.18, if $\angle A = 23.3°$, then

$$23.3° + \angle B = 90°$$
$$23.3° - \angle B - 23.3° = 90° - 23.3°$$
$$\angle B = 66.7°$$

**FIGURE 7.18**

A second important property of triangles refers to right triangles. This property is called the **Pythagorean rule**.

> **In a right triangle, the square of the length of the hypotenuse is equal to the sum of the squares of the lengths of the other two sides.**
> $$c^2 = a^2 + b^2$$

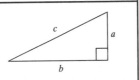

**EXAMPLE 5** In right triangle *ABC* in Figure 7.19, if $AC = 8.7$ ft and $BC = 5.0$ ft, then

$$(AB)^2 = 8.7^2 + 5.0^2 = 100.69$$
$$AB = \sqrt{100.69} = 10.034441$$

To the nearest tenth, $AB = 10.0$ ft.

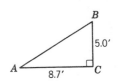

**FIGURE 7.19**

## ■ Applications of the Pythagorean Rule

Many practical problems involve right triangles in which two sides are known and the third side is to be found. The Pythagorean rule can be used to solve such problems.

**EXAMPLE 6** A V-block is formed by cutting a V-shaped groove in a metal block. For the V-block shown in Figure 7.20, compute the width of the cut (dimension *A*) to the nearest tenth.

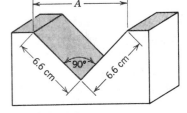

**FIGURE 7.20**

**Solution** Draw a sketch of the right triangle formed by the cut (Figure 7.21), and note that the width of the cut is the hypotenuse of the right triangle. By the Pythagorean rule,

$$A^2 = 6.6^2 + 6.6^2$$
$$A = \sqrt{6.6^2 + 6.6^2} = 9.3338095$$

To the nearest tenth, the cut is 9.3 cm wide.

**FIGURE 7.21**

**EXAMPLE 7** The bottom of a 25-ft ladder is placed 8 ft from the base of a building, as shown in Figure 7.22. To the nearest foot, how high up the side of the building will the ladder reach?

**Solution** Draw a sketch of the right triangle formed by the ladder, the ground, and the side of the building, where *h* is the required height (Figure 7.23). By the Pythagorean rule,

$$h^2 + 8^2 = 25^2$$
$$h = \sqrt{25^2 - 8^2} = 23.685439$$

To the nearest whole foot, the ladder will reach 24 ft up the side.

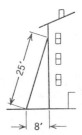 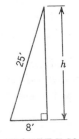

**FIGURE 7.22**      **FIGURE 7.23**

## Exercises for Section 7.2

*Classify each triangle according to the lengths of its sides. See Example 1.*

**1.**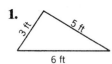

3 ft, 5 ft, 6 ft

**2.**

4 in., 4 in., 3 in.

**3.**

6 m, 6 m, 6 m

**4.**

5 mi, 3 mi, 4 mi

**5.**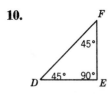

10 cm, 14 cm, 10 cm

**6.**

4 yd, 13 yd, 12 yd

*Classify each triangle according to the measure of its angles. If a triangle is a right triangle, name the hypotenuse. See Example 2.*

**7.**

C 100°, A 60°, B 20°

**8.**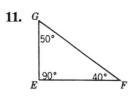

D 40°, B 75°, C 65°

**9.**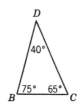

E 60°, C 60°, D 60°

**10.**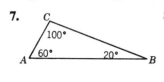

F 45°, D 45°, E 90°

**11.**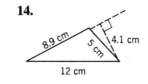

G 50°, E 90°, F 40°

**12.**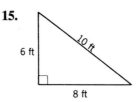

H 40°, F 110°, G 30°

*For each triangle, specify the length h of the altitude and the length b of the related base. See Example 3.*

**13.**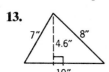

7″, 8″, 4.6″, 10″

**14.**

8.9 cm, 5 cm, 4.1 cm, 12 cm

**15.**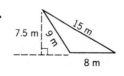

6 ft, 10 ft, 8 ft

**16.**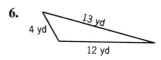

12 mi, 5 mi, 13 mi

**17.**

7.5 m, 9 m, 15 m, 8 m

**18.**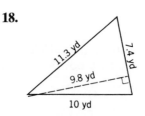

11.3 yd, 7.4 yd, 9.8 yd, 10 yd

*Given the measures of two angles of a triangle, say △BCD, find the measure of the third angle. See Example 4a.*

**19.** ∠B = 60°;   ∠C = 20°          **20.** ∠B = 80°;   ∠C = 40°

**21.** $\angle B = 100°$; $\angle D = 10°$ **22.** $\angle B = 115°$; $\angle D = 15°$

**23.** $\angle C = 10°40'$; $\angle D = 20°10'$ **24.** $\angle C = 25°35'$; $\angle D = 35°20'$

**25.** $\angle B = 51.7°$; $\angle D = 32.8°$ **26.** $\angle B = 90.6°$; $\angle D = 72.6°$

*Given the measure of one acute angle of right $\triangle ABC$, find the measure of the other acute angle. See Example 4b.*

**27.** $\angle A = 20°$ **28.** $\angle A = 35°$ **29.** $\angle B = 19°20'$

**30.** $\angle B = 70°25'$ **31.** $\angle A = 73.7°$ **32.** $\angle A = 57.6°$

*In $\triangle ADF$ in Figure 7.24, if $\angle A = \angle D$, find the measures of the other two angles, given the measure of one angle.*

**33.** $\angle A = 70°10'$ **34.** $\angle D = 85°20'$

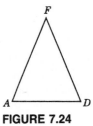

**FIGURE 7.24**

**35.** $\angle F = 20°40'$ **36.** $\angle F = 30°36'$

*Refer to Figure 7.25 for Exercises 37 to 42. Given the lengths of two sides, find the length of the third side, to the nearest tenth. See Example 5.*

**37.** $AC = 4$ cm; $BC = 5$ cm **38.** $AC = 9$ ft; $BC = 4$ ft

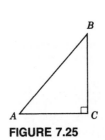

**FIGURE 7.25**

**39.** $AC = 5.8$ km;   $BC = 2.2$ km        **40.** $AC = 4.6$ mi;   $BC = 7.6$ mi

**41.** $AB = 84.6$ mi;   $BC = 22.3$ mi        **42.** $AB = 70.5$ cm;   $BC = 36.8$ cm

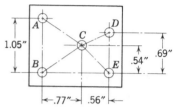

*Five holes are drilled in a rectangular steel plate, as shown in Figure 7.26. Find the indicated distances, to the nearest hundredth.*

**43.** BC        **44.** CE        **45.** AC        **46.** DC

**FIGURE 7.26**

**47.** A roof truss has the dimensions shown in Figure 7.27. Find the length of a rafter, to the nearest inch.

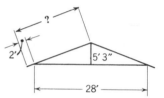

**FIGURE 7.27**

**48.** In Figure 7.28, the head of a bolt is a square 2 in. on a side and must pass through a hole with $\frac{1}{4}$-in. clearance at each corner. To the nearest tenth of an inch, what is the diameter of the hole?

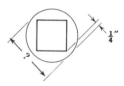

**FIGURE 7.28**

**49.** A thin strip of sheet aluminum is to be bent into the shape shown in Figure 7.29. To the nearest hundredth, find the approximate length of the strip before it is bent.

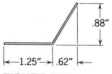

**FIGURE 7.29**

**50.** Find the length along the centerline of the piece of work shown in Figure 7.30. Round off to the nearest hundredth.

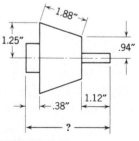

## 7.3  *Polygons and Circles*

In Section 7.2 we investigated the properties of triangles. Triangles and other closed plane figures formed by three or more segments are called **polygons**. Each of the segments is called a **side** of the polygon and each of the points at which the sides meet is called a **vertex**. Figure 7.31 shows a four-sided polygon with sides *AB*, *BC*, *CD*, and *DA*. Its vertices are *A*, *B*, *C*, and *D*.

**FIGURE 7.30**

Polygons with four, five, six, and eight sides are commonly called **quadrilaterals, pentagons, hexagons**, and **octagons**, respectively (Figure 7.32).

**FIGURE 7.31**

Quadrilateral   Pentagon   Hexagon   Octagon

**FIGURE 7.32**

A **parallelogram** is a quadrilateral whose opposite sides are parallel and equal in length. In parallelogram *ABCD* (Figure 7.33), sides *AB* and *CD* are parallel and equal, sides *AD* and *BC* are parallel and equal. A **rectangle** is a parallelogram with four right angles, as indicated in Figure 7.34. Because a rectangle is a parallelogram, opposite sides are equal, and

$$GD = EF \quad \text{and} \quad DE = GF$$

A **square** is a rectangle with all four sides equal. Thus, in square *ABCD* of Figure 7.34*b* we have

$$AB = BC = CD = DA$$

A **trapezoid** is a quadrilateral with exactly one pair of opposite sides parallel. Each of the parallel sides is called a **base**. Figure 7.35 shows trapezoid *STUV* with bases *ST* and *UV*.

**FIGURE 7.33**

**FIGURE 7.34**

(*a*)   (*b*)

**EXAMPLE 1**   A portion of a steel truss is shown in Figure 7.36. (Dimensions are in meters.)
**(a)** In parallelogram *ABEF*:  *AB* = *FE* = 3 m;  *AF* = *BE* = $3\sqrt{2}$ m.
**(b)** In rectangle *ACEG*:  *AC* = *GE* = 6 m;  *AG* = *CE* = 3 m.
**(c)** In square *ABFG*:  *AB* = *BF* = *FG* = *GA* = 3 m.
**(d)** In trapezoid *ADEF*:  base *AD* = 9 m;  base *FE* = 3 m.

**FIGURE 7.35**

## ■ Regular Polygons

A **regular polygon** is a polygon with all sides equal and all angles equal. Examples of such polygons are shown in Figure 7.37.

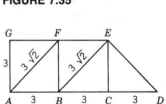

**FIGURE 7.36**

Equilateral   Square   Regular   Regular   Regular
triangle                 pentagon  hexagon  octagon

**FIGURE 7.37**

## ■ Perimeters

The **perimeter** of a plane geometric figure is the distance around the figure.

**EXAMPLE 2**

**(a)**  23 ft   8 in.
    13 ft   5 in.
    11 ft   2 in.
    10 ft
    ————————
    57 ft   15 in. = 58 ft   3 in.

In Figure 7.38, the perimeter is 58 ft 3 in.

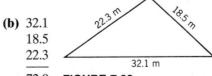

**FIGURE 7.38**

**(b)**  32.1
    18.5
    22.3
    ————
    72.9   **FIGURE 7.39**

In Figure 7.39, the perimeter is 72.9 m.

When computing with denominate numbers, we will always indicate units as part of the answer. However, we will show units in a computation only when more than one unit is involved, as in Example 2a.

A perimeter can often be computed by definition, as in Example 2. On the other hand, for certain special polygons, formulas are convenient. For example, the perimeter $P$ of a rectangle as shown in Figure 7.40 can be computed by the simple formula

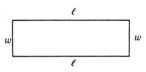

**FIGURE 7.40**

$$P = 2\ell + 2w$$

An alternate form, more convenient for calculator use, is

$$P = (\ell + w) \times 2 \qquad (1)$$

**EXAMPLE 3**   Find the perimeter of a rectangle 118.3 ft long and 52.9 ft wide.

**Solution**   Substitute 118.3 for $\ell$ and 52.9 for $w$ in Formula 1.

$$P = (\ell + w) \times 2$$
$$= (118.3 + 52.9) \times 2 = 342.4$$

The perimeter is 342.4 ft.

**EXAMPLE 4**   Find the width of a rectangle with length 205.4 m if the perimeter is 595.4 m.

**Solution**   Substitute 595.4 for $P$ and 205.4 for $\ell$ in Formula 1.

$$P = (\ell + w) \times 2$$
$$595.4 = (205.4 + w) \times 2$$

Divide each side by 2.

$$297.7 = 205.4 + w$$

Subtract 205.4 from each side.

$$92.3 = w$$

The width of the rectangle is 92.3 m.

All the sides of a regular polygon are equal in length. Hence, if $s$ represents the length of a side and $n$ is the number of sides, the perimeter $P$ of a regular polygon is given by

$$P = ns \qquad (2)$$

**EXAMPLE 5**   Find the side of a regular octagon if its perimeter is 29 yd 1 ft.

**Solution**   Change 29 yd 1 ft to 88 ft. Substitute 88 for $P$ and 8 for $n$ in Formula 2.

$$88 = 8s$$

Divide each side by 8.

$$11 = s$$

Each side is 11 ft (or 3 yd 2 ft) long.

## ■ *Circles*

A **circle** is a plane curve, every point of which is the same distance from a given point, called the **center** (point $Q$ in Figure 7.41). Following are some of the terms associated with circles (see Figure 7.41).

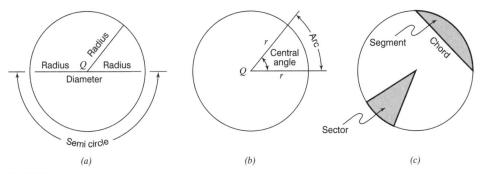

**FIGURE 7.41**

- **Radius** A line segment (or the length of the segment) between the center and any point of the circle. (Plural: radii.)
- **Diameter** A line segment with endpoints on the circle and including the center.
- **Chord** A line segment with endpoints on the circle and not including the center.
- **Segment** Area of a circle between a chord and the circle.
- **Arc** A part of a circle.
- **Semicircle** An arc of a circle whose endpoints determine a diameter.
- **Circumference** The perimeter of (distance around) a circle.
- **Central angle** An angle whose sides are two radii.
- **Sector** Area of a circle intercepted by the radii of a central angle.

In Figure 7.41*a* observe that a diameter contains two radii. Thus, for a circle of radius $r$ and diameter $d$, we have

$$d = 2r \qquad\qquad\qquad \textbf{(3)}$$

For at least 2000 years it has been known that if the circumference $C$ of a circle is divided by its diameter $d$, the quotient $C/d$ is always the same. This quotient is the nonterminating decimal approximately equal to 3.14159 and is traditionally represented by the Greek letter $\pi$ ("pi"). This remarkable fact leads to two formulas. The circumference $C$ of a circle is given by

$$C = 2\pi r \qquad\qquad\qquad \textbf{(4)}$$

or

$$C = \pi d \qquad\qquad\qquad \textbf{(5)}$$

**EXAMPLE 6** To the nearest tenth:
**(a)** Find the circumference of a circle with radius $r = 10.3$ ft.
**(b)** Find the diameter of a circle with circumference 412.8 km.

**Solutions**
**(a)** Substitute 10.3 for $r$ in Formula 4.

$$C = 2\pi r$$
$$= 2\pi(10.3) = 64.716809$$

To the nearest tenth, $C = 64.7$ ft.

**(b)** Substitute 412.8 for $C$ in Formula 5.

$$C = \pi d$$

$$412.8 = \pi d$$

$$\frac{412.8}{\pi} = d$$

$$d = 131.39832$$

To the nearest tenth,   $d = 131.4$ km.

## ■ Length of an Arc

The length $L$ of the arc associated with a central angle of $D°$ in a circle with radius $r$ (see Figure 7.42) is given by

$$L = \frac{\pi D°r}{180} \tag{6}$$

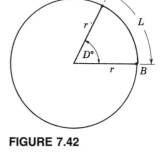

**FIGURE 7.42**

**EXAMPLE 7**   The length of an arc of a sector with a 43° central angle in a circle with a 14-in. radius is given by

$$L = \frac{\pi D°r}{180} = \frac{\pi \cdot 43 \cdot 14}{180} = 10.50688$$

To the nearest tenth,   $L = 10.5$ in.

**EXAMPLE 8**   Find the measure of the central angle associated with an arc 22.6 cm long in a circle with a radius of 32 cm.

**Solution**   Substitute 22.6 for $L$ and 32 for $r$ in Formula 6.

$$22.6 = \frac{\pi \cdot D° \cdot 32}{180}$$

Multiply each side by 180.

$$4068 = 32\pi D°$$

Divide each side by $32\pi$.

$$\frac{4068}{32\pi} = D°$$

$$D° = 40.46514$$

To the nearest tenth,   $D = 40.5°$.

## ■ Perimeters of Composite Figures

We sometimes need to use more than one formula to find the perimeter of a given figure.

**EXAMPLE 9**   A semicircle with radius 0.70 in. is cut out of a rectangle $ABCD$ as shown in Figure 7.43. To the nearest hundredth, find the perimeter of the resulting figure.

**Solution**

$$\text{perimeter} = AB + CD + DA + \text{arc length } BC \tag{7}$$

Because $ABCD$ is a rectangle,

$$CD = AB = 3.52 \quad \text{and} \quad DA = 1.40$$

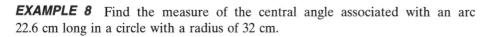

**FIGURE 7.43**

*BC* is the diameter of the semicircle with center at *P*, arc length *BC* is one-half the circumference of the circle with center at *P*, and diameter equal to 1.40. Hence,

$$\text{arc length } BC = \tfrac{1}{2}(\pi \times 1.40) = 2.1991149$$

The perimeter can now be computed from Equation 7:

$$\text{perimeter} = 3.52 + 3.52 + 1.40 + 2.1991149 = 10.639115$$

The perimeter is 10.64 in., to the nearest hundredth.

**EXAMPLE 10**  In Figure 7.44, *Q* is the center of a circle with radius 3.92 cm, and ∠*PQR* is equal to 135.8°. Find the perimeter of the figure to the nearest hundredth.

**Solution**

$$\text{perimeter} = AB + BR + PA + \text{arc length } PR \qquad \textbf{(8)}$$

Use Formula 6 to find arc length *PR*.

$$\text{arc length } PR = \frac{\pi \times 135.8 \times 3.92}{180} = 9.2910159$$

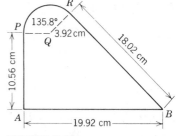

**FIGURE 7.44**

The perimeter can now be found from Equation 8:

$$\text{perimeter} = 19.92 + 18.02 + 10.56 + 9.2910159 = 57.791016$$

The perimeter is 57.79 cm, to the nearest hundredth.

## Exercises for Section 7.3

*Figure 7.45 shows a portion of a structure that will support a building extending from the top of a low cliff. Find the length of the indicated piece. See Example 1.*

1. Side *BD* of parallelogram *ABDE*.   2. Side *BE* of rectangle *BCDE*.

3. Base *FD* of trapezoid *ABDF*.   4. Side *AG* of square *GHBA*.

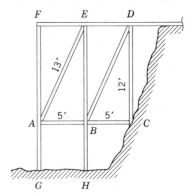

**FIGURE 7.45**

*Find the perimeter of each polygon. See Example 2.*

5.

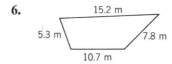

6.

15.2 m
5.3 m   7.8 m
10.7 m

7.

9.1 cm
5.0 cm
6.7 cm   10.8 cm
9.4 cm

8.

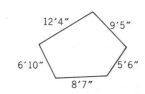

**9.**

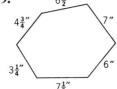

**10.**

100.0 km
87.1 km          95.7 km
88.4 km          101.2 km
98.3 km

*Find the perimeter of each rectangle having the given dimensions. See Example 3.*

**11.** $\ell = 11.1$ in.;   $w = 6.8$ in.          **12.** $\ell = 9.1$ m;   $w = 4.9$ m

**13.** $\ell = 71.6$ cm;   $w = 49.0$ cm          **14.** $\ell = 9$ ft 6 in.;   $w = 2$ ft 7 in.

**15.** $\ell = 5$ ft 8 in.;   $w = 2$ ft 6 in.          **16.** $\ell = 9.4$ mm;   $w = 3.2$ mm

*Find the missing dimension of each rectangle having the given perimeter. See Example 4. Round off to the nearest tenth.*

**17.** $P = 25.2$ m;   $\ell = 6.9$ m;   $w = ?$          **18.** $P = 21.2$ km;   $\ell = 7.8$ km;   $w = ?$

**19.** $P = 336.6$ km;   $\ell = ?$;   $w = 78.1$ km          **20.** $P = 129.4$ ft;   $\ell = ?$;   $w = 12.9$ ft

*Find the perimeter of (a) an equilateral triangle, (b) a square, (c) a regular hexagon, and (d) a regular octagon having sides of the given length. See Example 5.*

**21.** $s = 5$ ft 9 in.          **22.** $s = 9.2$ mm

**23.** $s = 8.9$ km          **24.** $s = 6$ ft 5 in.

*Find the length of a side of (a) an equilateral triangle, (b) a square, (c) a regular hexagon, and (d) a regular octagon having the given perimeter. See Example 5. Round off to the nearest tenth.*

**25.** $P = 69.6$ km      **26.** $P = 70.4$ ft

**27.** $P = 28.4$ in.      **28.** $P = 376.4$ m

*Find the circumference of a circle with the given radius (r) or diameter (d). See Example 6a. Round off to the nearest tenth.*

**29.** $r = 2.9$ yd      **30.** $r = 1.7$ in.      **31.** $r = 8.4$ m

**32.** $d = 37.9$ cm      **33.** $d = 84.0$ km      **34.** $d = 72.1$ ft

*Find the diameter of a circle with the given circumference. See Example 6b. Round off to four significant digits.*

**35.** $C = 4.248$ ft      **36.** $C = 4.676$ ft      **37.** $C = 32.37$ m

**38.** $C = 86.59$ m      **39.** $C = 385.3$ mm      **40.** $C = 132.8$ mm

*Find the length of an arc associated with a central angle of measure D°, in a circle with radius r. See Example 7. Round off to three significant digits.*

**41.** $D = 45.2°$;   $r = 18.2$ in.      **42.** $D = 16.3°$;   $r = 12.0$ in.

**43.** $D = 120°$;   $r = 27.4$ cm       **44.** $D = 105°$;   $r = 19.3$ cm

*Find the measure of the central angle associated with an arc of length L, in a circle of radius r. See Example 8. Round off to the nearest tenth.*

**45.** $L = 31.3$ in.;   $r = 8.0$ in.       **46.** $L = 4.5$ in.;   $r = 3.4$ in.

**47.** $L = 80.7$ cm.;   $r = 21.5$ cm       **48.** $L = 59.4$ cm;   $r = 17.6$ cm

*Find the perimeter of each figure to the nearest hundredth. See Examples 9 and 10.*

**49.**
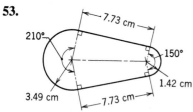
25 m

25 m

100 m

**50.**
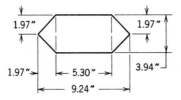
2.64 cm

9.43 cm

**51.**
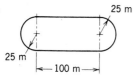
2.40″

.53″   1.64″

4.80″

**52.**
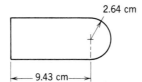
69.29″   45.79″

**53.**
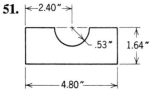
7.73 cm

210°

150°

1.42 cm

3.49 cm

7.73 cm

**54.**
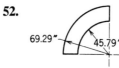
1.97″   1.97″

1.97″   5.30″   3.94″

9.24″

**55.** The top of a rectangular workbench measures 4 ft 8 in. long and 3 ft 2 in. wide. A strip of brass is to be fastened around the table top. Find the length of the brass strip.

**56.** The four sides of a plot of ground measure 90 ft, 125 ft, 92 ft, and 130

ft, respectively. How much fencing
is needed to enclose the plot?

57. Rain gutters are to be installed on
the four sides of a rectangular
structure that is 10.4 m long and 7.8
m wide. Find the cost of the rain
gutter if it sells for $1.56 per meter.

58. A certain type of fencing costs
$23.15 for a piece 8 ft long. (*a*) How
many such 8-ft pieces are required
for a fence around a rectangular lot
120 ft long and 96 ft wide? (*b*) What
is the total cost of the required
fencing?

59. It costs $18 per meter to apply a
certain kind of trim to the side of a
swimming pool. What is the total
cost (to the nearest cent) of
applying this trim completely
around a rectangular pool 8.2 m
wide and 16.5 m long?

60. Flexible plastic pipe is to be
installed completely around each of
two rectangular buildings. Each
building is 75 ft long and 25 ft wide.
(*a*) How many feet of pipe are
needed? (*b*) If a coil of 100 ft of
pipe sells for $16.70, what will be
the total cost of the pipe?

## 7.4 Area

In Section 7.3 we associated a measure called *perimeter* (or *circumference*) with
certain geometric figures. In this section we consider another measure, called *area*,
that can also be associated with such figures.

### ■ Square Units

A square with each side 1 unit long is called a *square unit* and is used as a unit of
measurement. The **area** of a surface is the number of square units that it contains.
If the unit of length is 1 inch, then the area of the square is 1 square inch (sq in.),

as in Figure 7.46*a*. If the unit of length is 1 centimeter, then the area of the square is 1 square centimeter (sq cm), as in Figure 7.46*b*. In general, to each unit of length, there is a corresponding square (sq) unit of area.

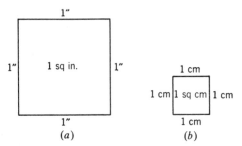

**FIGURE 7.46**

Recall from Section 7.2 that the segment from any vertex of a triangle, perpendicular to the opposite side, is an *altitude* of the triangle. A segment that measures the distance between a pair of opposite sides of a parallelogram is called an **altitude** of the parallelogram; a segment that measures the distance between the two bases of a trapezoid is called the **altitude** of the trapezoid. In general, as shown in Figure 7.47, we shall use the symbol *h* to name an altitude.

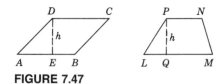

**FIGURE 7.47**

## ■ *Area Formulas for Polygons*

Table 7.2 lists some of the commonly used formulas for finding areas (*A*) of polygons that often appear in technical applications.

**TABLE 7.2**

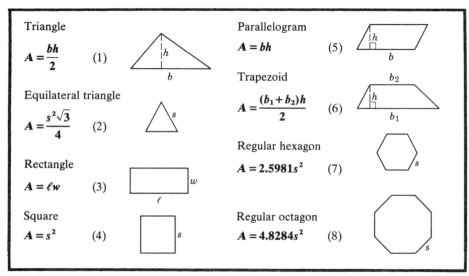

**EXAMPLE 1** Find the area of each triangle in Figure 7.48.

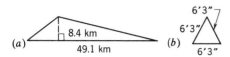

(a)

(b)

**FIGURE 7.48**

*Solutions*

**(a)** Substitute 49.1 for $b$ and 8.4 for $h$ in Formula 1.

$$A = \frac{bh}{2}$$

$$A = \frac{49.1 \times 8.4}{2} = 206.22$$

The area is 206.2 sq km, to the nearest tenth.

**(b)** First   6 ft 3 in = 6.25 ft.   Next, because the triangle is equilateral, substitute 6.25 for $s$ in Formula 2.

$$A = \frac{s^2\sqrt{3}}{4}$$

$$A = \frac{(6.25)^2\sqrt{3}}{4} = 16.914559$$

The area is 16.9 sq ft, to the nearest tenth.

**EXAMPLE 2** The area of a trapezoid is 157.5 sq ft. If the altitude is 5 ft and one base is 36 ft, find the length of the second base.

*Solution*   Substitute 157.5 for $A$, 5 for $h$, and 36 for $b_1$, in Formula 6.

$$A = \frac{(b_1 + b_2)h}{2}$$

$$157.5 = \frac{(36 + b_2) \times 5}{2}$$

Multiply each side by 2.

$$315 = (36 + b_2) \times 5$$

Divide each side by 5.

$$63 = 36 + b_2$$

Subtract 36 from each side.

$$27 = b_2$$

The second base is 27 ft long.

■ *Area Formulas for Circles*

Either of the following formulas can be used to find the area of a circle:

$$A = \pi r^2 \qquad\qquad\qquad (9)$$

or

$$A = \frac{\pi d^2}{4} \qquad (10)$$

where $r$ is the radius and $d$ is the diameter of the circle. If the central angle of a sector of a circle is $D°$, then the area of the sector is given by

$$A = \frac{\pi D° r^2}{360} \qquad (11)$$

**EXAMPLE 3**  The area of a circle with radius 19.2 yd is given by

$$A = \pi(19.2)^2 = 1158.1167$$

The area is 1158.1 sq yd, to the nearest tenth.

**EXAMPLE 4**  To the nearest tenth, find the diameter of a circle with area 22.9 sq m.

**Solution**  Substitute 22.9 for $A$ in formula 10.

$$22.9 = \frac{\pi d^2}{4}$$

Multiply each side by 4.

$$91.6 = \pi d^2$$

Divide each side by $\pi$.

$$29.157186 = d^2$$

Find the square root of 29.157186.

$$d = 5.3997394$$

The diameter is 5.4 m, to the nearest tenth.

**EXAMPLE 5**  To the nearest tenth, find the area of the sector shown in Figure 7.49.

**Solution**  Substitute 51° for $D°$ and 8 for $r$ in Formula 11.

$$A = \frac{\pi D° r^2}{360}$$

$$A = \frac{\pi \times 51° \times 8^2}{360} = 28.48377$$

**FIGURE 7.49**

The area is 28.5 sq in. to the nearest tenth.

**EXAMPLE 6**  To the nearest tenth, find the area of Figure 7.50.

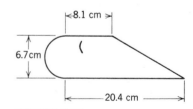

**FIGURE 7.50**

**Solution**  As shown in Figure 7.51, draw in the dashed segments to divide the figure into three regions:

$A_1$:  One-half of a circle with diameter 6.7 cm.
$A_2$:  A rectangle with length 8.1 cm and width 6.7 cm.
$A_3$:  A right triangle with base 12.3 cm and altitude 6.7 cm.

Compute the three separate areas and add the results.

$$A_1 = \frac{1}{2} \times \frac{\pi(6.7)^2}{4} = 17.62826$$

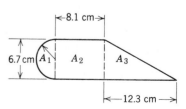

**FIGURE 7.51**

$$A_2 = 8.1 \times 6.7 = 54.27$$

$$A_3 = \frac{12.3 \times 6.7}{2} = 41.205$$

$$A_1 + A_2 + A_3 = 17.628261 + 54.27 + 41.205 = 113.10326$$

The total area is 113.1 sq cm, to the nearest tenth.

## Exercises for Section 7.4

*For Exercises 1 to 24, refer to Examples 1 to 3 and Table 7.2.*

*Find the area of each triangle with the given base and altitude to three significant digits.*

**1.** 26.1 in;  12.4 in.  **2.** 64.5 ft;  47.8 ft

**3.** 16.9 km;  29.0 km  **4.** 41.7 m;  57.2 m

*Find the area of each rectangle with the given length and width. Round off to two significant digits.*

**5.** 16 ft;  4.2 ft  **6.** 62 in.;  18.9 in.

**7.** 73.4 cm;  11 cm  **8.** 5.7 m;  4.9 m

*Find the area of (a) an equilateral triangle, (b) a square, (c) a regular hexagon, and (d) a regular octagon, with each side the given length. Round off to the nearest tenth.*

**9.** 14.1 mm  **10.** 38.0 m  **11.** 7.3 ft  **12.** 4.0 ft

*Find the area of each circle with the given radius. Round off to the nearest tenth.*

**13.** 5 in.  **14.** 9 ft  **15.** 11.6 m  **16.** 15.1 cm

*Find the area of each parallelogram with the given dimensions. Round off to three significant digits.*

**17.** $b = 9.64$ m;   $h = 2.64$ m       **18.** $b = 66.4$ in.;   $h = 26.4$ in.

**19.** $b = 94.3$ ft;   $h = 77.3$ ft       **20.** $b = 5.52$ cm;   $h = 8.86$ cm

*Find the area of each trapezoid with the given dimensions. Round off to three significant digits.*

**21.** $b_1 = 7.06$ in;   $b_2 = 18.73$ in.;   $h = 5.68$ in.

**22.** $b_1 = 84.3$ in.;   $b_2 = 94.5$ in.;   $h = 57.7$ in.

**23.** $b_1 = 38.8$ cm;   $b_2 = 56.8$ cm;   $h = 18.6$ cm

**24.** $b_1 = 36.3$ m;   $b_2 = 67.6$ m;   $h = 47.5$ m

*Find the diameter of each circle with the given area. See Example 4. Round off to the nearest hundredth.*

**25.** 81.52 sq cm       **26.** 29.67 sq cm

**27.** 536.64 sq in.       **28.** 919.21 sq in.

*Find the missing dimension for each triangle, to the nearest tenth.*

**29.** $A = 62.4$ sq m;   $b = 29$ m;   $h = ?$

**30.** $A = 352$ sq ft;   $b = 30$ ft;   $h = ?$

**31.** $A = 102.85$ sq km;   $b = ?$;   $h = 37.6$ km

**32.** $A = 11.22$ sq cm;   $b = ?$;   $h = 3.7$ cm

*Find the missing dimension for each rectangle, to the nearest tenth.*

**33.** $A = 1472$ sq in.;   $\ell = 63$ in.;   $w = ?$

**34.** $A = 990$ sq mi;   $\ell = 56$ mi;   $w = ?$

**35.** $A = 1462$ sq m;   $\ell = ?$;   $w = 19$ m

**36.** $A = 2464$ sq cm;   $\ell = ?$;   $w = 42$ cm

*Find the area of each sector of a circle with the given radius and central angle. See Example 5. Round off to four significant digits.*

**37.** $r = 90.72$ in.;   $D = 25.6°$     **38.** $r = 64.36$ in.;   $D = 64.1°$

**39.** $r = 5.012$ cm;    $D = 168.3°$      **40.** $r = 7.311$ cm;    $D = 115.2°$

**41.** $r = 3.040$ m;    $D = 120°$      **42.** $r = 1.633$ m;    $D = 110°$

*Find the area of each figure. See Example 6. Round off to three significant digits.*

**43.**

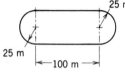

**44.**

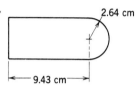

**45.**

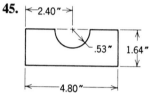

**46.**

**47.**

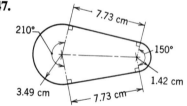

**48.**

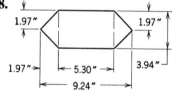

*Exercises 49 to 54 refer to the floor plan shown in Figure 7.52.*

**49.** Find (*a*) the total floor area in square feet, and (*b*) the monthly rental at 19¢ per square foot.

**50.** Find (*a*) the office area in square feet, and (*b*) the cost of carpeting the office with carpeting that costs $1.396 per square foot.

**51.** The shop area is to be covered with 9-in.-square floor tiles. Find (*a*) the number of tiles required, and (*b*) to the nearest cent, the total cost if the tiles sell at 45 for $22.05.

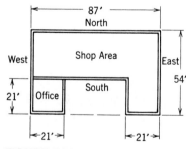

**FIGURE 7.52**

**52.** The east wall is 12 ft high and is to be faced with one layer of bricks. (*a*) If 616 bricks are needed to face 100 sq ft, to the nearest whole number, find the total number of bricks needed. (*b*) Find the total cost if the price of bricks is $45 for 300.

**53.** All interior walls of the shop area are 12 ft high. Each wall, except the east wall, is to be painted with waterproofing paint costing $19.84 per gallon. If each gallon covers 225 sq ft, find (*a*) to the nearest whole number the number of gallons needed, and (*b*) the total cost.

**54.** Repeat Exercise 53 if the paint costs $18.85 per gallon and 1 gallon covers 150 sq ft.

*A copper bus bar, used to carry heavy electric current, must sometimes be replaced by copper wire. To the nearest hundredth of an inch, find the diameter of wire having the same cross-sectional area as the given rectangular bus bar.*

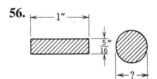

**55.** $\frac{3}{4}''$   $\frac{1}{4}''$   ?

**56.** $1''$   $\frac{5}{16}''$   ?

**57.** $\frac{7}{8}''$   $\frac{1}{2}''$   ?

**58.** $1\frac{1}{8}''$   $\frac{5}{8}''$   ?

## 7.5 Prisms and Pyramids; Volume

In Sections 7.3 and 7.4 we considered some measures associated with plane figures. In this section we consider some three-dimensional figures called **solids** and a measure called *volume* that can be associated with such figures.

### ■ Prisms

Figure 7.53*a* shows a solid bounded by seven faces. The upper and lower faces are pentagons that are parallel to each other and are called **bases** of the solid. The

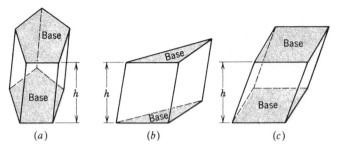

**FIGURE 7.53**

remaining faces are parallelograms and are referred to as the **lateral faces** of the solid. In general, if two of the faces of a solid are polygons that are parallel to each other, and the lateral faces are parallelograms, the solid is called a **prism**. The distance ($h$) between bases is called the **altitude** (or **height**) of the prism. The name of the polygon that forms the base can be used to name the prism. For example, a **pentagonal prism** is shown in Figure 7.53a; a **triangular prism** and a **rectangular prism** are shown in Figure 7.53b and c.

If the lateral faces of a prism are rectangles, as shown in Figure 7.54, the prism is a **right prism**. A six-faced right rectangular prism, as shown in Figures 7.54a and b, is a **rectangular prism**; if each of the faces is a square, as shown in Figure 7.54b, the prism is called a **cube**. Figure 7.55 shows a **right triangular prism**, a **right hexagonal prism**, and a **right octagonal prism**.

Rectangular prisms

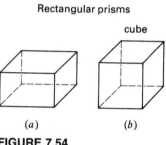

**FIGURE 7.54**

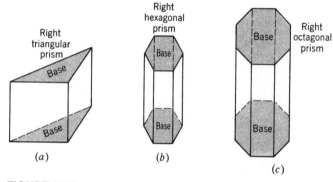

**FIGURE 7.55**

# ■ *Pyramids*

If one of the faces of a solid is a polygon and the other faces are triangles with a common vertex, the solid is called a **pyramid**. The common vertex of the triangular faces is called the **apex** of the pyramid; the polygonal face opposite the apex is called the **base** of the pyramid. The distance from the apex to the base is the **altitude** (or **height**) of the pyramid. A **rectangular pyramid** and a **triangular pyramid** are shown in Figure 7.56.

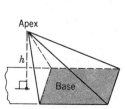

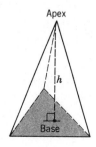

**FIGURE 7.56**

If a pyramid is cut at any point below the apex by a plane parallel to the base, the resulting solid that does *not* include the apex is called a **frustum** of the pyramid. Thus, a frustum has two bases. A frustum of a pyramid with each base a parallelogram and a frustum of a pyramid with each base a triangle are shown in Figure 7.57.

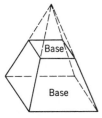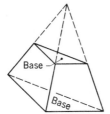

**FIGURE 7.57**

## ■ *Cubic Units*

A cube with each edge 1 unit long is called a *cubic unit* and is used as a unit of measurement. The **volume** of a solid is the number of cubic units that it contains. If the unit of length is 1 inch, then the volume of the cube is 1 cubic inch (cu in.), as in Figure 7.58*a*. If the unit of length is 1 centimeter, then the volume of the cube is 1 cubic centimeter (cu cm), as in Figure 7.58*b*. In general, to each unit of length there is a corresponding cubic (cu) unit of volume.

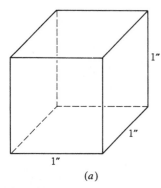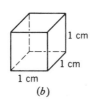

**FIGURE 7.58**

## ■ *Volume of a Prism*

The volume $V$ of any prism is given by

$$V = \textbf{area of base} \times \textbf{altitude} \qquad \textbf{(1)}$$

**EXAMPLE 1**  Find the volume of the triangular prism shown in Figure 7.59.

**Solution**  Because the base is an equilateral triangle, the area of the base is equal to

$$\frac{(13.1)^2\sqrt{3}}{4}$$

Substitute this value for "area of base" and 26.8 for "altitude" in Formula 1.

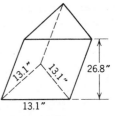

**FIGURE 7.59**

$$V = \frac{(13.1)^2\sqrt{3}}{4} \times 26.8 = 1991.4895$$

The volume is 1991.5 cu in., to the nearest tenth.

The base of a rectangular prism is a rectangle with an area equal to $\ell \times w$ (Figure 7.60). Hence, from Formula 1,

$$V = \ell wh \qquad (2)$$

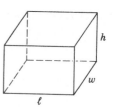

**FIGURE 7.60**

Thus, the volume of a rectangular prism is simply the product of the three dimensions—length, width, and height.

**EXAMPLE 2**   A tank in the form of a rectangular prism is 40 ft by 25 ft by 35 ft. If 1 cu ft is approximately equal to 7.48 gal, how many gallons of liquid can the tank contain?

**Solution**   First, compute the volume of the tank. By Formula 2,

$$V = \ell wh$$
$$V = 40 \times 25 \times 35 = 35,000$$

Next, convert 35,000 cu ft to gallons. If 1 cu ft equals 7.48 gal, and $G$ is the total number of gallons, then by the unit-product rule,

$$G = 35,000 \times 7.48 = 261,800$$

Hence, the tank can contain approximately 261,800 gal of liquid.

Because each of the three dimensions of a cube is the same (Figure 7.61), the volume of the cube is given by

$$V = s^3 \qquad (3)$$

**FIGURE 7.61**

**EXAMPLE 3**   The volume of a cube with an edge of length 7.6 m is given by

$$V = 7.6^3 = 438.976$$

The volume of the cube is 439.0 cu m, to the nearest tenth.

## ■ *Volume of a Pyramid*

The volume of a pyramid is given by the formula

$$V = \tfrac{1}{3} \times \textbf{area of base} \times \textbf{altitude} \qquad (4)$$

Note that the volume of a pyramid is equal to one-third the volume of the prism with the same base and altitude.

**EXAMPLE 4**   A pyramid has a square base 35 yd on each edge (Figure 7.62). Find the volume of the pyramid if the altitude is 23 yd.

**Solution**   The area of the base is $35^2$. Substitute $35^2$ for "area of base" and 23 for "altitude" in Formula 4.

$$V = \tfrac{1}{3} \times 35^2 \times 23 = 9391.6666$$

The volume of the pyramid is 9392 cu yd, to the nearest whole number.

The volume of a frustum of a pyramid is given by the formula

$$V = \tfrac{1}{3}h(B + b + \sqrt{Bb}) \qquad (5)$$

where $h$ is the altitude, $B$ is the area of the *larger* base, and $b$ is the area of the *smaller* base.

**EXAMPLE 5**   To the nearest tenth, find the volume of the frustum of a rectangular pyramid with one base 12.2 ft by 8.6 ft, the other base 6.1 ft by 4.3 ft, and the altitude 5.9 ft (Figure 7.63).

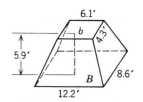

**FIGURE 7.63**

**Solution**   The areas of the bases are

$$B = 12.2 \times 8.6 = 104.92 \quad \text{and} \quad b = 6.1 \times 4.3 = 26.23$$

Substitute these values for $B$ and $b$, and 5.9 for $h$, in Formula 5.

$$V = \tfrac{1}{3}h(B + b + \sqrt{Bb})$$
$$V = \tfrac{1}{3} \times 5.9(104.92 + 26.23 + \sqrt{104.92 \times 26.23})$$
$$= 361.09967$$

The volume is 361.1 cu ft, to the nearest tenth.

**EXAMPLE 6**   To the nearest whole number, find the altitude of the frustum of a triangular pyramid with volume 450 cu m and bases 126 sq m and 14 sq m in area (Figure 7.64).

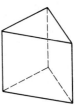

**FIGURE 7.64**

**Solution**   Substitute 450 for $V$, 126 for $B$, and 14 for $b$ in Formula 5.

$$V = \tfrac{1}{3}h(B + b + \sqrt{Bb})$$
$$450 = \tfrac{1}{3} \times h(126 + 14 + \sqrt{126 \times 14})$$
$$450 = \tfrac{1}{3} \times h(182)$$

Multiply each side by 3.

$$1350 = h \times 182$$

Divide each side by 182.

$$h = 7.4175824$$

The altitude is 7 m, to the nearest whole number.

## Exercises for Section 7.5

*Find the volume of a prism whose base is an equilateral triangle, having the given dimensions. See Example 1 and Figure 7.65. Round off to the nearest tenth.*

**1.** $s = 7.8$ cm;   $h = 2.4$ cm       **2.** $s = 1.6$ cm;   $h = 7.3$ cm

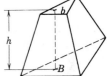

**FIGURE 7.65**

**3.** $s = 17.5$ ft;   $h = 29.7$ ft       **4.** $s = 26.5$ ft;   $h = 37.2$ ft

*Find the volume of a prism whose base is a right triangle, having the given dimensions (see Figure 7.66). Round off to the nearest tenth.*

**5.** $a = 3.7$ cm;   $b = 7.4$ cm;   $h = 17$ cm

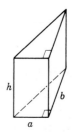

**FIGURE 7.66**

**6.** $a = 4.4$ cm;   $b = 5.7$ cm;   $h = 8.3$ cm

**7.** $a = 7.4$ in.;   $b = 5.8$ in.;   $h = 4.7$ in.

**8.** $a = 8.2$ in.;   $b = 7.9$ in.;   $h = 4.3$ in.

*Find the volume of a rectangular prism with the given dimensions. See Example 2. Round off to two significant digits.*

**9.** $\ell = 13.1$ m;   $w = 5.2$ m;   $h = 3.6$ m

**10.** $\ell = 9.6$ m;   $w = 7.1$ m;   $h = 8.1$ m

**11.** $\ell = 11.6$ cm;   $w = 9.0$ cm;   $h = 5.3$ cm

**12.** $\ell = 9.2$ cm;   $w = 4.9$ cm;   $h = 2.8$ cm

*Find the number of gallons capacity of a tank in the form of a rectangular prism with the given dimensions. See Example 2. Round off to the nearest whole number.*

**13.** 20 ft × 15.6 ft × 9.8 ft       **14.** 36.1 ft × 22 ft × 12.2 ft

**15.** 10 ft × 8 ft × 4 ft 3 in       **16.** 15 ft 8 in. × 10 ft 4 in. × 6 ft

*Find the volume of a cube with the given length of one edge. See Example 3. Round off to three significant digits.*

**17.** 8.00 ft     **18.** 12.0 ft     **19.** 6.10 m

**20.** 9.30 m     **21.** 0.420 in.     **22.** 0.750 in.

*Find the volume of a pyramid with a rectangular base, having the given dimensions. See Example 4. Round off to three significant digits.*

**23.**

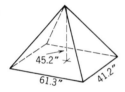

**24.**

**25.**

**26.**

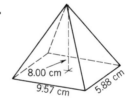

*Find the volume of a pyramid with a base in the shape of an equilateral triangle, having sides of length s. Round off to the nearest tenth.*

**27.** $s = 12.0$ cm;   $h = 3.0$ cm     **28.** $s = 13.0$ cm;   $h = 4.1$ cm

**29.** $s = 21.8$ in.;   $h = 12.3$ in.     **30.** $s = 35.2$ in.;   $h = 13.6$ in.

*In Exercises 31 to 36, find the volume of a frustum of a pyramid, having the given dimensions. See Example 5. Round off to three significant digits.*

| | Area Larger Base | Area Smaller Base | Altitude | Volume |
|---|---|---|---|---|
| **31.** | 73.5 sq m | 33.4 sq m | 4.00 m | ____ |
| **32.** | 32.6 sq m | 22.9 sq m | 9.00 m | ____ |
| **33.** | 59.6 sq ft | 20.2 sq ft | 4.70 ft | ____ |
| **34.** | 98.7 sq ft | 68.4 sq ft | 13.4 ft | ____ |

**35.**
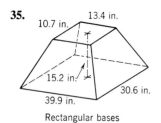
Rectangular bases

**36.**
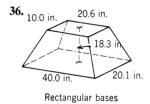
Rectangular bases

*Find the altitude of a frustum of a pyramid, having the given dimensions. See Example 6. Round off to the nearest tenth.*

| | Area Larger Base | Area Smaller Base | Volume | Altitude |
|---|---|---|---|---|
| **37.** | 58.2 sq cm | 47.1 sq cm | 363 cu cm | ____ |
| **38.** | 97.4 sq cm | 66.2 sq cm | 744 cu cm | ____ |
| **39.** | 24.4 sq in. | 13.5 sq in. | 234 cu in. | ____ |
| **40.** | 32.2 sq in. | 21.9 sq in. | 252 cu in. | ____ |

**41.** If 1 cu in. of iron weighs 0.284 lb, what is the weight, to the nearest whole number, of an iron plate 6 ft long, 3 ft wide, and 2 in. thick?

**42.** If 1 cu cm of aluminum weighs 2.58 g, find the weight of an aluminum bar 10 cm square and 2 m long.

*Each of the following right prisms is machined from aluminum weighing 2.58 g per 1 cu cm. Find (a) the volume, and (b) the weight of each prism. Round off each answer to three significant digits.*

**43.**

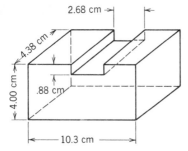

**44.**

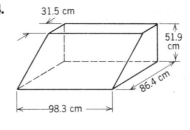

**45.** A concrete retaining wall 88 ft long has the given cross section. Find (*a*) the number of cubic feet of concrete needed, and (*b*) the total cost if concrete costs $1.09 per cubic foot.

**46.** Repeat Exercise 45 if concrete costs $1.14 per cubic foot.

**FIGURE 7.67**

*Exercises 47 and 48 refer to the floor plan of a building shown in Figure 7.68.*

**47.** The ceiling is to be 10 ft high. If an air-conditioning unit can move 1000 cu ft of air per minute, find, to the nearest tenth of a minute, how long it will take to move all of the air in the building.

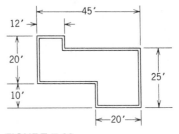

**FIGURE 7.68**

**48.** The building is to have a concrete floor 4 in. thick. (*a*) To the nearest tenth, how many cubic feet of concrete will be required? (*b*) At $1.18 per cubic foot, what will be the cost of pouring the floor?

## 7.6 Cylinders, Cones, and Spheres

In Section 7.5 we considered solids bounded by planes. In this section we consider solids bounded by curved surfaces.

## ■ *Cylinders*

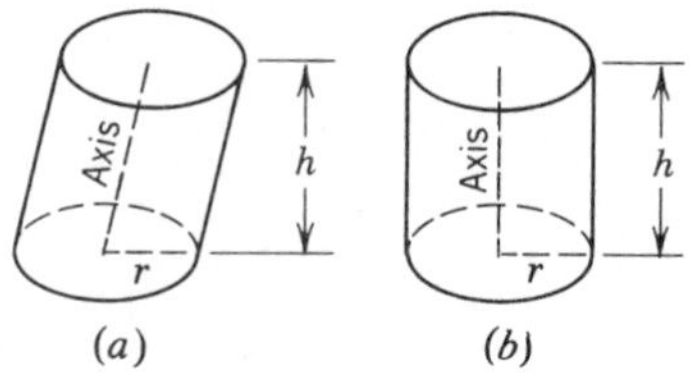

A solid with two parallel circular bases of equal area is called a **circular cylinder** (Figure 7.69a). The distance between the bases is the **altitude** (*h*), and the segment between the centers of the bases is the **axis** of the cylinder. If the axis is perpendicular to the bases, the cylinder is a **right circular cylinder**, and the altitude *h* is equal to the length of the axis (Figure 7.69b).

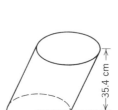

**FIGURE 7.69**

The volume of a cylinder can be computed in the same way that we computed the volume of a prism (page 255)—the product of the area of the base ($\pi r^2$) and the altitude (*h*). For a circular cylinder, we have the formula

$$V = \pi r^2 h \tag{1}$$

where *r* is the radius of the base.

**EXAMPLE 1**   Find the volume of the circular cylinder shown in Figure 7.70.

**Solution**   Substitute 16.8 for *r* and 35.4 for *h* in Formula 1.

$$V = \pi r^2 h$$
$$= \pi (16.8)^2 \times 35.4 = 31{,}388.582$$

The volume is 31,388.6 cu cm, to the nearest tenth.

**FIGURE 7.70**

## ■ *Cones*

A solid with a circular base and a curved lateral surface that comes to a point (the **vertex**), as in Figure 7.71, is called a **circular cone**. The distance from the vertex to the base is the **altitude** (*h*), and the segment from the vertex to the center of the base is the **axis** of the cone. If the axis is perpendicular to the base, the cone is a **right circular cone**, and the altitude *h* is equal to the length of the axis. The distance $\ell$ between the vertex and any point of the circular edge of the base of a right circular cone (Fig. 7.71b) is called the **slant height**.

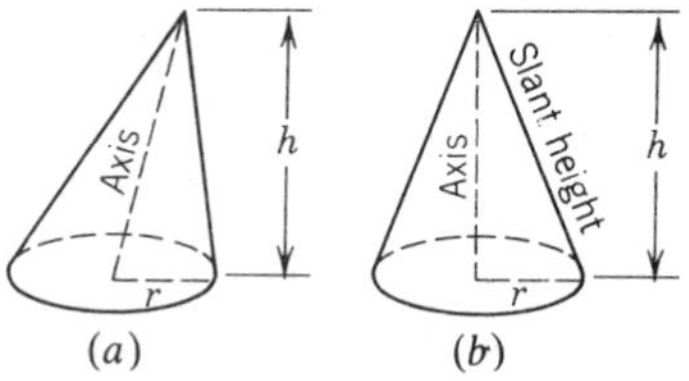

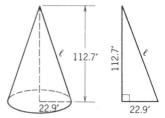

**FIGURE 7.71**

**EXAMPLE 2**   Find the slant height of the right circular cone in Figure 7.72.

**Solution**   The segment of length $\ell$ is the hypotenuse of a right triangle formed by the axis and a radius of the base. By the Pythagorean rule,

$$\ell^2 = (112.7)^2 + (22.9)^2 = 13{,}225.7$$
$$\ell = \sqrt{13{,}225.7} = 115.00304$$

The slant height is 115.0 ft, to the nearest tenth.

The volume of a cone is equal to one-third the volume of the cylinder with the same base and altitude. Hence, for a circular cone, we have the formula

$$V = \tfrac{1}{3}\pi r^2 h \tag{2}$$

**FIGURE 7.72**

**EXAMPLE 3**   Find the volume of the circular cone of Example 2.

**Solution**   Substitute 22.9 for *r* and 112.7 for *h* in Formula 2.

$$V = \tfrac{1}{3}\pi r^2 h$$
$$= \tfrac{1}{3}\pi (22.9)^2 \times 112.7 = 61{,}890.43$$

The volume is 61,980.4 cu ft, to the nearest tenth.

## ■ *Frustum of a Cone*

If a cone is cut at any point below the vertex by a plane parallel to the base, the resulting solid that does *not* include the vertex is called a **frustum** of the cone (Figure 7.73). The variable $r$ represents a radius of the smaller base, $R$ represents a radius of the larger base, and $h$ represents the height of the frustum. The slant height $\ell$ of a frustum of a right circular cone (Figure 7.73*b*) can be computed as shown in the next example.

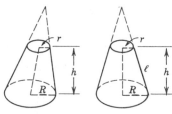

**FIGURE 7.73**

**EXAMPLE 4** Find the slant height of the frustum of the right circular cone with $r = 5.0$ m, $R = 7.0$ m, and $h = 8.0$ m (Figure 7.74).

**Solution** Note that $\triangle ABC$ (Figure 7.75) is a right triangle, with $\ell$ as the hypotenuse. Side $AB$ is equal in length to the altitude $h$; hence, $AB = 8.0$ m. Side $BC$ is equal in length to $R - r$, or 2.0 m. By the Pythagorean rule,

$$\ell^2 = (8.0)^2 + (2.0)^2 = 68.0$$
$$\ell = \sqrt{68.0} = 8.246211$$

The slant height is 8.2 m, to the nearest tenth.

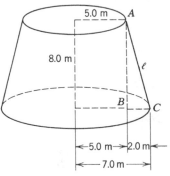

**FIGURE 7.74**

The volume of a frustum of a cone is given by the formula

$$V = \tfrac{1}{3}\pi h(R^2 + r^2 + Rr) \tag{3}$$

where $R$ is the radius of the larger base, and $r$ is the radius of the smaller base (Figure 7.76).

**FIGURE 7.75**

**EXAMPLE 5** To the nearest tenth, find the volume of the frustum of a cone if the frustum is 12.0 in. high, one base radius is 6.2 in., and the other base radius is 3.8 in.

**Solution** Substitute 12.0 for $h$, 6.2 for $R$, and 3.8 for $r$ in Formula 3.

$$V = \tfrac{1}{3}\pi h(R^2 + r^2 + Rr)$$
$$= \tfrac{1}{3}\pi \times 12.0(6.2^2 + 3.8^2 + 6.2 \times 3.8)$$
$$= 960.57337$$

The volume is 960.6 cu in., to the nearest tenth.

**FIGURE 7.76**

## ■ *Spheres*

A solid bounded by a curved surface, each point of which is the same distance from a fixed point (the **center**), is called a **sphere**. The distance from the center to any point of the sphere is called a **radius** of the sphere (Figure 7.77). The word "radius" is also used to name any segment between the center and any point of the sphere.

The volume of a sphere is given by the formula

$$V = \tfrac{4}{3}\pi r^3 \tag{4}$$

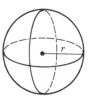

**FIGURE 7.77**

**EXAMPLE 6** To the nearest whole number, find the volume of a sphere with radius 14 cm.

**Solution** Substitute 14 for $r$ in Formula 4.

$$V = \tfrac{4}{3}\pi r^3$$
$$= \tfrac{4}{3}\pi(14)^3 = 11{,}494.04$$

The volume is 11,494 cu cm, to the nearest whole number.

## Exercises for Section 7.6

*Find the volume of each circular cylinder. See Example 1. Round off to the near-est hundredth.*

**1.**

6.74"

2.21"→

**2.**

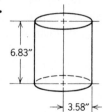

6.83"

→ 3.58" ←

**3.**

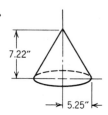

53.91 cm

9.70 cm→

**4.**

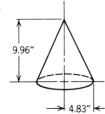

5.56 cm

→ 8.37 cm ←

*Find the slant height of each cone. See Example 2. Round off to three significant digits.*

**5.**

7.22"

→ 5.25" ←

**6.**

9.96"

→ 4.83" ←

**7.**

8.25 m

→ 3.89 m ←

**8.**

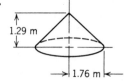

1.29 m

→ 1.76 m ←

*Find the volume of each circular cone. See Example 3. Round off to three significant digits.*

**9.**

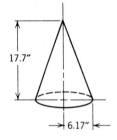

**10.**

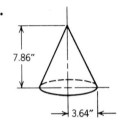

**11.**

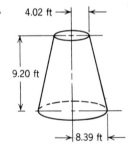

**12.**

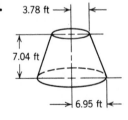

*Find the slant height of each frustum of a right circular cone. See Example 4. Round off to the nearest hundredth.*

**13.**

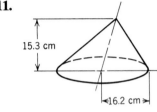

**14.**

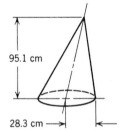

**15.**

**16.**

*Find the volume of each frustum of a circular cone. See Example 5. Round off to the nearest hundredth.*

**17.**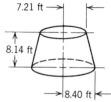

**18.**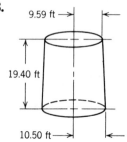

**19.**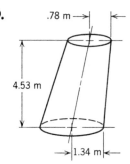

**20.**

*Find the volume of a sphere having the given radius. See Example 6. Round off to four significant digits.*

**21.** 4 ft 9 in.  **22.** 9 ft 2 in.  **23.** 39.04 cm  **24.** 57.62 cm

**25.** A cylindrical storage tank is 6 ft in diameter and 15 ft long. To the nearest gallon, how many gallons of liquid will it hold? (One cubic foot is approximately 7.48 gal.)

**26.** A cylindrical storage tank for natural gas has a diameter of 56 ft and a height of 40 ft. Find the volume to the nearest cubic foot.

**27.** A storage tank (Figure 7.78) is in the shape of a right circular cylinder with half of a sphere (a hemisphere)

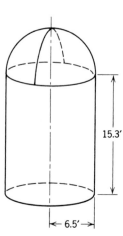

**FIGURE 7.78**

on top. (*a*) To the nearest tenth, what is the volume of the tank? (*b*) If 1 cu ft contains 7.48 gal, how many gallons will the tank hold (to the nearest gallon)?

**28.** Assuming that the Louisiana Superdome is a hemisphere with a diameter of 226 yd, find the volume (in cubic yards) of the air inside the dome. Round off to the nearest whole unit.

**29.** If aluminum weighs 2.58 g/cu cm,* find the weight, to the nearest gram, of an aluminum rod 10 cm in diameter and 2 m long.

**30.** If iron weighs 0.284 lb/cu in., find the weight, to the nearest pound, of an iron bar 1 ft in diameter and 12 ft long.

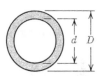

**FIGURE 7.79**

*Circular tubing has a cross section as shown in Figure 7.79. Find the volume of material in the wall of tubing having the given dimensions. Round off to the nearest tenth of a unit.*

|  | *d* | *D* | Length | Volume |
|---|---|---|---|---|
| **31.** | 0.50 in. | 0.75 in. | 8 ft | ____ cu in. |
| **32.** | 0.375 in. | 0.50 in. | 10 ft | ____ cu in. |
| **33.** | 1.5 cm | 2.0 cm | 3 m | ____ cu cm |
| **34.** | 5.0 cm | 5.5 cm | 2 m | ____ cu cm |

*If the tubing in Exercises 31 to 34 is made from copper weighing 0.322 lb/cu in. or, equivalently, 8.94 g/cu cm, find the weight of the indicated tubing, to the nearest tenth.*

**35.** Exercise 31 (pounds)    **36.** Exercise 32 (pounds)

---

*In the notation "g/cu cm," the symbol / is read as "per."

**37.** Exercise 33 (grams)          **38.** Exercise 34 (grams)

**39.** At a certain seaport, liquefied natural gas (l.n.g.) is pumped ashore at the rate of 4 cu m/sec. To the nearest second, how long will it take to fill a spherical tank 10 m in diameter?

**40.** Solve Exercise 39 for a cylindrical tank 10 m in diameter and 10 m high.

**41.** A round steel bar 12 in. long is turned on a lathe until its diameter is reduced from 3 in to $2\frac{1}{2}$ in. To the nearest tenth, how many cubic inches of steel are removed?

**42.** Find the weight of the bronze pin shown in Figure 7.80, if bronze weighs 8.79 g/cu cm. Round off to the nearest gram.

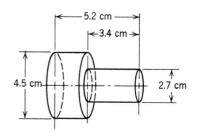

**FIGURE 7.80**

## 7.7 Surface Area of Solids

In this section we consider a measure called *surface area* that can be associated with solids.

### ■ Prisms

The **lateral surface area** ($A$) of a prism is the sum of the areas of the lateral faces of the prism. The **total surface area** ($S$) of a prism is the sum of the areas of the bases and the areas of the lateral faces.

**EXAMPLE 1**  Find the lateral surface area of the triangular prism shown in Figure 7.81.

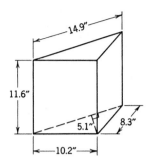

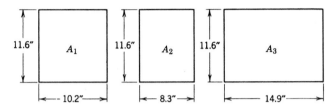

**FIGURE 7.81**

**Solution**  There are three lateral faces, each a rectangle, as shown. First find the area of each lateral face.

$$A_1 = 10.2 \times 11.6 = 118.32$$
$$A_2 = 8.3 \times 11.6 = 96.28$$
$$A_3 = 14.9 \times 11.6 = 172.84$$

The sum of the areas of the lateral faces is

$$A_1 + A_2 + A_3 = 118.32 + 96.28 + 172.84 = 387.44$$

The lateral surface area is 387.4 sq in., to the nearest tenth.

**EXAMPLE 2**  Find the total surface area of the prism of Example 1.

**Solution**  There are two triangular bases, each with the same area ($h = 5.1$ in., and $b = 14.9$ in.). Hence the sum of the areas of the bases equals twice the area of either base.

$$B_1 + B_2 = 2 \times \left( \frac{1}{2} \times 14.9 \times 5.1 \right) = 75.99$$

The total surface area $S$ is the sum of the lateral surface area (from Example 1) and the areas of the bases.

$$S = 387.44 + 75.99 = 463.43$$

The total surface area is 463.4 sq in., to the nearest tenth.

**EXAMPLE 3**  Find the total surface area of the rectangular prism shown in Figure 7.82.

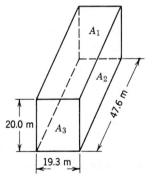

**FIGURE 7.82**

**Solution**  Although there are six rectangular faces, there are only three "different" rectangles. Hence, compute each of the three areas, find their sum, and multiply by 2.

$$A_1 = 19.3 \times 47.6 = 918.68$$
$$A_2 = 47.6 \times 20.0 = 952.0$$
$$A_3 = 19.3 \times 20.0 = 386.0$$
$$S = 2(918.68 + 952.0 + 386.0) = 4513.36$$

The total surface area is 4513.4 sq m, to the nearest tenth.

**EXAMPLE 4**  Find the total surface area of the cube shown in Figure 7.83.

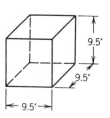

**Solution**  Each face of the cube is a square with area equal to $9.5^2$ sq ft. Because there are six such faces, the total surface area of the cube is given by

$$S = 6 \times 9.5^2 = 541.5 \text{ sq ft}$$

**FIGURE 7.83**

## ■ *Right Circular Cylinders*

If we picture the lateral surface of a right circular cylinder "unrolled" into a rectangle, as indicated in Figure 7.84, the area of the rectangle ($\ell \times w$) is equal to the product of the circumference of the base of the cylinder ($2\pi r$) and the altitude ($h$). Hence, the lateral surface area of a right circular cylinder is given by

$$A = 2\pi rh \qquad\qquad (1)$$

where $r$ is a radius of the base, and $h$ is the altitude of the cylinder. The total surface area equals the sum of the lateral surface area and twice the area of a base:

$$S = 2\pi rh + 2\pi r^2 \qquad\qquad (2)$$

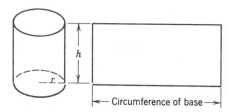

**FIGURE 7.84**

**EXAMPLE 5**  For the cylinder shown in Figure 7.85, find (*a*) the lateral surface area, and (*b*) the total surface area.

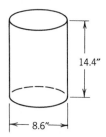

**Solutions**
(a) The radius of a base is

$$\frac{1}{2} \times 8.6 = 4.3$$

Substitute 4.3 for $r$ and 14.4 for $h$ in Formula 1.

$$A = 2\pi rh$$
$$= 2\pi(4.3)(14.4) = 389.05484$$

**FIGURE 7.85**

The lateral surface area is 389.1 sq. in., to the nearest tenth.

**(b)** Substitute 389.05484 for $2\pi rh$ and 4.3 for $r$ in Formula 2.

$$S = 389.05484 + 2\pi(4.3)^2 = 505.2309$$

The total surface area is 505.2 sq in., to the nearest tenth.

## ■ *Right Circular Cones*

The lateral surface area of a right circular cone is given by the formula

$$A = \pi r \ell \qquad (3)$$

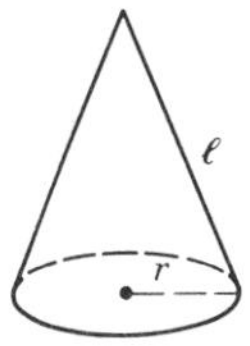

**FIGURE 7.86**

where $r$ is a radius of the base, and $\ell$ is the slant height (Figure 7.86). The total surface area is found by adding the area of the base ($\pi r^2$) to the lateral surface area ($\pi r l$), as expressed by

$$S = \pi r^2 + \pi r \ell \qquad (4)$$

**EXAMPLE 6**  Find the total surface area of the right circular cone in Figure 7.87 with base radius 14.8 cm and altitude 38.1 cm.

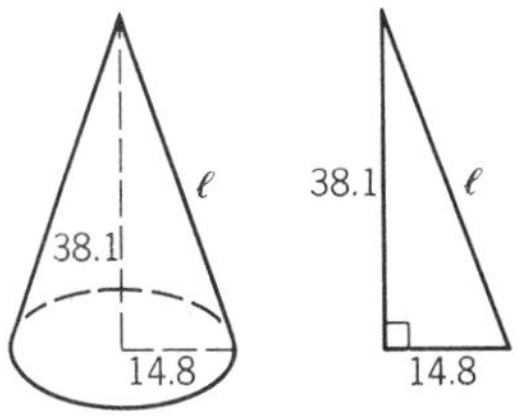

**Solution**  First, find the slant height, $\ell$. By the Pythagorean rule,

$$\ell = \sqrt{(14.8)^2 + (38.1)^2} = 40.873586$$

**FIGURE 7.87**

Next, substitute 14.8 for $r$ and 40.873586 for $\ell$ in Formula 4.

$$S = \pi r^2 + \pi r \ell$$
$$= \pi(14.8)^2 + \pi(14.8)(40.873586) = 2588.575$$

The total surface area is 2588.6 sq cm, to the nearest tenth.

The lateral surface area of a frustum of a right circular cone is given by

$$A = \pi(R + r)\ell \qquad (5)$$

**FIGURE 7.88**

where $R$ is the radius of the larger base, $r$ is the radius of the smaller base, and $\ell$ is the slant height of the frustum (Figure 7.88).

**EXAMPLE 7**  Find the lateral surface area of the frustum of the right circular cone shown in Figure 7.89.

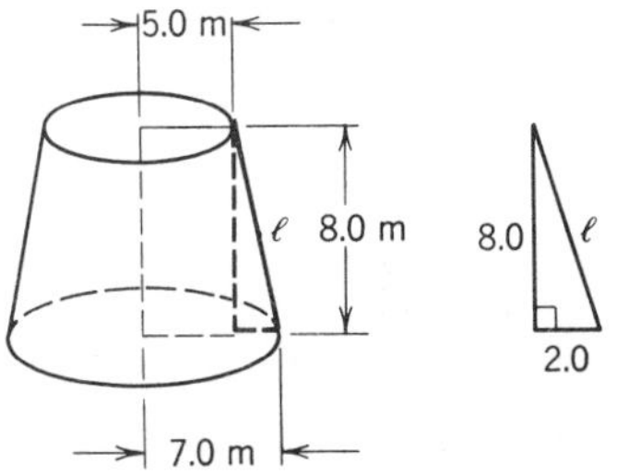

**Solution**  First, find the slant height, $\ell$. By the Pythagorean rule,

$$\ell = \sqrt{8^2 + 2^2} = 8.2462113$$

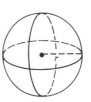

**FIGURE 7.89**

Next, substitute 7 for $R$, 5 for $r$, and 8.2462113 for $\ell$ in Formula 5.

$$A = \pi(R + r)\ell$$
$$= \pi(7 + 5)(8.2462113) = 310.87484$$

The lateral surface area is 310.9 sq m, to the nearest tenth.

**EXAMPLE 8**  Find the total surface area of the frustum of Example 7.

**Solution**  Add the area of the bases to the lateral surface area in Example 7.

$$S = \pi \cdot 7^2 + \pi \cdot 5^2 + 310.87484 = 543.3527$$

The total surface area is 543.4 sq m, to the nearest tenth.

## ■ *Spheres*

The (total) surface area of a sphere is given by the formula

$$S = 4\pi r^2 \qquad (6)$$

**FIGURE 7.90**

where $r$ is the radius of the sphere (Figure 7.90).

**EXAMPLE 9**  Find the radius of a sphere with a surface area of 4332 sq yd.

**Solution**  Substitute 4332 for $S$ in Formula 6.

$$S = 4\pi r^2$$
$$4332 = 4\pi r^2$$

Divide each side by $4\pi$.

$$344.7296 = r^2$$
$$r = \sqrt{344.7296} = 18.566895$$

The radius is 18.6 yd, to the nearest tenth.

## Exercises for Section 7.7

*Find the lateral surface area of each right prism. See Example 1.*

**1.**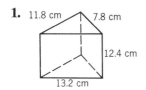

**2.**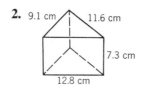

**3.**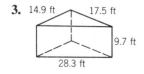

**4.**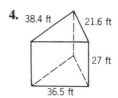

*Find the total surface area of each right prism. (Each base is a right triangle.) See Example 2. Round off to two significant digits.*

**5.**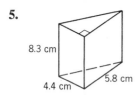

**6.**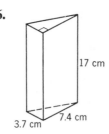

**7.**
4.3 in.
8.2 in.   7.9 in.

**8.**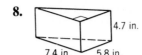
4.7 in.
7.4 in.   5.8 in.

*Find the total surface area of each rectangular prism. See Example 3. Round off to three significant digits.*

**9.**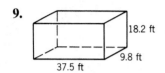
18.2 ft
9.8 ft
37.5 ft

**10.**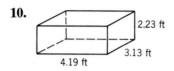
2.23 ft
3.13 ft
4.19 ft

**11.**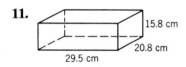
15.8 cm
20.8 cm
29.5 cm

**12.**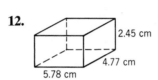
2.45 cm
4.77 cm
5.78 cm

*Find the total surface area of a cube with the given length of one edge. See Example 4. Round off to three significant digits.*

**13.** 9.30 m     **14.** 6.13 m     **15.** 7.42 in.     **16.** 4.25 in.

*Find (a) the lateral surface area, and (b) the total surface area of a right cylinder, having the given dimensions. See Example 5. Round off to the nearest hundredth.*

**17.** $r = 3.58$ in;   $h = 6.83$ in.     **18.** $r = 2.21$ in.;   $h = 6.74$ in.

**19.** $r = 8.37$ m;   $h = 5.56$ m     **20.** $r = 90.7$ m;   $h = 53.9$ m

*Find (a) the lateral surface area, and (b) the total surface area of a right circular cone, having the given dimensions. See Example 6. Round off to three significant digits.*

**21.** $r = 3.64$ ft;   $h = 7.86$ ft     **22.** $r = 6.17$ ft;   $h = 17.7$ ft

**23.** $r = 28.3$ cm;   $h = 95.1$ cm       **24.** $r = 16.2$ cm;   $h = 15.3$ cm

*Find (a) the lateral surface area, and (b) the total surface area of each frustum of a right circular cone. See Examples 7 and 8. Round off to the nearest hundredth.*

**25.**
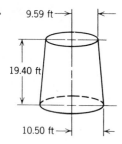
9.59 ft

19.40 ft

10.50 ft

**26.**
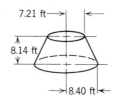
7.21 ft

8.14 ft

8.40 ft

**27.**
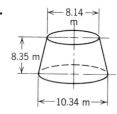
8.14 m

8.35 m

10.34 m

**28.**

1.56 m

4.53 m

2.68 m

*Find the surface area of a sphere having the given radius. Round off to four significant digits.*

**29.** 19.10 mm       **30.** 26.23 mm       **31.** 9 ft 2 in.       **32.** 4 ft 9 in.

*Find the radius of a sphere having the given surface area. See Example 9. Round off to the nearest tenth.*

**33.** 907.2 sq in.       **34.** 643.6 sq in.       **35.** 98.6 sq m       **36.** 95.0 sq m

**37.** If 1 gal of paint covers 240 sq ft, how many gallons of paint (to the nearest gallon) are needed to cover the total surface of a cylindrical tank 10 ft in diameter and 20 ft long?

**38.** Solve Exercise 37 if the tank is spherical with a diameter of 20 ft.

*To the nearest square inch, how many square inches of sheet metal are needed to manufacture each of the items shown?*

**39.**

10.5″
2″
12.2″
2″
6.5″
Open ends

**40.**

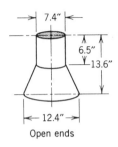

7.4″
6.5″
13.6″
12.4″
Open ends

**41.**

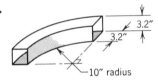

3.2″
3.2″
10″ radius

**42.**

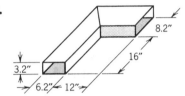

8.2″
16″
3.2″
6.2″ 12″

## ■ *Review Exercises*

### Section 7.1

*Name the vertex and sides of each angle (see Figure 7.91).*

**1.** ∠2    **2.** ∠4

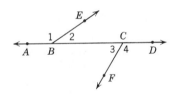

**FIGURE 7.91**

**3.** Write 24°45′ in decimal form.

**4.** Write 56.4° in degree-minute form.

**5.** Specify whether each angle is acute, right, or obtuse.
   **(a)** 47°    **(b)** 147°    **(c)** 90°

**6.** In Figure 7.92, if ∠1 = 45° and ∠3 = 35°, find the measure of ∠2.

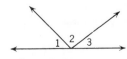

**FIGURE 7.92**

## Section 7.2

**7.** Classify △*ACD* in all possible ways according to its sides and its angles. (See Figure 7.93.)

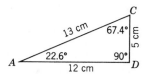

**FIGURE 7.93**

**8.** In △*BFG*, name the altitude and the related base. (See Figure 7.94.)

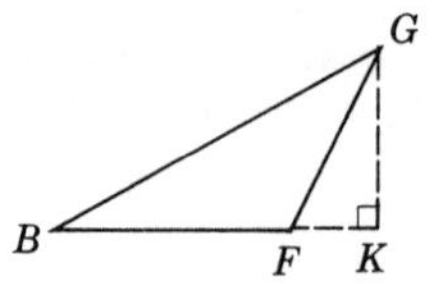

**FIGURE 7.94**

**9.** In a triangle, one angle measures 41.3° and another measures 28.4°. Find the measure of the third angle.

**10.** In an isosceles triangle, the vertex angle measures 46° and the other two angles are equal. Find the measures of the other two angles.

**11.** The hypotenuse of a right triangle is 12 mm long, and one side is 6 mm long. Find the length of the third side, correct to the nearest tenth.

## Section 7.3

**12.** Find the perimeter of a rectangle that is 18 ft long and 6 ft wide.

**13.** Find the length of a rectangle that is 8.3 m wide if its perimeter is 30.7 m. Round off to the nearest tenth.

**14.** Find the perimeter of (*a*) an equilateral triangle, (*b*) a square, (*c*) a regular hexagon, and (*d*) a regular octagon, given that each side is 32.9 cm long. Round off to the nearest tenth.

**15.** To the nearest tenth, find the area of a circle with a diameter of 16.2 in.

**16.** Find the perimeter of Figure 7.95.

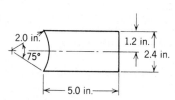

**FIGURE 7.95**

## Section 7.4

*In Exercises 17 to 22, find the area of each figure.*

**17.** A triangle with base 10.2 mi and altitude 6 mi.

**18.** A rectangle 47 km long and 15 km wide.

**19.** (*a*) An equilateral triangle, (*b*) a square, (*c*) a regular hexagon, and (*d*) a regular octagon, if a side is 14.3 in. Round off to three significant digits.

**20.** A circle with radius 5.1 m. Answer to the nearest hundredth.

**21.** A parallelogram with $b = 4.69$ m and $h = 6.42$ m. Round off to three significant digits.

**22.** A trapezoid with $b_1 = 9.61$ in., $b_2 = 4.06$ in., and $h = 3.52$ in. Round off to the nearest hundredth.

**23.** Find the diameter of a circle to four significant digits if the area is 67.92 sq cm.

**24.** To the nearest tenth, find the area of a sector of a circle having a radius of 4.3 in. and a central angle of 88.5°.

**25.** Find the area of Figure 7.96 to two significant digits.

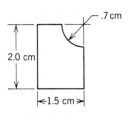

**FIGURE 7.96**

## Section 7.5

**26.** To the nearest tenth, find the volume of a prism with an altitude of 18 in. and a base that is an equilateral triangle with a side 6 in. long.

**27.** To the nearest gallon, how many gallons of water can be contained in a tank built as a rectangular prism 32 ft by 16 ft by 12 ft?

**28.** What is the volume of a cube with edges 15 m long?

**29.** What is the volume of a pyramid with an altitude of 21 ft if its base is a rectangle 35 ft by 70 ft?

**30.** Find the volume of a frustum of a pyramid given that the areas of the larger base and the smaller base are 62.3 sq ft and 46.8 sq ft, respectively, and the altitude is 7.00 ft. Answer to three significant digits.

## Section 7.6

*In Exercises 31 to 34, find the volume of each solid. Round off to three significant digits.*

**31.** A circular cylinder 12.0 cm in diameter and 15.0 cm long.

**32.** A circular cone with base radius 4.32 in. and altitude 9.21 in.

**33.** A frustum of a cone if the radius of the larger base is 7.05 m, the radius of the smaller base is 3.88 m, and the altitude is 5.34 m.

**34.** A sphere with diameter 26 ft.

## Section 7.7

*In Exercises 35 to 40, find the total surface area of the given solid, rounded off to the nearest whole number.*

**35.** A prism 16 cm high with a base that is an equilateral triangle with a side 6 cm long.

**36.** A rectangular prism 10 ft high with a base that is 8 ft by 4 ft.

**37.** A cube with edges 3.9 m long.

**38.** A right circular cylinder 8.2 ft in diameter and 10 ft long.

**39.** A right circular cone with a base 6 cm in diameter and altitude 8 cm.

**40.** A frustum of a cone with radius of the larger base 10 in., radius of the smaller base 6 in., and altitude 5 in.

# Essentials of Trigonometry

Trigonometry is the branch of mathematics that deals with measurement of angles and triangles. In this chapter we consider methods that can be used either to find particular parts of triangles (sides or angles) or to completely *solve a triangle*. A triangle is said to be solved when all three sides and all three angles are known.

## 8.1 Trigonometric Ratios

In Section 7.2 we used symbols such as *AB*, *BC*, *CD*, and so forth to name sides of triangles. Sometimes it is convenient to use lowercase letters *a*, *b*, and *c* to indicate the lengths of the sides (of $\triangle ABC$) opposite $\angle A$, $\angle B$, and $\angle C$, respectively (see Fig. 8.1).

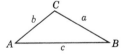

**FIGURE 8.1**

In any triangle, the two sides that form a given angle are said to be *adjacent* to the angle, and the third side that does not form the given angle is said to be *opposite* the angle. For example, in $\triangle RST$ of Fig. 8.2, we have:

| ANGLE | SIDES ADJACENT | SIDE OPPOSITE |
|-------|----------------|---------------|
| $\angle R$ | *RS* and *RT* | *ST* |
| $\angle S$ | *RS* and *ST* | *RT* |
| $\angle T$ | *RT* and *ST* | *RS* |

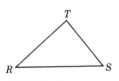

**FIGURE 8.2**

In a right triangle, we consider adjacent sides and opposite sides for the acute angles only. Furthermore, for each acute angle of a right triangle, only the side of the triangle that is *not* the hypotenuse is referred to as the side *adjacent* to the angle. For example, in right triangle *ABC* in Fig. 8.3, *AC* is the side adjacent to $\angle A$, and *BC* is the side opposite $\angle A$; *BC* is the side adjacent to $\angle B$, and *AC* is the side opposite $\angle B$.

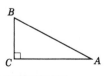

**FIGURE 8.3**

**EXAMPLE 1**  For each triangle in Figure 8.4, name the hypotenuse and the side opposite and the side adjacent to each angle.

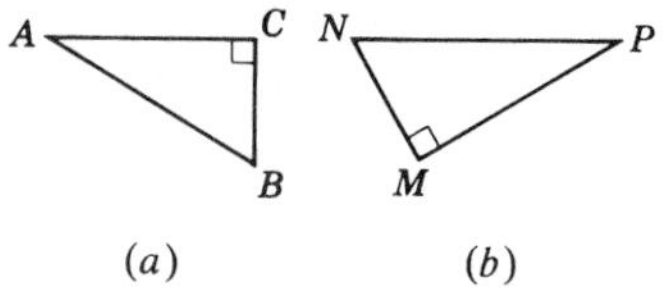

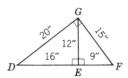

**FIGURE 8.4**

**Solutions**

| | HYPOTENUSE | ACUTE ANGLE | SIDE OPPOSITE | SIDE ADJACENT |
|---|---|---|---|---|
| **(a)** | AB | ∠A | BC | AC |
| | | ∠B | AC | BC |
| **(b)** | NP | ∠N | MP | NM |
| | | ∠P | NM | MP |

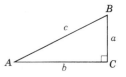

**FIGURE 8.5**

**EXAMPLE 2**  For each triangle in Figure 8.5, specify the length of the hypotenuse and the lengths of the side opposite and the side adjacent to each acute angle.

**(a)** △DEG     **(b)** △GEF     **(c)** △DGF

**Solutions**

| | HYPOTENUSE | ANGLE | SIDE OPPOSITE | SIDE ADJACENT |
|---|---|---|---|---|
| **(a)** △DEG | 20 in. | ∠D | 12 in. | 16 in. |
| | | ∠DGE | 16 in. | 12 in. |
| **(b)** △GEF | 15 in. | ∠F | 12 in. | 9 in. |
| | | ∠EGF | 9 in. | 12 in. |
| **(c)** △DGF | 25 in. | ∠D | 15 in. | 20 in. |
| | | ∠F | 20 in. | 15 in. |

## ■ Sine, Cosine, and Tangent

Three ratios that involve the sides and the hypotenuse of a right triangle, which are useful for solving triangles, are given special names in terms of the acute angles of the triangle. For ∠A in Figure 8.6:

**FIGURE 8.6**

$$\left.\begin{array}{l} \text{sine of } \angle A = \dfrac{\text{side opposite } \angle A}{\text{hypotenuse}} \\[2mm] \text{cosine of } \angle A = \dfrac{\text{side adjacent to } \angle A}{\text{hypotenuse}} \\[2mm] \text{tangent of } \angle A = \dfrac{\text{side opposite } \angle A}{\text{side adjacent to } \angle A} \end{array}\right\} \qquad \textbf{(1)}$$

These ratios are called **trigonometric ratios**; the names sine, cosine, and tangent are commonly abbreviated as sin, cos, and tan, respectively. Thus, using $a$, $b$, and $c$ to name the sides and hypotenuse, Equations 1 are customarily given as

$$\sin A = \frac{a}{c}, \qquad \cos A = \frac{b}{c}, \qquad \text{and} \qquad \tan A = \frac{a}{b}$$

Referring to ∠B in Figure 8.6:

$$\text{sine of } \angle B = \frac{\text{side opposite } \angle B}{\text{hypotenuse}}$$

$$\text{cosine of } \angle B = \frac{\text{side adjacent to } \angle B}{\text{hypotenuse}}$$

$$\text{tangent of } \angle B = \frac{\text{side opposite } \angle B}{\text{side adjacent to } \angle B}$$

Using sides $a$, $b$, and $c$, we have

$$\sin B = \frac{b}{c}, \qquad \cos B = \frac{a}{c}, \qquad \text{and} \qquad \tan B = \frac{b}{a}$$

**FIGURE 8.7**

**EXAMPLE 3** In Figure 8.7:

**(a)** $\sin A = \dfrac{3}{5} = 0.6$      **(b)** $\sin B = \dfrac{4}{5} = 0.8$

$\cos A = \dfrac{4}{5} = 0.8$            $\cos B = \dfrac{3}{5} = 0.6$

$\tan A = \dfrac{3}{4} = 0.75$         $\tan B = \dfrac{4}{3} = 1.3333333$

**EXAMPLE 4**

In $\triangle DEG$ of Figure 8.8:      In $\triangle DFG$ of Figure 8.8:

$\sin D = \dfrac{12}{20} = 0.6$           $\sin D = \dfrac{15}{16 + 9} = 0.6$

$\cos D = \dfrac{16}{20} = 0.8$           $\cos D = \dfrac{20}{16 + 9} = 0.8$

$\tan D = \dfrac{12}{16} = 0.75$        $\tan D = \dfrac{15}{20} = 0.75$

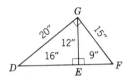

**FIGURE 8.8**

In Example 4, note that the trigonometric ratios associated with $\angle D$ are the same, respectively, even though they were computed for two different "size" triangles. In fact, *the sine, cosine, and tangent of any angle depend only on the measure of the angle and not on the "size" of the triangle that includes the angle.*

## ■ Finding Values for Trigonometric Ratios

Approximate values for each of the three trigonometric ratios for *any* angle are readily found on a calculator by use of the *trigonometric keys* $\boxed{\textbf{SIN}}$, $\boxed{\textbf{COS}}$, and $\boxed{\textbf{TAN}}$. The calculator sequence calls for entering the angle *before* pressing the appropriate trigonometric key.

**EXAMPLE 5** (Set the D-R switch* to D, if applicable.)

$$\sin 73° = 0.95630; \qquad \cos 73° = 0.29237; \qquad \tan 73° = 3.27085$$

as computed by the following sequences:

$$73 \ \boxed{\textbf{SIN}} \ \longrightarrow 0.95630476$$
$$73 \ \boxed{\textbf{COS}} \ \longrightarrow 0.29237171$$
$$73 \ \boxed{\textbf{TAN}} \ \longrightarrow 3.2708526$$

**EXAMPLE 6** Find tan 58°42′ to five places.

**Solution** First, change 58°42′ to decimal form.

$$58°42' = \left(58\frac{42}{60}\right)^{\circ} = 58.7°$$

---

*Most scientific calculators have a D-R switch (or key) that allows the calculator to operate with different types of angle measures. For our purposes, in this section we will use only the D (for "degree") setting.

Next,   tan 58.7° = 1.64471   (to five places) as computed by the sequences

$$58.7 \; \boxed{\text{TAN}} \longrightarrow 1.6447111$$

### ■ Finding the Acute Angle Associated with a Given Ratio

We can use the calculator keys $\boxed{\text{sin}^{-1}}$, $\boxed{\text{cos}^{-1}}$, and $\boxed{\text{tan}^{-1}}$ to find the acute angle associated with a given trigonometric ratio. (Your calculator may use other labels for these keys.) We will refer to these keys as "inverse" keys.

**EXAMPLE 7**   To the nearest tenth of a degree, find $\angle A$ such that:

**(a)** sin $A$ = 0.64312       **(b)** tan $A$ = 1.4371

### Solutions
**(a)** $0.64312 \; \boxed{\text{sin}^{-1}} \longrightarrow 40.024866$

To the nearest tenth, 40.0° is the angle whose sine is 0.64312.

**(b)** $1.4371 \; \boxed{\text{tan}^{-1}} \longrightarrow 55.168035$

To the nearest tenth, 55.2° is the angle whose tangent is 1.4371.

### Exercises for Section 8.1

*For each triangle, name the hypotenuse and the side opposite and the side adjacent to each acute angle. See Example 1.*

**1.**

**2.**

**3.**

**4.**

**5.**

**6.**

*For each right triangle, specify the length of the hypotenuse and the lengths of the side opposite and the side adjacent to each acute angle. See Example 2.*

**7.**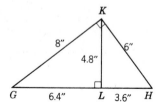

**8.**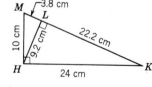

**9.**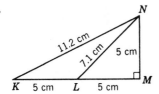

**10.**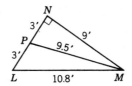

**11.**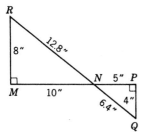

**12.**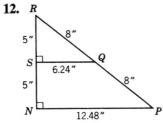

*Refer to the related drawing to find the sine, cosine, and tangent of each angle. Round off to five decimal places. See Examples 3 and 4.*

**13.** ∠*B*          **14.** ∠*C*

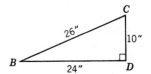

**15.** ∠*G*          **16.** ∠*F*

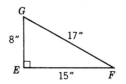

**17.** ∠*H*          **18.** ∠*K*

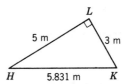

**19.** ∠*DBC*        **20.** ∠*BDC*

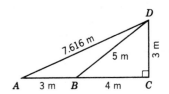

**21.** ∠*A*          **22.** ∠*ADC*

**23.** ∠F        **24.** ∠H

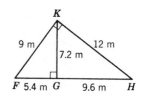

**25.** ∠GKH      **26.** ∠FKG

*Use a calculator to find the sine, cosine, and tangent of each angle, rounded off to five decimal places. See Examples 5 and 6.*

**27.** 68°        **28.** 76°        **29.** 49°        **30.** 52°

**31.** 44.9°      **32.** 61.4°      **33.** 26.2°      **34.** 45.8°

**35.** 27°43′     **36.** 79°25′     **37.** 10°13′     **38.** 30°30′

*Find the angle, to the nearest tenth of a degree, given the sine, cosine, or tangent. See Example 7.*

**39.** $\sin A = 0.38324$     **40.** $\sin A = 0.13240$     **41.** $\cos A = 0.53257$

**42.** $\cos A = 0.28373$     **43.** $\tan A = 0.37755$     **44.** $\tan A = 8.7516$

**45.** $\sin B = 0.98919$     **46.** $\sin B = 0.56550$     **47.** $\cos B = 0.20628$

**48.** $\cos B = 0.49084$     **49.** $\tan B = 5.9244$      **50.** $\tan B = 0.47481$

*Find the indicated angle, to the nearest tenth of a degree.*

**51.** ∠*A* (first find tan *A*)

**52.** ∠*B* (first find tan *B*)

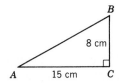

**53.** ∠*D* (first find sin *D*)

**54.** ∠*F* (first find sin *F*)

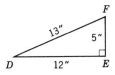

## 8.2 Calculations with Trigonometric Ratios

In this section we consider calculator sequences for combined operations that involve trigonometric ratios. For simple computations the keystroke sequences are direct.

**EXAMPLE 1**  Compute to five decimal places.
**(a)** 8.64 cos 73° = 2.52609;   this is computed by

$$8.64 \;\boxed{\times}\; 73 \;\boxed{\textbf{COS}}\; \boxed{=} \longrightarrow 2.5206915$$

**(b)** $\dfrac{24.6}{\tan 59°}$ = 14.78117;   this is computed by

$$24.6 \;\boxed{\div}\; 59 \;\boxed{\textbf{TAN}}\; \boxed{=} \longrightarrow 14.781171$$

Examples 1*a* and 1*b* involve either a single multiplication or division. Computations involving trigonometric ratios frequently require more than a single operation.

**EXAMPLE 2**
**(a)** $\dfrac{84.9 \sin 78.3°}{62.5}$ = 1.33018;   this is computed by

$$84.9 \;\boxed{\times}\; 78.3 \;\boxed{\textbf{SIN}}\; \boxed{\div}\; 62.5 \;\boxed{=} \longrightarrow 1.3301763$$

**(b)** $\dfrac{204 \sin 83°}{\sin 25°}$ = 479.10711;   this is computed by

$$204 \;\boxed{\times}\; 83 \;\boxed{\textbf{SIN}}\; \boxed{\div}\; 25 \;\boxed{\textbf{SIN}}\; \boxed{=} \longrightarrow 479.10711$$

## ■ Sequences with $\boxed{sin^{-1}}$, $\boxed{cos^{-1}}$, $\boxed{tan^{-1}}$ Keys

Recall from Section 8.1 that you can use the inverse keys on your calculator to find the angle associated with a given trigonometric ratio. Sequences that include the inverse keys operate only on the number in the display and, if that number is the result of one or more operations, all operations must be completed first.

**EXAMPLE 3**  To the nearest tenth of a degree, find $\angle B$ such that

$$\tan B = \frac{64.8}{7.9}$$

**Solution**  Because $\angle B$ is the angle whose tangent is $64.8 \div 7.9$, first compute the quotient. Use the sequence:

$$64.8 \boxed{\div} 7.9 \boxed{=} \boxed{\tan^{-1}} \longrightarrow 83.049168$$

To the nearest tenth,   $\angle B = 83.0°$.

**EXAMPLE 4**  To the nearest tenth of a degree, find $\angle C$ such that

$$\sin C = \frac{12.6 \sin 42.5°}{14.4}$$

**Solution**  Because $\angle C$ is the angle whose sine is equal to

$$12.6 \times \sin 42.5° \div 14.4$$

first compute the combined product and quotient. Use the sequence:

$$12.6 \boxed{\times} 42.5 \boxed{\text{SIN}} \boxed{\div} 14.4 \boxed{=} \boxed{\sin^{-1}} \longrightarrow 36.23805$$

To the nearest tenth,   $\angle C = 36.2°$.

Trigonometric computations often include squares of numbers. In some computations you may want to write intermediate results if you are not comfortable using a single keying sequence.

**EXAMPLE 5**  To the nearest tenth, find $b$ if

$$b^2 = 2.4^2 + 3.8^2 - 2(2.4)(3.8) \cos 42°$$

**Solution**  First compute

$$2.4^2 + 3.8^2 \longrightarrow 20.2$$

Record this result. Then compute

$$2(2.4)(3.8) \cos 42° \longrightarrow 13.554962$$

Then,

$$\sqrt{20.2 - 13.554962} \longrightarrow 2.5777971$$

To the nearest tenth,   $b = 2.6$.

## Exercises for Section 8.2

*Compute. See Example 1. Round off to three significant digits.*

**1.** $16.9 \cos 48°$    **2.** $11.1 \cos 28°$    **3.** $71.6 \tan 24°$

**4.** $72.7 \tan 54°$    **5.** $1.76 \sin 7.2°$    **6.** $2.04 \sin 5.1°$

**7.** $\dfrac{9.95}{\cos 6.2°}$    **8.** $\dfrac{9.63}{\cos 3.1°}$    **9.** $\dfrac{89.5}{\tan 4.7°}$

**10.** $\dfrac{85.4}{\tan 9.1°}$       **11.** $\dfrac{2.89}{\sin 47°5'}$       **12.** $\dfrac{6.35}{\sin 55°34'}$

*Compute. See Example 2. Round off to four significant digits.*

**13.** $\dfrac{94.29 \sin 9.3°}{96.95}$       **14.** $\dfrac{10.36 \sin 56.1°}{11.29}$

**15.** $\dfrac{7.119 \sin 73.3°}{6.710}$       **16.** $\dfrac{5.108 \sin 5.12°}{2.765}$

**17.** $\dfrac{2.368 \sin 21.3°}{\sin 82.5°}$       **18.** $\dfrac{1.011 \sin 54.0°}{\sin 23.3°}$

**19.** $\dfrac{70.56 \sin 9°7'}{\sin 6°9'}$       **20.** $\dfrac{52.16 \sin 25°39'}{\sin 16°46'}$

*Find each angle, to the nearest tenth of a degree. See Examples 3 and 4.*

**21.** $\sin A = \dfrac{3.26}{3.93}$       **22.** $\sin A = \dfrac{29.33}{42.70}$

**23.** $\cos B = \dfrac{81.5}{99.6}$       **24.** $\tan B = \dfrac{2.643}{4.608}$

**25.** $\sin C = \dfrac{6.09 \sin 52.6°}{7.28}$       **26.** $\sin C = \dfrac{6.65 \sin 66.1°}{7.78}$

**27.** $\sin A = \dfrac{89.7 \sin 68.8°}{88.5}$       **28.** $\sin A = \dfrac{92.4 \sin 77.6°}{95.6}$

**29.** $\cos B = \dfrac{16.3^2 + 8.9^2 - 9.6^2}{2(16.3)(8.9)}$       **30.** $\cos B = \dfrac{17.0^2 + 8.7^2 - 10.3^2}{2(17.0)(8.7)}$

*Find a, b, or c, to the nearest tenth. See Example 5.*

**31.** $a^2 = 6.4^2 + 2.7^2 - 2(6.4)(2.7) \cos 70.0°$

**32.** $b^2 = 8.3^2 + 3.5^2 - 2(8.3)(3.5) \cos 23.5°$

**33.** $c^2 = 8.8^2 + 8.2^2 - 2(8.8)(8.2) \cos 47.1°$

**34.** $a^2 = 12.5^2 + 4.44^2 - 2(12.5)(4.44) \cos 10.3°$

## 8.3  Solving Right Triangles

One or more of the following basic concepts can be helpful when solving right triangles:

**1.** The trigonometric ratios
**2.** The fact that the sum of the acute angles of a right triangle is 90°
**3.** The Pythagorean rule

In general, the concepts to be used in specific problems are chosen on the basis of the given data. Unless specified otherwise, we shall find angles to the nearest tenth of a degree and find sides and the hypotenuse to no more places than the smallest number of significant digits in the given data.

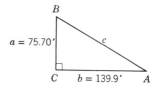

**FIGURE 8.9**

**EXAMPLE 1**  Find $\angle A$ in $\triangle ABC$ (Figure 8.9)

**Solution**  Note that the 75.70-ft side is *opposite* $\angle A$, and the 139.9-ft side is *adjacent* to $\angle A$. Hence, use the tangent ratio.

$$\tan A = \frac{a}{b} = \frac{75.70}{139.9}$$
$$= 0.54110079$$

from which, to the nearest tenth,   $\angle A = 28.4°$.   Note that all the calculations can be completed by the sequence

$$75.70 \boxed{\div} 139.9 \boxed{=} \boxed{\tan^{-1}} \longrightarrow 28.417856$$

**EXAMPLE 2**  In $\triangle ABC$ of Example 1, find $\angle B$.

**Solution**  The sum of the acute angles is 90° and, from Example 1, $\angle A = 28.4°$. Hence,

$$\angle B = (90 - 28.4)° = 61.6°$$

If two sides (or one side and the hypotenuse) of a right triangle are given, then the hypotenuse (or the second side) can be found by the Pythagorean rule. For instance, in $\triangle ABC$ of Example 1,

$$c = \sqrt{75.70^2 + 139.9^2} = 159.0676$$

and   $c = 159.1$ ft,   to four significant digits.

If only one side (or the hypotenuse) and an acute angle of a right triangle are given, we can use trigonometric ratios to find the second side and the hypotenuse (or the other two sides).

**EXAMPLE 3**  Find sides $a$ and $b$ in $\triangle ABC$ (Figure 8.10).

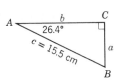

**FIGURE 8.10**

**Solution**  In this case the hypotenuse is given ($c = 15.5$).
**1.** Because $a$ is the side *opposite* $\angle A$, use the sine ratio.

$$\sin 26.4° = \frac{a}{15.5}$$

Multiply each side by 15.5.

$$15.5 \sin 26.4° = \frac{a}{15.5} \times 15.5$$
$$a = 6.8918453$$

Thus,  $a = 6.89$ cm,  to three significant digits.

**2.** Because $b$ is the side *adjacent* to $\angle A$, use the cosine ratio.

$$\cos 26.4° = \frac{b}{15.5}$$

Multiply each side by 15.5.

$$15.5 \cos 26.4° = \frac{b}{15.5} \times 15.5$$
$$b = 13.883532$$

Thus,  $b = 13.9$ cm,  to three significant digits.

If a problem does not include a figure, it is helpful to sketch and label a figure before proceeding to solve the triangle. If the problem refers to "right triangle $ABC$," it is to be understood that $\angle C$ is the right angle.

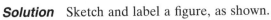

**FIGURE 8.11**

**EXAMPLE 4**  Solve right triangle $ABC$ if  $\angle B = 74.0°$  and  $a = 8.7$ yd (Figure 8.11).

**Solution**  Sketch and label a figure, as shown.
**1.** Note that $b$ is *opposite* $\angle B$ and $a$ is *adjacent* to $\angle B$. Hence, use the tangent ratio.

$$\tan 74.0° = \frac{b}{8.7}$$

Multiply each side by 8.7 to obtain

$$8.7 \tan 74.0° = b$$
$$b = 30.340506$$

Thus,  $b = 30$ yd,  to two significant digits.

**2.** Because $c$ is the *hypotenuse* and $a$ is *adjacent* to $\angle B$, use the cosine ratio.

$$\cos 74.0° = \frac{8.7}{c}$$

Multiply each side by $c$ to obtain

$$c \cos 74.0° = 8.7$$

Divide each side by $\cos 74.0°$ to obtain

$$c = \frac{8.7}{\cos 74.0°} = 31.563212$$

Thus,  $c = 32$ yd,  to two significant digits.

**3.** The sum of $\angle A$ and $\angle B$ is 90°. Hence,

$$\angle A = (90 - 74.0)° = 16.0°$$

## ■ Applications

Many practical problems that involve right triangles can be solved by the methods shown in the examples above. Of course, any angle measures given in degree-minute form should be changed to decimal form in order to use your calculator.

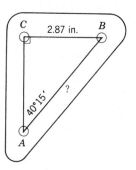

**EXAMPLE 5**   Three holes are to be drilled in a triangular plate at points $A$, $B$, and $C$, as shown in Figure 8.12. To the nearest hundredth, find the dimension $AB$.

**FIGURE 8.12**

**Solution**   First, change 40°15′ to 40.25°. Next, note that $\angle A$, $BC$, and $AB$ are related by

$$\sin A = \frac{BC}{AB}$$

$$\sin 40.25° = \frac{2.87}{AB}$$

Multiply each side by $AB$ to obtain

$$AB \sin 40.25° = 2.87$$

Divide each side by $\sin 40.25°$ to obtain

$$AB = \frac{2.87}{\sin 40.25°} = 4.441872$$

Thus,   $AB = 4.44$ in.,   to the nearest hundredth.

**EXAMPLE 6**   An anchor cable is to be fastened from the top of a transmission tower to a point ($P$) 582 ft from the center ($Q$) of the base of the tower, as shown in Figure 8.13. Find the length of the cable if the angle at $P$ measured from the ground to the cable is 43.9°.

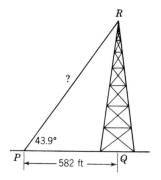

**FIGURE 8.13**

**Solution**   Note that $\angle P$, $PR$, and $PQ$ are related by

$$\cos P = \frac{PQ}{PR}$$

Hence,

$$\cos 43.9° = \frac{582}{PR}$$

Multiply each side by $PR$ to obtain

$$PR \cos 43.9° = 582$$

Divide each side by $\cos 43.9°$ to obtain

$$PR = \frac{582}{\cos 43.9°} = 807.71508$$

To three significant digits, the cable is 808 ft long.

## Exercises for Section 8.3

Find the measures of the acute angles of a right triangle having the given dimensions. See Examples 1 and 2. Round off to the nearest tenth of a degree. For Exercises 1 to 16, refer to Figure 8.14.

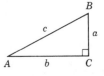

**FIGURE 8.14**

**1.** $a = 4$ yd;   $b = 7$ yd

**2.** $a = 5$ yd;   $b = 8$ yd

**3.**  $a = 9.7$ in.;   $b = 8.7$ in.

**4.** $a = 7.9$ in.;   $b = 5.4$ in.

**5.** $a = 14$ cm;   $c = 55$ cm

**6.** $a = 12$ cm;   $c = 44$ cm

**7.** $b = 12.2$ km;   $c = 61.3$ km

**8.** $b = 75.7$ km;   $c = 90.4$ km

Solve each right triangle having the dimensions given. See Examples 3 and 4. Round off angles to the nearest tenth of a degree and lengths to three significant digits.

**9.** $\angle A = 56.2°$;   $c = 19.2$ cm

**10.** $\angle A = 20.3°$;   $c = 76.4$ m

**11.** $\angle B = 34°17'$;   $b = 7.70$ yd

**12.** $\angle B = 53°26'$;   $b = 91.0$ yd

**13.** $a = 10.4$ cm;   $b = 22.3$ cm

**14.** $a = 24.1$ cm;   $b = 37.5$ cm

**15.** $a = 77.9$ mi;   $c = 99.5$ mi

**16.** $a = 89.5$ mi;   $c = 96.3$ mi

*See Examples 5 and 6 for the following exercises. Find all angles correct to the nearest tenth of a degree and all lengths to the number of significant digits used in the drawing.*

**17.** Find the measure of ∠A for machining the steel block (Figure 8.15).

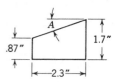

**FIGURE 8.15**

**18.** Find the height of the tower (Figure 8.16).

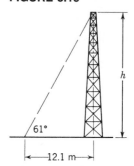

**FIGURE 8.16**

**19.** Find the dimensions *a* and *b* between the two holes (Figure 8.17).

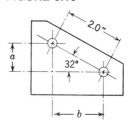

**FIGURE 8.17**

**20.** A strip of thin metal is to be bent as shown in Figure 8.18. Find ∠A.

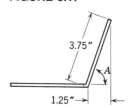

**FIGURE 8.18**

**21.** A fitting is formed as shown in Figure 8.19. Find ∠A.

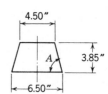

**FIGURE 8.19**

**22.** Three holes are to be located as shown in Figure 8.20. Find dimension *a*.

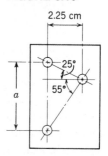

**FIGURE 8.20**

**23.** A steel block is machined as shown in Figure 8.21. Find the indicated dimension.

**FIGURE 8.21**

**24.** Three holes are to be drilled in a plate as shown in Figure 8.22. Find the indicated dimension.

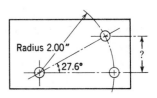

**FIGURE 8.22**

**25.** A roof truss has the dimensions given in Figure 8.23. Find the length of girder *AB*.

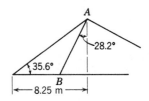

**FIGURE 8.23**

**26.** A burglar alarm mounted on the wall of a building (Figure 8.24) detects motion within an angle of 80°. Find the length of the interval protected on the property line 50 ft away.

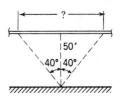

**FIGURE 8.24**

**27.** A tower (Figure 8.25) is supported in part by two guy wires forming angles of 54° and 62° with the ground. Find the distance between the wires at the ground level.

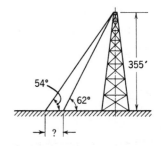

**FIGURE 8.25**

**28.** Five holes are to be drilled on a semicircle with a 12.25-cm radius such that the four angles shown at the center are equal (see Figure 8.26). Find the indicated dimension.

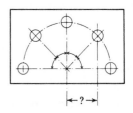

**FIGURE 8.26**

## 8.4 Solving Oblique Triangles: Law of Sines

A triangle that does not include a right angle is called an **oblique** triangle. In this section we introduce a formula that can be used to solve some oblique triangles.

### ■ Sine of an Obtuse Angle

In Section 8.1 we obtained trigonometric ratios associated with angles from 0° to 90°. It is also possible to assign values to sin $A$, cos $A$, and tan $A$ for *any* $\angle A$. For our purposes in this section we shall need only to consider values for sin $A$ for angles from 0° to 180°. Calculator sequences for computing such values are the same as those used for acute angles.

### EXAMPLE 1
**(a)** sin 107.4° = 0.95424;  this is computed by the sequence

$$107.4 \boxed{\text{SIN}} \longrightarrow 0.95424033$$

**(b)** sin 180° = 0;  this is computed by the sequence

$$180 \boxed{\text{SIN}} \longrightarrow 0$$

Use your calculator to verify that

$$\sin 60° = 0.86603 \quad \text{and} \quad \sin 120° = 0.86603$$

Note that the sum of 60° and 120° is 180° and that sin 60° is equal to sin 120°. In fact:

---

**The sine of any angle ($\angle A$) is equal to the sine of (180° − $\angle A$).**

---

Thus the following procedure can be used to find an *obtuse* $\angle A$ when sin $A$ is given.

**1.** Use $\boxed{\text{SIN}^{-1}}$ to find the associated acute angle.

**2.** Subtract the acute angle from 180° to find the obtuse angle.

**EXAMPLE 2**  Find the obtuse $\angle A$ such that  sin $A$ = 0.70527.

**Solution**  To the nearest tenth, the *acute* angle whose sine is 0.70527 is 44.9°. The required *obtuse* angle is

$$A = (180 - 44.9)° = 135.1°$$

A convenient calculator sequence for these computations is

$$180 \boxed{-} 0.70527 \boxed{\text{sin}^{-1}} \boxed{=} \longrightarrow 135.14864$$

### ■ The Law of Sines

The ratio of any side of a triangle to the sine of the angle opposite that side is the same for each angle. That is, if $\angle A$, $\angle B$, and $\angle C$ are the angles of a triangle and $a$, $b$, and $c$ are the respective opposite sides (Figure 8.27), then

**FIGURE 8.27**

$$\frac{a}{\sin A} = \frac{b}{\sin B} = \frac{c}{\sin C}$$

This relationship is called the **law of sines**. When applying this law to solve a triangle, we choose whichever two of the three ratios form a proportion in which three of the four terms are known. Thus, in order to use this law, we must be given at least *one side and two angles* or *two sides and the angle opposite one of them*.

## ■ *One Side and Two Angles*

As noted in Section 8.3, we are sometimes given problems that do not include figures. In such cases it is helpful to sketch and label a figure.

**EXAMPLE 3**   Solve $\triangle ABC$, given that   $\angle A = 20.2°$,   $\angle B = 83.9°$,   and   $c = 43.2$ m

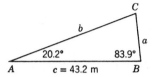

**FIGURE 8.28**

**Solution**   Sketch and label a figure (Figure 8.28). Then, find $\angle C$, $a$, and $b$.

**1.** Because the sum of the angles is $180°$,

$$\angle C = (180 - 20.2 - 83.9)° = 75.9°$$

**2.** To find $a$, note that $\angle A$, $\angle C$ (from part 1), and $c$ are known. Hence, from the law of sines, choose

$$\frac{a}{\sin A} = \frac{c}{\sin C}$$

$$\frac{a}{\sin 20.2°} = \frac{43.2}{\sin 75.9°}$$

By the cross-multiplication rule,

$$a \sin 75.9° = 43.2 \sin 20.2°$$

Divide each side by $\sin 75.9°$ to obtain

$$a = \frac{43.2 \sin 20.2°}{\sin 75.9°} = 15.380258$$

Thus,   $a = 15.4$ m,   to three significant digits.

**3.** To find $b$, note that $\angle B$, $\angle C$, and $c$ are known. Hence, from the law of sines, choose

$$\frac{b}{\sin B} = \frac{c}{\sin C}$$

$$\frac{b}{\sin 83.9°} = \frac{43.2}{\sin 75.9°}$$

By the cross-multiplication rule,

$$b \sin 75.9° = 43.2 \sin 83.9°$$

Divide each side by $\sin 75.9°$ to obtain

$$b = \frac{43.2 \sin 83.9°}{\sin 75.9°} = 44.289761$$

Thus,   $b = 44.3$ m,   to three significant digits.

## ■ *Two Sides and the Angle Opposite One of Them*

Given two sides of a triangle and the angle opposite the first given side, we begin to solve the triangle by using the law of sines to find the angle opposite the second given side. Then we proceed to find the third angle and the third side.

**EXAMPLE 4**   Solve the given *acute* triangle in Figure 8.29.

**Solution**

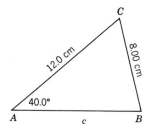

**FIGURE 8.29**

1. Because $\angle A$, $a$, and $b$ are known, use the law of sines to find $\angle B$. Choose

$$\frac{a}{\sin A} = \frac{b}{\sin B}$$

$$\frac{8.00}{\sin 40.0°} = \frac{12.0}{\sin B}$$

By the cross-multiplication rule,

$$8.00 \sin B = 12.0 \sin 40.0°$$

Divide each side by 8.00.

$$\sin B = \frac{12.0 \sin 40.0°}{8.00} = 0.96418141$$

Use the $\boxed{\sin^{-1}}$ key to obtain $\angle B$.

$$\angle B = 74.618568$$

To the nearest tenth,   $\angle B = 74.6°$.

2. Because the sum of the angles is 180°,

$$\angle C = (180 - 40.0 - 74.6)° = 65.4°$$

3. Use the law of sines to find $c$. Choose

$$\frac{a}{\sin A} = \frac{c}{\sin C}$$

$$\frac{8.00}{\sin 40.0°} = \frac{c}{\sin 65.4°}$$

By the cross-multiplication rule,

$$8.00 \sin 65.4° = c \sin 40.0°$$

Divide each side by $\sin 40.0°$.

$$c = \frac{8.00 \sin 65.4°}{\sin 40.0°} = 11.316162$$

Thus,   $c = 11.3$ cm,   to three significant digits.

**EXAMPLE 5**   Solve the *obtuse* triangle shown in Figure 8.30.

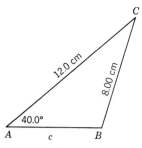

**FIGURE 8.30**

**Solution**

1. Because $\angle A$, $a$, and $b$ are known, use the law of sines to find $\angle B$. Choose

$$\frac{a}{\sin A} = \frac{b}{\sin B}$$

$$\frac{8.00}{\sin 40.0°} = \frac{12.0}{\sin B}$$

From Example 4, *acute* $\angle B$ is equal to 74.6°. Find *obtuse* $\angle B$.

$$\angle B = (180 - 74.6)° = 105.4°$$

2. Because the sum of the angles of the triangle is 180°,

$$\angle C = (180 - 40.0 - 105.4)° = 34.6°$$

**3.** Use the law of sines to find $c$. Choose

$$\frac{a}{\sin A} = \frac{c}{\sin C}$$

$$\frac{8.00}{\sin 40.0°} = \frac{c}{\sin 34.6°}$$

By the cross-multiplication rule,

$$8.00 \sin 34.6° = c \sin 40.0°$$

Divide each side by $\sin 40.0°$.

$$c = \frac{8.00 \sin 34.6°}{\sin 40.0°} = 7.0672643$$

Thus,   $c = 7.07$ cm,   to three significant digits.

## ◾ Applications of the Law of Sines

Many practical problems that involve triangles other than right triangles can be solved by using the law of sines.

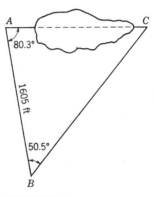

**FIGURE 8.31**

**EXAMPLE 6**   In Figure 8.31, points $A$ and $C$ are at opposite ends of a lake. A surveyor determines that   $\angle A = 80.3°$,   $\angle B = 50.5°$,   and the distance from $A$ to $B$ is 1605 ft. Find the distance $AC$ across the lake.

**Solution**   Let   $AC = b$,   and   $c = 1605$ ft.   First, find $\angle C$, the angle opposite side $c$.

$$\angle C = (180 - 80.3 - 50.5)° = 49.2°$$

Next, to find $b$, note that $\angle B$, $\angle C$, and $c$ are known. Hence, from the law of sines, choose

$$\frac{b}{\sin B} = \frac{c}{\sin C}$$

$$\frac{b}{\sin 50.5°} = \frac{1605}{\sin 49.2°}$$

By the cross-multiplication rule,

$$b \sin 49.2° = 1605 \sin 50.5°$$

Divide each side by $\sin 49.2°$ to obtain

$$b = \frac{1605 \sin 50.5°}{\sin 49.2°} = 1636.0179$$

The distance across the lake is 1636 ft, to four significant digits.

**EXAMPLE 7**   In Figure 8.32, microwave relay transmitters are erected at $A$, $B$, and $C$, so that a signal can be beamed from $A$ to $C$ and then from $C$ to $B$. Through what angle must the transmitter at $A$ be turned so that a signal can be beamed directly from $A$ to $B$?

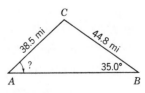

**FIGURE 8.32**

**Solution** From the figure: $a = 44.8$; $b = 38.5$; $\angle B = 35.0°$. Because $a$, $b$, and $\angle B$ are known and $\angle A$ is required, choose the law of sines in the form

$$\frac{a}{\sin A} = \frac{b}{\sin B}$$

$$\frac{44.8}{\sin A} = \frac{38.5}{\sin 35.0°}$$

By the cross-multiplication rule,

$$44.8 \sin 35.0° = 38.5 \sin A$$

Divide each side by 38.5.

$$\sin A = \frac{44.8 \sin 35.0°}{38.5} = .66743442$$

Use the $\boxed{\sin^{-1}}$ key to obtain

$$\angle A = 41.9°$$

to the nearest tenth.

### Exercises for Section 8.4

*Find the sine of each angle. Round off to five places. See Example 1.*

**1.** 104°        **2.** 150°        **3.** 102°        **4.** 118°

**5.** 164.7°      **6.** 91.6°       **7.** 141.9°      **8.** 162.5°

**9.** 124°13′     **10.** 161°56′    **11.** 93°36′     **12.** 160°46′

*Find the obtuse angle whose sine is given. See Example 2. Round off to the nearest tenth of a degree.*

**13.** 0.42488    **14.** 0.78077    **15.** 0.69882

**16.** 0.61657    **17.** 0.34136    **18.** 0.79180

*Refer to Figure 8.33 for Exercises 19 to 24. Solve each triangle having the given dimensions. See Example 3. Round off angles to the nearest tenth of a degree and lengths to three significant digits.*

**19.** $\angle A = 24°$; $\angle B = 74°$; $c = 13.2$ in.

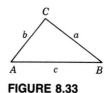

**FIGURE 8.33**

**20.** $\angle A = 35°$;   $\angle B = 53°$;   $a = 21.2$ in.

**21.** $\angle C = 125.2°$;   $\angle A = 27.5°$;   $c = 97.4$ km

**22.** $\angle C = 10.8°$;   $\angle A = 109.2°$;   $a = 35.2$ km

**23.** $\angle A = 72°30'$;   $\angle B = 19°20'$;   $b = 50.6$ m

**24.** $\angle A = 50°25'$;   $\angle B = 24°15'$;   $c = 21.9$ m

*See Examples 4 and 5 for Exercises 25 to 28.*

**25.** Find an acute angle $C$ if   $\angle A = 29°$,   $a = 42.2$ in.,   and   $c = 78.7$ in.

**26.** Find an acute angle $C$ if   $\angle B = 32°$,   $b = 17.3$ yd,   and   $c = 28.9$ yd.

**27.** Find an obtuse angle $A$ if   $\angle C = 35.6°$,   $c = 55.2$ m,   and   $a = 75.9$ m.

**28.** Find an obtuse angle $A$ if   $\angle C = 23.4°$,   $c = 45.6$ m,   and   $a = 93.8$ m.

*For the following exercises, round off angles to the nearest tenth of a degree and lengths to the smallest number of significant digits used in the given dimensions. Assume that the final zeros on a whole number are significant. See Examples 6 and 7.*

**29.** In Figure 8.34, points $A$, $B$, and $C$ are on the banks of a river. Find $AC$.

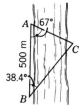

**FIGURE 8.34**

**30.** In the figure for Exercise 29, find $BC$ if   $AB = 750$ yd,   $\angle A = 52.4°$, and   $\angle B = 67.9°$.

**31.** In Figure 8.35, two observation points ($P$ and $R$) are 1000 yd apart on shore. A ship at sea is located so that   $\angle P = 46.8°$   and   $\angle R = 52.1°$. (*a*) Find $PS$. (*b*) Find $h$.

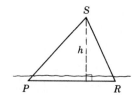

**FIGURE 8.35**

**32.** In the figure for Exercise 31, (*a*) find $RS$, and (*b*) find $h$ if   $PR = 1500$ m, $\angle P = 38.7°$,   and   $\angle R = 49.6°$.

**33.** The holes in a fitting are located as shown in Figure 8.36. (*a*) Find *AB*. (*b*) Find dimension *x*. (*c*) Find dimension *y*.

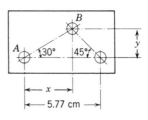

**FIGURE 8.36**

**34.** Two sightings to the top of a cliff were made as shown in Figure 8.37. (*a*) Find *AB*. (*b*) Find the height of the cliff.

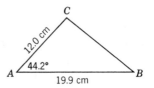

**FIGURE 8.37**

## 8.5 Solving Oblique Triangles: Law of Cosines

In Section 8.4 we used the law of sines to solve oblique triangles where the given data consisted of two angles and one side, or two sides and the angle opposite one of the given sides. In this section we introduce a second formula that can be used to solve triangles in which the given data consist of two sides and the angle between the two sides, or the three sides.

### ■ The Law of Cosines

If $\angle A$, $\angle B$, and $\angle C$ are the angles of a triangle, and *a*, *b*, and *c* are the respective opposite sides (Figure 8.38), then the three sides of the triangle are related by the formulas

$$a^2 = b^2 + c^2 - 2bc \cos A$$
$$b^2 = c^2 + a^2 - 2ca \cos B$$
$$c^2 = a^2 + b^2 - 2ab \cos C$$

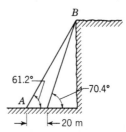

**FIGURE 8.38**

Each of these equations is called the **law of cosines**.

### ■ Side-Angle-Side

If two sides and the angle *between* the two sides of an oblique triangle are given, the law of cosines can be used to find the third side.

**EXAMPLE 1** Find side *BC* of $\triangle ABC$ (Figure 8.39).

**Solution** Let $a = BC$. Then, choose the law of cosines in the form

$$a^2 = b^2 + c^2 - 2bc \cos A$$

Substitute 12.0 for *b*, 19.9 for *c*, 44.2° for *A*, and solve for *a*.

$$a^2 = 12.0^2 + 19.9^2 - 2(12.0)(19.9) \cos 44.2°$$
$$a = \sqrt{12.0^2 + 19.9^2 - 2(12.0)(19.9) \cos 44.2°}$$
$$a = 14.057507$$

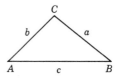

**FIGURE 8.39**

Thus, $a = BC = 14.1$ cm, to three significant digits.

Sometimes it is necessary to use the law of cosines for solving triangles in which one of the angles is an obtuse angle. Calculator sequences for finding the cosine of such an angle are the same as those used for finding the cosine of an acute angle.

### EXAMPLE 2
**(a)** cos 136° = −0.71934;   this is computed by

$$136 \boxed{\text{COS}} \longrightarrow -0.7193398$$

**(b)** cos 163°45′ = cos 163.75° = −0.96005;   this is computed by

$$163.75 \boxed{\text{COS}} \longrightarrow -0.96004986$$

Note that each of the results in Example 2 is a negative number. In fact:

> *The cosine of any angle from 90° up to and including 180° is a negative number.*

Given the cosine of an obtuse angle, calculator sequences for finding the angle are the same as those used to find an acute angle, except that the sign change key $\boxed{^{+}/_{-}}$ is used.

**EXAMPLE 3**   Find the $\angle A$ such that   cos $A$ = −0.25687.

**Solution**   To the nearest tenth,   $\angle A$ = 104.9°;   this is computed by

$$0.25687 \boxed{^{+}/_{-}} \boxed{\cos^{-1}} \longrightarrow 104.88442$$

**EXAMPLE 4**   Find side $AB$ of $\triangle ABC$ (Figure 8.40).

**Solution**   Let   $c = AB$.   Then, choose the law of cosines in the form

$$c^2 = a^2 + b^2 - 2ab \cos C$$

Substitute 19.1 for $a$, 8.90 for $b$, 119.6° (equal to 119°36′) for $\angle C$, and solve for $c$.

$$c^2 = 19.1^2 + 8.90^2 - 2(19.1)(8.90) \cos 119.6°$$
$$c = \sqrt{19.1^2 + 8.90^2 - 2(19.1)(8.90) \cos 119.6°}$$
$$c = 24.73763$$

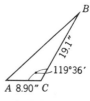

**FIGURE 8.40**

Thus,   $c = 24.7$ in.,   to three significant digits.

After the side opposite one of the angles has been found, the law of sines can be used to find a second angle. However, because the sine of an angle ($\angle A$) is equal to the sine of (180° − $\angle A$), when using the law of sines it is not always clear whether the required angle is acute or obtuse. To avoid this difficulty, we *use the law of sines to find the angle that is opposite the shorter given side*—this angle will always be an acute angle. Of course, after two of the angles are known, the third angle of the triangle can be found by subtraction from 180°.

**EXAMPLE 5**   Complete the solution of $\triangle ABC$ of Example 4. (Find $\angle A$ and $\angle B$.)

**Solution**   From Example 4,   $c = 24.7$ in.   First, find $\angle B$ because it is the angle opposite the shorter given side (8.90 in.). Use the law of sines in the form

$$\frac{b}{\sin B} = \frac{c}{\sin C}$$

$$\frac{8.90}{\sin B} = \frac{24.7}{\sin 119.6°}$$

By the cross-multiplication rule,

$$8.90 \sin 119.6° = 24.7 \sin B$$

Divide each side by 24.7.

$$\sin B = \frac{8.90 \sin 119.6°}{24.7} = 0.31329979$$

Use the $\boxed{\sin^{-1}}$ key to obtain $\quad \angle B = 18.3°.\quad$ Next,

$$\angle A = (180 - 119.6 - 18.3)° = 42.1°$$

## ■ Side-Side-Side

If the three sides of a triangle are given, the law of cosines can also be used to find one of the angles. It is best to *find the angle opposite the longest given side first.* This angle will be the *largest angle* and will be obtuse if the cosine is negative, or acute if the cosine is positive. In either case, the other two angles must be acute, and we can thus proceed to solve for the other angles of the triangle.

**EXAMPLE 6** In Figure 8.41, find the largest angle of $\triangle ABC$, given $\quad a = 13.6$ ft, $b = 23.0$ ft, and $c = 24.0$ ft.

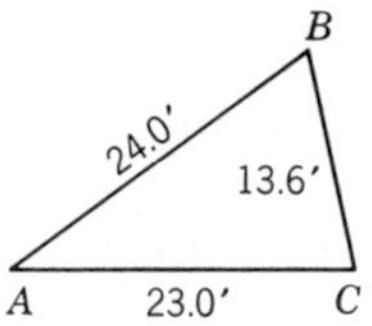

**Solution** Sketch and label a figure. $AB$ is the longest side. Hence, $\angle C$ (the angle opposite $AB$) is the largest angle. Choose the law of cosines in the form

$$c^2 = a^2 + b^2 - 2ab \cos C$$

**FIGURE 8.41**

Make the appropriate substitutions and solve for $\cos C$.

$$24.0^2 = 13.6^2 + 23.0^2 - 2(13.6)(23.0) \cos C$$
$$576 = 713.96 - 625.6 \cos C$$

Add $625.6 \cos C$ to, and subtract 576 from, each side:

$$625.6 \cos C = 137.96$$

Divide each side by 625.6.

$$\cos C = \frac{137.96}{625.6} = 0.2205243$$

Because $\cos C$ is positive, $\angle C$ is acute and we use the $\boxed{\cos^{-1}}$ key to obtain $\angle C = 77.3°$.

**EXAMPLE 7** Complete the solution of $\triangle ABC$ of Example 6. (Find $\angle A$ and $\angle B$.)

**Solution** To find $\angle A$, use the law of sines in the form

$$\frac{a}{\sin A} = \frac{c}{\sin C}$$
$$\frac{13.6}{\sin A} = \frac{24}{\sin 77.3°}$$

By the cross-multiplication rule,

$$13.6 \sin 77.3° = 24 \sin A$$

Divide each side by 24.

$$\sin A = \frac{13.6 \sin 77.3°}{24} = 0.55280292$$

Use the $\boxed{\sin^{-1}}$ key to obtain   $\angle A = 33.6°$.   Next

$$\angle B = (180 - 77.3 - 33.6)° = 69.1°$$

## ■ Applications of the Law of Cosines

The law of cosines can be used to solve many practical problems that involve oblique triangles.

**EXAMPLE 8**   A triangular roof span is to be built as shown in Figure 8.42. If $AC = 24$ ft 6 in.   and   $CB = 28$ ft 9 in., find dimension $AB$ to the nearest tenth.

**Solution**   As in Figure 8.43, sketch $\triangle ABC$ showing the given information:

24 ft 6  in. = 24.5 ft;   28 ft 9 in. = 28.75 ft;   $115°24' = 115.4°$

Because $c$ is required, use the law of cosines in the form

$$c^2 = a^2 + b^2 - 2ab \cos C$$

Make the appropriate substitutions and solve for $c$.

$$c^2 = 28.75^2 + 24.5^2 - 2(28.75)(24.5) \cos 115.4°$$
$$c = \sqrt{28.75^2 + 24.5^2 - 2(28.75)(24.5) \cos 115.4°}$$
$$c = 45.067448$$

To the nearest tenth,   $AB = 45.1$ ft.

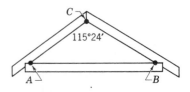

**FIGURE 8.42**

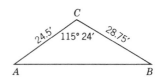

**FIGURE 8.43**

**EXAMPLE 9**   A portion of a bridge truss is shown in Figure 8.44. Find $\angle 1$ and $\angle 2$, as formed by the brace and the beams of the truss.

**Solution**   Draw a simplified figure, as shown in Figure 8.45. Use $\angle B$ for $\angle 1$ and $\angle C$ for $\angle 2$.

**1.** Note that $\angle B$ is the largest angle of $\triangle ABC$. Thus, first use the law of cosines to solve for $\angle B$.

$$b^2 = a^2 + c^2 - 2ac \cos B$$

Make the appropriate substitutions and solve for $\cos B$.

$$18.4^2 = 11.2^2 + 9.6^2 - 2(11.2)(9.6) \cos B$$
$$338.56 = 217.6 - 215.04 \cos B$$

Add 215.04 cos $B$ to, and subtract 338.56 from, each side.

$$215.04 \cos B = -120.96$$

Divide each side by 215.04.

$$\cos B = \frac{-120.96}{215.04} = -0.5625$$

Use the $\boxed{\cos^{-1}}$ key to obtain   $\angle B = \angle 1 = 124.2°$.

**2.** Next use the law of sines to find $\angle C$.

$$\frac{b}{\sin B} = \frac{c}{\sin C}$$

$$\frac{18.4}{\sin 124.2°} = \frac{9.6}{\sin C}$$

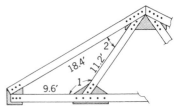

**FIGURE 8.44**

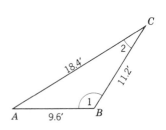

**FIGURE 8.45**

By the cross-multiplication rule,

$$18.4 \sin C = 9.6 \sin 124.2°$$

Divide each side by 18.4.

$$\sin C = \frac{9.6 \sin 124.2°}{18.4} = 0.4315203$$

Use the $\boxed{\sin^{-1}}$ key to obtain   $\angle C = \angle 2 = 25.6°.$

## Exercises for Section 8.5

*Find the cosine of each angle. Round off to five places. See Example 2.*

**1.** 162°        **2.** 165°        **3.** 126°        **4.** 99°

**5.** 161.9°      **6.** 136.9°      **7.** 146.4°      **8.** 91.5°

**9.** 132°6′      **10.** 107°27′    **11.** 96°10′     **12.** 97°12′

*Find the angle having the given cosine. See Example 3. Round off to the nearest tenth.*

**13.** −0.46920   **14.** −0.99378   **15.** −0.66092   **16.** −0.16834

**17.** 0.06004    **18.** 0.21597    **19.** −0.92532   **20.** −0.73572

**21.** 0.50501    **22.** 0.85065    **23.** −0.70925   **24.** −0.07896

*Refer to Figure 8.46 for Exercises 25 to 28. Find the length of the required side for a triangle having the given dimensions. See Examples 1 and 4. Round off to three significant digits.*

**25.** $a$ = 42.4 in.;   $b$ = 46.7 in.;   $\angle C$ = 32°;   $c$ = ?

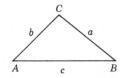

**FIGURE 8.46**

**26.** $a$ = 71.0 in.;   $b$ = 11.3 in.;   $\angle C$ = 96°;   $c$ = ?

**27.** $b = 37.8$ m;  $c = 81.3$ m;  $\angle A = 117.4°$;  $a = ?$

**28.** $b = 10.4$ m;  $c = 38.8$ m;  $\angle A = 139.4°$;  $a = ?$

*Complete the solution of the triangle in the indicated exercise. See Example 5. Round off angles to the nearest tenth of a degree, and lengths to three significant digits.*

**29.** Exercise 25       **30.** Exercise 26       **31.** Exercise 27       **32.** Exercise 28

*Solve each triangle with the given dimensions. See Examples 6 and 7. Round off results to the nearest tenth.*

**33.** $a = 5.8$ in;  $b = 3.4$ in.;      **34.** $a = 4.3$ in.;  $b = 7.5$ in.;
     $c = 7.1$ in.                      $c = 8.7$ in.

**35.** $a = 16.7$ m;  $b = 29.2$ m;     **36.** $a = 27.2$ m;  $b = 13.5$ m;
     $c = 14.7$ m                      $c = 15.9$ m

*In each of the following problems, round off angles to the nearest tenth of a degree and lengths to the smallest number of significant digits used in the given dimensions.*

**37.** Four holes are to be drilled to form a parallelogram as shown in Figure 8.47. Find the distance between holes $E$ and $C$.

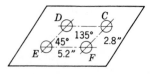

**FIGURE 8.47**

**38.** Find the distance between holes $D$ and $F$ in the plate shown for Exercise 37.

**39.** The boom of a derrick is to be supported by a brace as shown in Figure 8.48. Find the length, $b$, of the brace.

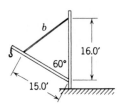

**FIGURE 8.48**

**40.** A vertical pole on a slope is braced by a guy wire as shown in Figure 8.49. Find the length of the guy wire.

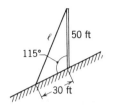

**FIGURE 8.49**

**41.** A bridge is built with beams having the lengths given in Figure 8.50.
(a) Find the measure of ∠1.
(b) Find the length, $h$, of the vertical support.

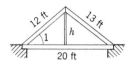

**FIGURE 8.50**

**42.** Three pulleys are mounted as shown in Figure 8.51. Find the measure of ∠A.

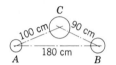

**FIGURE 8.51**

## 8.6  *Radians*

Recall that, in Section 7.1, we defined *degree*. One degree is the measure of an angle which, if one of its sides is rotated $\frac{1}{360}$ of one complete rotation about its endpoint, it will coincide with the other side (Figure 8.52*a*). Another useful unit for measuring angles can be defined in a similar manner. An angle has a measure of one **radian** if one of the sides of the angle coincides with the other after a rotation of $\frac{1}{2\pi}\left(\text{approximately } \frac{1}{6.28}\right)$ of one complete rotation about its endpoint (Figure 8.52*b*).

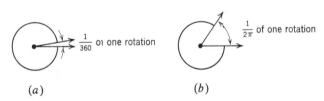

(*a*)                (*b*)

**FIGURE 8.52**

### ■ *Degree-Radian Conversions*

Because 1 radian is $\frac{1}{2\pi}$ of a complete rotation, there are $2\pi$ (approximately 6.28) radians, or 360°, in one complete rotation, and $\pi$ radians in 180°. Figure 8.53 shows the degree-radian relationship of other common angle measures, where the symbol ᴿ is used to denote radian measure.

Because  $\pi$ radians = 180°,  the following proportion can be used to convert *any* radian measure ($R$) to degrees ($D$) or *any* degree measure to radians.

$$\frac{R}{\pi} = \frac{D^\circ}{180} \tag{1}$$

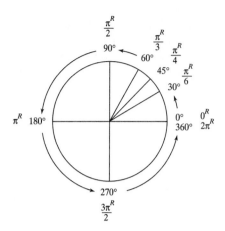

**FIGURE 8.53**

For example, we can change 1 radian to degrees by substituting 1 for $R$ and solving for $D$.

$$\frac{1^R}{\pi} = \frac{D°}{180}$$

$$180 = \pi \times D°$$

$$D° = \frac{180}{\pi} = 57.295779°$$

Thus, 1 radian = 57.3°, to the nearest tenth. It is sometimes useful to remember that 1 radian is approximately 57°.

**EXAMPLE 1**  2.3 radians = ? degrees   (to the nearest tenth).

**Solution**  In Formula 1, substitute 2.3 for $R$ and solve for $D$.

$$\frac{2.3^R}{\pi} = \frac{D°}{180}$$

$$2.3 \times 180 = \pi \times D°$$

$$D° = \frac{2.3 \times 180}{\pi} = 131.78029°$$

Thus, 2.3 radians = 131.8°, to the nearest tenth (see Figure 8.54).

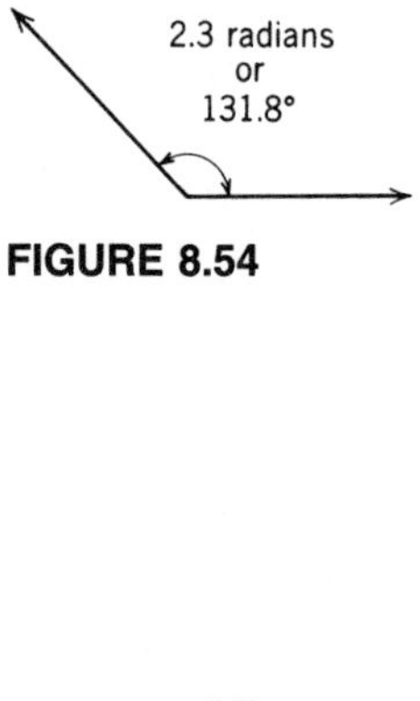

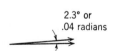

**FIGURE 8.54**

**EXAMPLE 2**  2.3° = ? radians   (to the nearest hundredth).

**Solution**  In Formula 1, substitute 2.3 for $D$ and solve for $R$.

$$\frac{R}{\pi} = \frac{2.3°}{180}$$

$$R \times 180 = \pi \times 2.3$$

$$R = \frac{\pi \times 2.3}{180} = 0.04014257$$

Thus, 2.3° = 0.04 radians,   to the nearest hundredth (see Figure 8.55).

2.3° or
.04 radians

**FIGURE 8.55**

## ■ Trigonometric Ratios Using Radians

When using trigonometric ratio notation in which *no unit of measure* is shown, such as

$$\sin 3, \quad \cos 0.02, \quad \tan 0.67$$

it is understood that the numbers refer to radian measure. A calculator must be set to radian units when such ratios are involved in calculator sequences. We shall use the phrase "radian mode" to remind you to set your calculator for this unit of measure. In general, a D-R switch (or key) is used for this purpose.

**EXAMPLE 3** Set to radian mode.

**(a)** sin 3 = 0.14112; this is computed by

$$3 \boxed{\text{SIN}} \longrightarrow 0.14111998$$

**(b)** cos 0.02 = 0.99980; this is computed by

$$0.02 \boxed{\text{COS}} \longrightarrow 0.99980001$$

**(c)** tan 0.67 = 0.79225; this is computed by

$$0.67 \boxed{\text{TAN}} \longrightarrow 0.79225417$$

Radian measure is frequently given in multiples of $\pi$. (See Figure 8.53.)

**EXAMPLE 4** Set to radian mode.

**(a)** $\sin \dfrac{\pi}{4} = 0.70711$; this is computed by

$$\boxed{\pi} \boxed{\div} 4 \boxed{=} \boxed{\text{SIN}} \longrightarrow 0.70710679$$

**(b)** $\cos \dfrac{5\pi}{6} = -0.86603$; this is computed by

$$5 \boxed{\times} \boxed{\pi} \boxed{\div} 6 \boxed{=} \boxed{\text{COS}} \longrightarrow -0.86602549$$

**(c)** $\tan \dfrac{4\pi}{3} = 1.73205$; this is computed by

$$4 \boxed{\times} \boxed{\pi} \boxed{\div} 3 \boxed{=} \boxed{\text{TAN}} \longrightarrow 1.7320508$$

Calculator sequences for finding angles in *radians* associated with given trigonometric ratios are similar to those we used for finding angles in *degrees* in Section 8.1.

**EXAMPLE 5** To the nearest hundredth of a radian, find the angle such that
**(a)** sin $A$ = 0.08493    **(b)** cos $B$ = 0.06745    **(c)** tan $C$ = 3.12309

**Solutions** Set to radian mode.

**(a)** $\angle A$ = 0.09 radian; this is computed by

$$0.08493 \boxed{\sin^{-1}} \longrightarrow 0.08503243$$

**(b)** $\angle B$ = 1.50 radians; this is computed by

$$0.06745 \boxed{\cos^{-1}} \longrightarrow 1.5032951$$

**(c)** $\angle C$ = 1.26 radians; this is computed by

$$3.12309 \boxed{\tan^{-1}} \longrightarrow 1.2609159$$

■ **Sectors: Arc Length and Area**

In Chapter 7 we introduced formulas for the length ($L$) of the arc of a sector and the area ($A$) of a sector of a circle (Figure 8.56) when the central angle is in degrees:

$$L = \frac{\pi D° r}{180} \quad \text{and} \quad A = \frac{\pi D° r^2}{360}$$

**FIGURE 8.56**

One advantage of the use of radian measure is the simplification of such formulas. In particular, for a sector with radius $r$ and central angle $\theta$* radians, the arc length formula becomes

$$L = r\theta \tag{2}$$

and the area formula becomes

$$A = \frac{r^2\theta}{2}, \tag{3}$$

where in each formula $\frac{\pi D^\circ}{180}$ has been replaced by $\theta$ in radians.

**EXAMPLE 6**  For a sector of a circle with radius 12.4 cm and central angle $\theta = 0.94$ radian,   find (*a*) the arc length, and (*b*) the area.

**Solutions**   In each case, substitute 12.4 for $r$ and 0.94 for $\theta$.

(**a**)                                   $L = r\theta$

$$L = 12.4 \times 0.94 = 11.656$$

The arc length is 11.7 cm, to the nearest tenth.

(**b**)                                   $A = \frac{r^2\theta}{2}$

$$A = \frac{(12.4)^2 \times 0.94}{2} = 72.2672$$

The area is 72.3 sq cm, to the nearest tenth.

### ■ Applications

Certain types of physical phenomena, such as an alternating electric current, the oscillation of a spring, and the motion of a pendulum, are called "periodic." The mathematical formulas used for the study of such phenomena usually involve trigo-nometric ratios using radians. However, the variables in such formulas frequently refer to such physical quantities as time and distance, rather than the measure of an angle. As a consequence, although angles may not be directly involved, calculator procedures for computations related to such formulas must be done with the calcula-tor set to radian mode.

**EXAMPLE 7**  The current $i$ in a 30-ohm resistor at a time $t$ can be computed by the formula

$$i = 12 \sin 118t$$

To the nearest tenth of an ampere, find the current in the resistor when (*a*) $t = 0.003$ sec,   and   (*b*) $t = 0.03$ sec.

**Solution**   Set to radian mode. Use the sequence 118 $\boxed{\times}$ $t$ $\boxed{=}$ $\boxed{\text{SIN}}$ $\boxed{\times}$ 12 $\boxed{=}$

(**a**)                       $i = 12 \sin (118 \times 0.003) = 4.1598306$

To the nearest tenth,   $i = 4.2$ amp.

(**b**)                       $i = 12 \sin (118 \times 0.03) = -4.655411$

To the nearest tenth,   $i = -4.7$ amp.

*The Greek letter $\theta$ (read "theta") is commonly used to represent the radian measure of an angle.

**EXAMPLE 8** At a time *t*, the position *d* of a weight that is oscillating (moving up and down) about the zero mark can be computed by

$$d = 5 \sin 96t$$

To the nearest tenth, find the position of the weight when (*a*)  *t* = 0.07 sec,  and (*b*) *t* = 0.7 sec.

**FIGURE 8.57**

**Solutions**  Set to radian mode.

**(a)**  $d = 5 \sin (96 \times 0.07) = 2.115277$

To the nearest tenth,  *d* = 2.1 cm  (the weight is 2.1 cm above the zero mark).

**(b)**  $d = 5 \sin (96 \times 0.7) = -4.7066576$

To the nearest tenth,  *d* = −4.7 cm  (the weight is 4.7 cm below the zero mark).

## Exercises for Section 8.6

*Convert each radian measure to degrees and each degree measure to radians. Round off degrees to the nearest tenth and radians to the nearest hundredth. See Examples 1 and 2.*

**1.** 1.80 radians     **2.** 2.95 radians     **3.** 1.57 radians

**4.** 1.31 radians     **5.** 4.01 radians     **6.** 3.89 radians

**7.** 97.7°     **8.** 77.5°     **9.** 29.5°

**10.** 92.4°     **11.** 197.3°     **12.** 284.9°

*Find each value to five decimal places. All angles are in radians. See Examples 3 and 4.*

**13.** sin 2.6     **14.** sin 1.5     **15.** cos 2.2

**16.** cos 0.87     **17.** tan 2.1     **18.** tan 0.12

**19.** $\tan \dfrac{7\pi}{6}$      **20.** $\tan \dfrac{3\pi}{4}$      **21.** $\sin \dfrac{\pi}{3}$

**22.** $\sin \dfrac{5\pi}{6}$      **23.** $\cos \dfrac{2\pi}{3}$      **24.** $\cos \dfrac{5\pi}{6}$

*Find $\angle A$ to the nearest hundredth of a radian. See Example 5.*

**25.** $\sin A = 0.56079$      **26.** $\sin A = 0.05431$      **27.** $\cos A = -0.41728$

**28.** $\cos A = 0.81605$      **29.** $\tan A = 3.30051$      **30.** $\tan A = 7.00024$

*To the nearest tenth, find (a) the arc length, and (b) the area of the sector described. See Example 6.*

**31.** $r = 2.87$ cm;   $\theta = 0.65$ radian      **32.** $r = 5.43$ cm;   $\theta = 1.6$ radians

**33.** $r = 35.1$ ft;   $\theta = 2.3$ radians      **34.** $r = 72.4$ ft;   $\theta = 1.3$ radians

*See Examples 7 and 8 for Exercises 35 to 40.*

**35.** The voltage $E$ in a 60-cycle-per-second electric generator, producing a maximum of 150 volts, is given by

$$E = 150 \sin 377t$$

To the nearest tenth, find the voltage for   $t = 0.007$ second.

**36.** Use the formula of Exercise 35 to find the voltage when   $t = 0.005$.

**37.** The speed $s$ at which the tip of an oscillating spring is moving is given by

$$s = 7.854 \sin 1.571t$$

where $s$ is in centimeters per second and $t$ is in seconds. To the nearest hundredth, find the speed of the tip when   $t = 1.3$ seconds.

**38.** Use the formula of Exercise 37 to find the speed when $t = 1.8$ seconds.

**39.** Under certain conditions, the displacement $d$ of a point on a vibrating spring is given by

$$d = \frac{1}{60} \sin 60t - t \cos 60t$$

where $d$ is in centimeters and $t$ is in seconds. To the nearest thousandth, find $d$ for $t = 0.4$ seconds.

**40.** Use the formula of Exercise 39 for $t = 0.8$ second.

## ■ *Review Exercises*

### Section 8.1

**1.** For each triangle in Figure 8.58, specify the length of the hypotenuse and the lengths of the side opposite and the side adjacent to each acute angle.

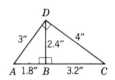

**FIGURE 8.58**

**2.** In Figure 8.59, find sin $F$, cos $F$, and tan $F$.

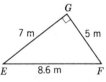

**FIGURE 8.59**

**3.** Find sin 16.4°, cos 16.4°, and tan 16.4°.

*Find the angle, correct to the nearest tenth of a degree.*

**4.** sin $A$ = 0.62820     **5.** cos $B$ = 0.55056     **6.** tan $C$ = 5.2375

### Section 8.2

*Compute. Round off to three significant digits.*

**7.** 96.1 cos 84°     **8.** $\dfrac{3.65}{\sin 55.3°}$     **9.** $\dfrac{2.49 \sin 6.5°}{1.93}$

**10.** Find $a$ if    $a = \sqrt{4.61^2 + 7.02^2 - 2(4.61)(7.02)\cos 12.8°}$.

*Find each angle to the nearest tenth of a degree.*

**11.** $\sin B = \dfrac{3.924}{4.702}$      **12.** $\cos C = \dfrac{3.45^2 + 4.53^2 - 5.12^2}{2(3.45)(4.53)}$

## Section 8.3

**13.** In Figure 8.60, find the measure of each acute angle of $\triangle ABC$ to the nearest tenth of a degree.

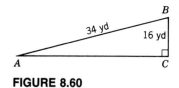

**FIGURE 8.60**

**14.** Solve right $\triangle ABC$ if    $a = 6.3$ in.    and    $b = 7.6$ in.    Round off angles to the nearest tenth and lengths to two significant digits.

## Section 8.4

**15.** $\sin 114.3° = ?$

**16.** If    $\sin A = 0.13634$    and $\angle A$ is obtuse, then    $\angle A = ?$    Round off to the nearest tenth.

**17.** Solve $\triangle ABC$ given    $a = 23.1$ in. $\angle A = 53°$,    and $\angle B = 17°$. Round off lengths to three significant digits.

**18.** In $\triangle DAB$ (Figure 8.61), find an obtuse $\angle D$ if    $d = 7.95$ m,    $\angle A =$ 20.7°,    and    $a = 5.46$ m.    Round off to the nearest tenth.

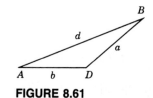

**FIGURE 8.61**

## Section 8.5

**19.** $\cos 100°20' = ?$ (To five decimal places.)

**20.** If $\cos C = -0.06585$, then $\angle C = ?$ Round off to the nearest tenth of a degree.

*In Exercises 21 and 22, round off angles to the nearest tenth of a degree and lengths to two significant digits.*

**21.** Solve $\triangle DEF$, given $\angle E = 70.9°$, $d = 6.4$ in., and $f = 4.2$ in.

**22.** Solve $\triangle ABC$, given $a = 1.4$ m, $b = 2.7$ m, and $c = 6.2$ m.

## Section 8.6

**23.** Convert:
**(a)** $1.73^R \rightarrow$ degrees (to the nearest tenth)
**(b)** $49.3° \rightarrow$ radians (to the nearest hundredth)

**24.** **(a)** Find $\cos 0.53^R$ to five decimal places.
**(b)** Find $\angle A$ such that $\tan A = 1.73214$, to the nearest hundredth of a radian.

**25.** Given that a voltage

$$E = 180 \sin 377t$$

find the voltage to the nearest tenth when $t = 0.006$.

**26.** Given that the displacement $d$ (in centimeters) of a point on a vibrating spring is given by

$$d = \frac{1}{40} \sin 40t - t \cos 40t$$

find the displacement to the nearest thousandth when $t = 0.3$ sec.

# *Graphing*

In earlier chapters we observed many examples in which pairs of related numbers were displayed in tables. Sometimes it is effective to show relationships between pairs of numbers using graphs.

## 9.1 Bar Graphs, Broken-Line Graphs, and Circle Graphs

### ■ Bar Graphs

In a **bar graph**, vertical or horizontal bars are used to represent numbers. The length of each bar is related to the number that it represents. As an example, the following table shows the number of electronic components inspected by each of six factory employees; Figure 9.1a shows a bar graph that illustrates the data of the table.

| Employee | 1 | 2 | 3 | 4 | 5 | 6 |
|---|---|---|---|---|---|---|
| Number of components | 225 | 230 | 175 | 200 | 190 | 250 |

Note that the employee numbers are entered at equally spaced intervals along the horizontal axis and that the vertical axis is scaled from 0 to 250. It is important to scale an axis of a bar graph so that the greatest number shown is greater than or equal to the greatest number in the table of values.

The data in the above table can also be illustrated by a bar graph with horizontal bars by interchanging the choice of scales for horizontal and vertical axes, as shown in Figure 9.1b.

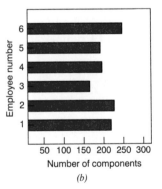

**FIGURE 9.1**

**EXAMPLE 1**    The following table shows the number of different operations needed to machine each of five engine parts from a casting. The vertical bar graph in Figure 9.2 illustrates the data.

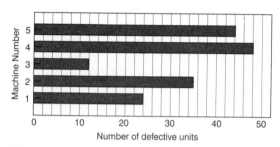

| Part Number | 1 | 2 | 3 | 4 | 5 |
|---|---|---|---|---|---|
| Number of operations | 8 | 5 | 15 | 16 | 10 |

**EXAMPLE 2**    Each of five machines in a die-stamping plant produces 3000 parts. The bar graph in Figure 9.3 shows the number of defective parts turned out by each machine. Find the percent of defective parts turned out by machine number 4.

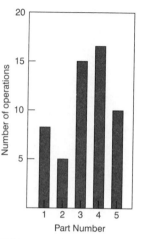

FIGURE 9.2

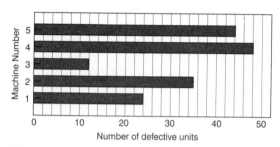

**FIGURE 9.3**

**Solution**    Note that machine number 4 turned out 48 defective parts out of 3000, and consider the question

$$48 \text{ is what percent of } 3000?$$

Let $P$ represent the required percent. Then

$$48 = P \times 3000$$

Divide each side by 3000.

$$\frac{48}{3000} = \frac{P \times 3000}{3000}$$
$$P = 0.016 = 1.6\%$$

Thus, 1.6% of the parts turned out by machine number 4 were defective.

**EXAMPLE 3**    Find the average number of defective parts turned out by the five machines of Example 2.

**Solution**    Note that the respective numbers of defective parts are: 24, 35, 12, 48, and 44. Let $A$ represent the average. Then

$$A = \frac{24 + 35 + 12 + 48 + 44}{5} = 32.6$$

The average number of defective parts is 32.6.

## ■ *Broken-Line Graphs*

In a **broken-line graph**, successive points are connected with segments. Such graphs are particularly appropriate for illustrating situations that are changing.

**EXAMPLE 4**   The following table shows the number of electric motors overhauled by a maintenance crew each month for 6 months. The broken-line graph in Figure 9.4 illustrates the data.

| Month | Jan | Feb | Mar | Apr | May | Jun |
|-------|-----|-----|-----|-----|-----|-----|
| Number of motors overhauled | 12 | 18 | 23 | 25 | 15 | 32 |

In some cases it is convenient to start the scale of an axis with a number other than zero.

**EXAMPLE 5**   To the nearest 10 units, the following table lists the number of hydraulic jacks sold by a company each year for 6 years. The broken-line graph in Figure 9.5 illustrates the data.

| Years | 1973 | 1974 | 1975 | 1976 | 1977 | 1978 |
|-------|------|------|------|------|------|------|
| Number sold | 420 | 500 | 750 | 720 | 840 | 890 |

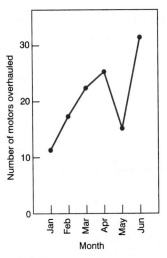

**FIGURE 9.4**

## ■ Circle Graphs

A **circle graph** is particularly useful for displaying how the parts of a quantity are related to the whole (100%) of the quantity.

**EXAMPLE 6**   A manufacturer of integrated circuits determines the following division of each manufacturing dollar into parts, expressed as a percent of each dollar spent.

| | | | |
|---|---|---|---|
| Labor: | 38.9% | Taxes: | 7.9% |
| Raw materials: | 19.8% | Fixed expenses: | 23.9% |
| Profit: | 9.5% | | |

Prepare a circle graph to display the given data.

**Solution**   Compute the corresponding percent of 360° for each part in the list.

$$0.389 \times 360° = 140.04°$$
$$0.198 \times 360° = 71.28°$$
$$0.095 \times 360° = 34.20°$$
$$0.239 \times 360° = 86.04°$$
$$0.079 \times 360° = 28.44°$$

Draw a circle and, using a protractor, measure the listed angles (rounded off to the nearest whole number) for each of the five parts. Label the resulting graph, as shown in Figure 9.6.

Bar graphs, broken-line graphs, and circle graphs are used in various technological and business fields to present data in a form that enables the viewer to readily see the overall picture of the relationship between sets of numbers. These graphs can be constructed in different ways with a variety of different choices for scales and size. For this reason, your graphs in the following exercises may differ from those in the answer section.

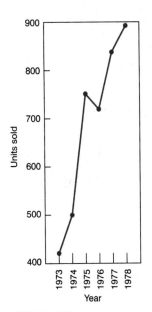

**FIGURE 9.5**

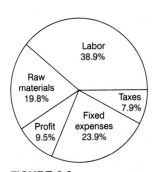

**FIGURE 9.6**

## Exercises for Section 9.1

*See Examples 1 to 3 for Exercises 1 to 14.*

*Construct a vertical bar graph for each of Exercises 1 to 4.*

**1.** The table shows the numbers of farm tractors sold in 1 year by different salespersons.

| Salesperson | 1 | 2 | 3 | 4 | 5 | 6 |
|---|---|---|---|---|---|---|
| Number sold | 32 | 18 | 34 | 12 | 21 | 40 |

To the nearest tenth:
   **(a)** Find the average yearly sales figure.
   **(b)** The number of sales made by salesperson number 6 is what percent of the total sales?

**2.** The table shows the number of persons employed at each of several branch offices of one company.

| Office | A | B | C | D | E | F |
|---|---|---|---|---|---|---|
| Number of employees | 10 | 6 | 12 | 18 | 24 | 16 |

To the nearest tenth:
   **(a)** Find the average number of people employed by a branch office.
   **(b)** What percent of the total number of employees are at office E?

**3.** The table shows the number of thousands of tons of alloy and stainless steel produced in the United States during certain years.

| Year | 1950 | 1955 | 1960 | 1965 | 1970 |
|---|---|---|---|---|---|
| Thousands of tons | 8570 | 10,660 | 8420 | 14,811 | 14,103 |

To the nearest whole number of thousands of pounds, find the average amount of metal produced over the 5 years.

**4.** The table shows the number of millions of pounds of cotton used in the United States during certain years.

| Year | 1950 | 1955 | 1960 | 1965 | 1970 |
|---|---|---|---|---|---|
| Millions of pounds | 4683 | 4382 | 4191 | 4477 | 3815 |

To the nearest whole number of millions of pounds, find the average amount of cotton used over the 5 years.

*Construct a horizontal bar graph for each of Exercises 5 to 8.*

**5.** The chart shows the average fuel consumption in miles per gallon of different automobiles.

| Automobile | 1 | 2 | 3 | 4 | 5 |
|---|---|---|---|---|---|
| Fuel consumption | 15.6 | 18.9 | 11.3 | 20.4 | 31.2 |

To the nearest tenth:
**(a)** Find the average fuel consumption.
**(b)** The fuel consumption of car 2 is what percent of the fuel consumption of car 5?

**6.** The chart shows the horsepower ratings of various automobile engines.

| Engine | A | B | C | D | E |
|---|---|---|---|---|---|
| Horsepower | 150 | 212 | 97 | 113 | 289 |

To the nearest tenth:
**(a)** Find the average horsepower rating.
**(b)** The horsepower of engine D is what percent of the horsepower of engine E?

**7.** The chart shows the value, in millions of dollars, of automobiles exported from the United States during certain years.

| Year | 1968 | 1969 | 1970 | 1971 | 1972 | 1973 |
|---|---|---|---|---|---|---|
| Value | 972 | 1010 | 822 | 1170 | 1304 | 1764 |

To the nearest whole number of millions of dollars, find the average value over the 6 years.

**8.** The chart shows the value, in millions of dollars, of automobiles imported into the United States during certain years.

| Year | 1968 | 1969 | 1970 | 1971 | 1972 | 1973 |
|---|---|---|---|---|---|---|
| Value | 2782 | 3355 | 3722 | 5085 | 7724 | 6479 |

To the nearest whole number of millions of dollars, find the average value over the 6 years.

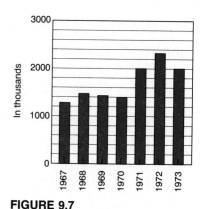

**FIGURE 9.7**

*The bar graph in Figure 9.7 shows the number of single-unit private homes started in the United States in certain years. Use the graph to answer Exercises 9 to 14.*

How many new homes were started in the following years?

**9.** 1967      **10.** 1968      **11.** 1971      **12.** 1972

**13.** Find the average number of new homes started per year in the period 1967–1970, inclusive.

**14.** Find the average number of new homes started per year in the period 1971–1973, inclusive.

*Construct a broken line graph for each of Exercises 15 to 18. See Examples 4 and 5.*

**15.** The chart shows the minimum hourly wage during various years for certain occupations in the United States.

| Year | 1975 | 1976 | 1977 | 1978 | 1979 | 1980 | 1981 |
|---|---|---|---|---|---|---|---|
| Wage | $2.10 | $2.30 | $2.30 | $2.65 | $2.90 | $3.10 | $3.35 |

**16.** The chart shows the value, in millions of dollars, of electrical apparatus imported into the United States during certain years.

| Year | 1967 | 1968 | 1969 | 1970 | 1971 | 1972 | 1973 |
|---|---|---|---|---|---|---|---|
| Value of apparatus | 133 | 168 | 196 | 247 | 263 | 356 | 458 |

**17.** The chart shows the number of units sold by a manufacturing company in each of several months.

| Month | Jan. | Feb. | Mar. | Apr. | May | June |
|---|---|---|---|---|---|---|
| Units sold | 80 | 65 | 65 | 75 | 91 | 106 |

**18.** The chart shows the average monthly temperatures in one city during several months.

| Month | July | Aug. | Sept. | Oct. | Nov. | Dec. |
|---|---|---|---|---|---|---|
| Aver. temp. (C°) | 18.9 | 17.2 | 12.2 | 5.6 | −5.6 | −13.9 |

*The broken-line graph in Figure 9.8 shows the value, in millions of dollars, of new houses started in California. Use the graph to answer Exercises 19 to 22.*

**19.** What was the value of new houses started in 1969?

**20.** What was the value of new houses started in 1971?

**21.** Between which two years was there the greatest increase in value?

**22.** Between which two years was there the least increase in value?

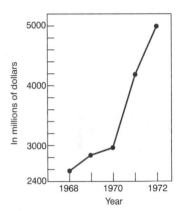

**FIGURE 9.8**

*Draw a circle graph for each of Exercises 23 to 26. State, to the nearest whole number, how many degrees are used for each item listed. See Example 6.*

**23.** In producing a bracket, the percent of the total time for each operation is:

| | |
|---|---|
| Milling | 36.3% |
| Drilling | 37.4% |
| Grinding | 18.5% |
| Bench work | 7.8% |

**24.** The percent of the total cost for each material required to build a work and storage building for a brake repair company was:

| | |
|---|---|
| Lumber | 28.6% |
| Concrete | 17.4% |
| Electrical | 16.9% |
| Plumbing | 15.1% |
| Roofing | 11.4% |
| Miscellaneous | 10.6% |

**25.** In one year in the United States, 4593 million barrels of crude oil were refined into the following products (in millions of barrels). (Hint: First calculate what percent of 4593 each amount is.)

|              |      |
|--------------|------|
| Gasoline     | 2399 |
| Fuel oil     | 1383 |
| Jet fuel     | 314  |
| Asphalt      | 168  |
| Liquefied gas| 128  |
| Miscellaneous| 201  |

**26.** In one year in the United States, shipments of a total of 3032 million pounds of castings were broken down as follows (in millions of pounds):

|           |      |
|-----------|------|
| Copper    | 751  |
| Aluminum  | 1506 |
| Zinc      | 696  |
| Magnesium | 34   |
| Lead      | 45   |

## 9.2  *Rectangular Coordinate System*

In Section 9.1 we constructed bar graphs and line graphs from tables with positive numbers only. In this section we construct graphs that also include *negative numbers*.

### ■ *Graphing Pairs of Numbers*

The construction of two reference lines, a **vertical axis** and a **horizontal axis**, intersecting at a point called the **origin**, as shown in Figure 9.9, makes it possible to associate a particular point with a given pair of numbers. The point is called the **graph** of the pair of numbers, and the numbers are called the **coordinates** of the point.

On the horizontal axis, positive numbers are located *to the right* of the origin, and negative numbers are located *to the left* of the origin. The arrowhead on the right side of the axis indicates that numbers (coordinates) increase in this direction. On the vertical axis, positive numbers are located *above* the origin and negative numbers are located *below* the origin. The arrowhead on the upper end of the axis indicates that numbers (coordinates) increase in this direction.

If $x$ and $y$ are the unknowns, it is standard practice to label the horizontal axis as the $x$-axis and the vertical axis as the $y$-axis, as in Figure 9.10. If other letters are used as unknowns, there is no standard rule—in such cases we shall specify a choice of axes.

We can graph a point with coordinates $x$ and $y$ in the following manner:

**1.** Count $x$ units from the origin—*to the right* if $x$ is positive and *to the left* if $x$ is negative.

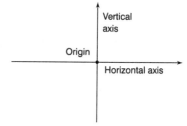

**FIGURE 9.9**

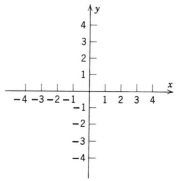

**FIGURE 9.10**

**2.** From the location reached on the *x*-axis, count *y* units *up* if *y* is positive and *down* if *y* is negative.

For example, to graph the pair of numbers  *x* = 2  and  *y* = −3,  start at the origin and count 2 units to the right. Then count 3 units down and mark the point, as in Figure 9.11. To graph the pair of numbers *x* = −3 and *y* = 2, start at the origin and count 3 units to the left, and then count 2 units up.

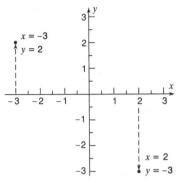

**FIGURE 9.11**

**EXAMPLE 1**  Graph each pair of numbers.

**(a)** *x* = 2,  *y* = 8            **(b)** *x* = −9,  *y* = 2

**(c)** *x* = −8,  *y* = −3         **(d)** *x* = 7,  *y* = −5

**Solutions**  The graph of each pair of numbers is shown in Figure 9.12.

The coordinates of a point are sometimes shown by (*x*, *y*). This symbol is called an **ordered pair**. It is understood that the first number is the *x*-coordinate and the second number is the *y*-coordinate. The coordinates in Example 1 could be shown as the ordered pairs (2, 8), (−9, 2), (−8, −3), and (7, −5).

In Section 9.1 we connected points with a straight line. Sometimes it is preferable to use a smooth curve to connect points on a graph.

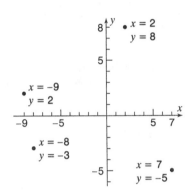

**EXAMPLE 2**  Graph the table of values, and connect the points with a smooth curve.

| *x* | −4 | −3 | −2 | 0 | 2 | 4 | 5 |
|-----|----|----|----|----|----|----|----|
| *y* | −4 | −1 | 1 | 2.5 | 3 | 5 | 8 |

**FIGURE 9.12**

Use one space equal to 1 unit on each axis.

**Solution**
**1.** Draw and label a horizontal *x*-axis and a vertical *y*-axis, as in part (*a*) of Figure 9.13, and indicate the scale on each axis.
**2.** Graph each pair of related numbers from the table, as in part (*b*).
**3.** Draw a smooth curve through the resulting points, as in part (*c*).

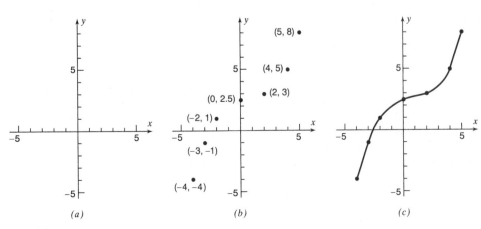

(*a*)                    (*b*)                    (*c*)

**FIGURE 9.13**

## ■ *Reading Values from Graphs*

It is often necessary to be able to read the value of one coordinate of a point on a graph when the other coordinate is given. As we noted in Section 9.1, in most cases, *numbers that are read from graphs are estimates only*, and therefore are not exact.

**EXAMPLE 3**   From the graph shown in Figure 9.14,
**(a)** Estimate the $y$ value paired with the $x$ value, $-4$.
**(b)** Estimate the $y$ value paired with the $x$ value, 4.
**(c)** Estimate the $y$ value paired with the $x$ value, $-9.5$.
**(d)** Estimate the $x$ value(s) paired with the $y$ value, zero.

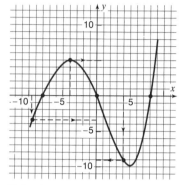

### *Solutions*
**(a)** From the point on the graph with $x$ value $-4$, estimate 5 as the $y$ value.
**(b)** From the point on the graph with $x$ value 4, estimate $-9$ as the $y$ value.
**(c)** From the point on the graph with (estimated) $x$ value $-9.5$, estimate $-3.5$ as the $y$ value.
**(d)** From the points on the graph with $y$ value zero, estimate $-8$, 0, and 8.2 as the $x$ values.

**FIGURE 9.14**

When drawing a graph we usually use the same scale on both axes. However, it is sometimes more convenient to use different scales on each of the axes.

## EXAMPLE 4
**(a)** Graph the following table, and connect the points with a smooth curve. Use a horizontal $t$-axis with one space equal to 5 units and a vertical $s$-axis with one space equal to 10 units.

| $t$ | $-20$ | $-10$ | 0 | 10 | 20 | 30 | 40 | 50 |
|---|---|---|---|---|---|---|---|---|
| $s$ | $-18$ | $-30$ | $-38$ | $-30$ | 0 | 58 | 84 | 92 |

**(b)** Estimate (to the nearest unit) the missing values in the following table:

| $t$ | $-15$ | 35 | ? | ? |
|---|---|---|---|---|
| $s$ | ? | ? | $-15$ | 35 |

### *Solutions*
**(a)** Draw and label a horizontal $t$-axis and a vertical $s$-axis, and scale each axis as shown in Figure 9.15. Graph each related pair of numbers from the table, and draw a smooth curve through the resulting points.
**(b)** Estimate the missing values from the graph and enter them in the table.

| $t$ | $-15$ | 35 | 17 | 26 |
|---|---|---|---|---|
| $s$ | $-25$ | 74 | $-15$ | 35 |

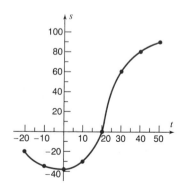

**FIGURE 9.15**

Some tables include only positive numbers. In such cases we need only show those parts of the axes where positive numbers are listed.

**EXAMPLE 5**   Graph the following table and connect the points with a smooth curve.

| $I$ | 15 | 20 | 25 | 30 | 35 | 40 | 45 |
|---|---|---|---|---|---|---|---|
| $R$ | 6.7 | 3.8 | 2.4 | 1.7 | 1.2 | 0.9 | 0.7 |

Use a horizontal *I*-axis with one space equal to 5 units and a vertical *R*-axis with two spaces equal to 1 unit.

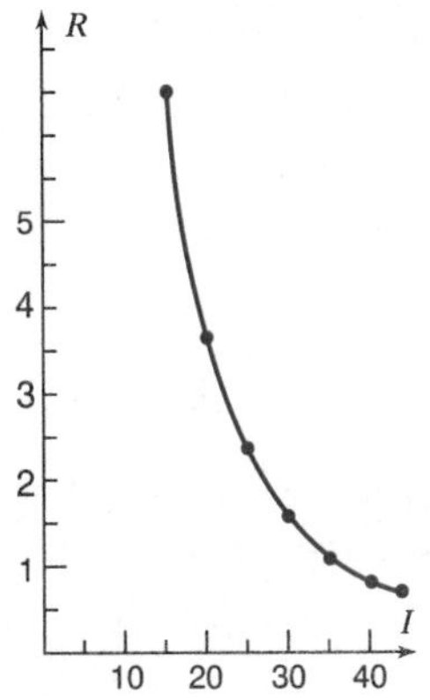

**Solution** Draw and label a horizontal *I*-axis from the origin to the right only; draw and label a vertical *R*-axis from the origin and up only.

Scale each axis as shown in Figure 9.16. Graph each pair of related numbers, and draw a smooth curve through the resulting points.

It may sometimes be necessary to "compress" the scale on one or both axes so that the resulting graph will be a reasonable size.

**EXAMPLE 6** Graph the following table, and connect the points with a smooth curve. Use a horizontal *t*-axis with two spaces equal to 5 units and a vertical *v*-axis with two spaces equal to 0.10 unit.

| $t$ | 0 | 10 | 20 | 30 | 40 |
|---|---|---|---|---|---|
| $v$ | 50.00 | 50.22 | 50.38 | 50.45 | 50.42 |

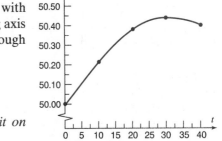

**FIGURE 9.16**

**Solution** Draw and label a horizontal *t*-axis as shown in Figure 9.17. It is not convenient to scale a vertical *v*-axis starting with zero. Hence, start the *v*-axis with a "wavy" line to indicate a "compression" of the axis, and scale the remaining axis as shown. Graph each pair of related numbers, and draw a smooth curve through the resulting points.

**FIGURE 9.17**

## Exercises for Section 9.2

*Graph each pair of numbers. See Example 1. Use one space equal to 1 unit on each axis.*

1. **(a)** $x = 1$, $y = 3$    **(b)** $x = -1$, $y = 0$
   **(c)** $x = -2$, $y = -3$    **(d)** $x = 0$, $y = -1$

2. **(a)** $x = 2$, $y = -2$    **(b)** $x = -1$, $y = 0$
   **(c)** $x = 0$, $y = 1$    **(d)** $x = 3$, $y = 4$

3. **(a)** $x = -4$, $y = -3$    **(b)** $x = 5$, $y = -2$
   **(c)** $x = 0$, $y = 5$    **(d)** $x = 2$, $y = 0$

4. (a) $x = 3, \quad y = 3$      (b) $x = -3, \quad y = -3$
   (c) $x = -3, \quad y = 3$      (d) $x = 3, \quad y = -3$

*Graph each table of values and connect the points with a smooth curve (or straight line). See Example 2. Use one space equal to 1 unit on each side.*

5.

| $x$ | -2 | -1 | 0 | 1 | 2 | 3 |
|---|---|---|---|---|---|---|
| $y$ | -3 | -1 | 1 | 3 | 5 | 7 |

6.

| $x$ | -3 | -2 | -1 | 0 | 1 | 2 |
|---|---|---|---|---|---|---|
| $y$ | 2 | 1 | 0 | -1 | -2 | -3 |

7.

| $x$ | -3 | -2 | -1 | 0 | 1 | 2 | 3 |
|---|---|---|---|---|---|---|---|
| $y$ | 3 | -2 | -5 | -6 | -5 | -2 | 3 |

8.

| $x$ | -3 | -2 | -1 | 0 | 1 | 2 | 3 |
|---|---|---|---|---|---|---|---|
| $y$ | -1 | 4 | 7 | 8 | 7 | 4 | -1 |

9.

| $x$ | -3 | -2 | -1 | 0 | 1 | 2 | 3 |
|---|---|---|---|---|---|---|---|
| $y$ | -9 | -4 | -1 | 0 | 1 | 4 | 9 |

**10.**

| x | -3 | -2 | -1 | 0 | 1 | 2 | 3 |
|---|----|----|----|---|---|---|---|
| y | -2 | 1 | 2 | 0 | -2 | -1 | 2 |

*From the graph in Figure 9.18, estimate the y value paired with each given x value and the x value paired with each given y value. Estimate to the nearest tenth. See Example 3.*

**11.** $x = -4$     **12.** $x = -1$     **13.** $x = 5$     **14.** $x = 4$

**15.** $y = 4$     **16.** $y = 5$     **17.** $y = -2$     **18.** $y = -3$

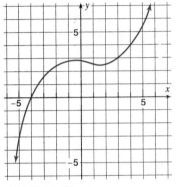

**FIGURE 9.18**

*In Exercises 19 to 22, (a) graph the entries in the first table, and connect the points with a smooth curve; (b) use the graph to estimate (to the nearest integer) the missing entries in the second table. See Example 4 for Exercises 19 and 20.*

**19. (a)**

| t | -10 | -5 | 0 | 5 | 10 | 15 | 20 |
|---|-----|----|---|---|----|----|----|
| s | 20 | 5 | 0 | 5 | 20 | 45 | 80 |

*t*-axis horizontal; one space = 5
*s*-axis vertical; one space = 10

**(b)**

| t | -7 | 12 | ? | ? |
|---|----|----|---|---|
| s | ? | ? | 30 | 65 |

**20. (a)**

| P | -20 | -15 | -10 | -5 | 0 | 5 | 10 |
|---|-----|-----|-----|----|---|---|----|
| I | 0 | 56 | 79 | 97 | 112 | 125 | 137 |

*P*-axis horizontal; one space = 5
*I*-axis vertical; one space = 10

**(b)**

| P | -12 | 3 | ? | ? |
|---|-----|---|---|---|
| I | ? | ? | 82 | 130 |

*See Example 5 for Exercises 21 and 22.*

**21. (a)**

| P | 10 | 20 | 30 | 40 | 50 | 60 |
|---|----|----|----|----|----|----|
| V | 60 | 30 | 20 | 15 | 12 | 10 |

*P*-axis horizontal; one space = 10
*V*-axis vertical; one space = 5

**(b)**

| P | 18 | 43 | ? | ? |
|---|----|----|----|----|
| V | ? | ? | 27 | 54 |

**22. (a)**

| d | 10 | 20 | 30 | 40 | 50 | 60 |
|---|----|----|----|----|----|----|
| Z | 35 | 118 | 167 | 200 | 228 | 250 |

*d*-axis horizontal; one space = 5
*Z*-axis vertical; one space = 25

**(b)**

| d | 25 | 57 | ? | ? |
|---|----|----|----|----|
| Z | ? | ? | 80 | 130 |

*Graph each table, and connect the points with a smooth curve. Compress the vertical axis as in Example 6.*

**23.**

| t | 10 | 15 | 20 | 25 | 30 | 35 | 40 |
|---|----|----|----|----|----|----|----|
| T | 70.20 | 70.40 | 70.51 | 70.53 | 70.49 | 70.35 | 70.00 |

*t*-axis horizontal; one space = 5
*T*-axis vertical; one space = 0.1

**24.**

| i | 2.0 | 2.5 | 3.0 | 3.5 | 4.0 | 4.5 | 5.0 |
|---|-----|-----|-----|-----|-----|-----|-----|
| r | 46.81 | 46.60 | 46.51 | 46.49 | 46.52 | 46.65 | 47.00 |

*i*-axis horizontal; one unit = 0.5
*r*-axis vertical; one unit = 0.1

*In Exercises 25 and 26, compress both axes.*

**25.**

| p | 50.5 | 51.0 | 51.5 | 52.0 | 52.5 | 53.0 | 53.5 |
|---|------|------|------|------|------|------|------|
| v | 38.78 | 38.49 | 38.24 | 38.05 | 37.90 | 37.80 | 37.75 |

*p*-axis horizontal; one space = 0.5
*v*-axis vertical; one space = 0.1

**26.**

| $d$ | 20.0 | 20.5 | 21.0 | 21.5 | 22.0 | 22.5 | 23.0 |
|---|---|---|---|---|---|---|---|
| $S$ | 42.63 | 42.93 | 43.18 | 43.38 | 43.52 | 43.61 | 43.65 |

$d$-axis horizontal; one space $= 0.5$
$S$-axis vertical; one space $= 0.1$

## 9.3 Preparing Tables from Equations

In Section 4.1 we considered equations that involved *one unknown*. Recall that if the unknown is replaced by a number such that the equation becomes a true statement, then the number is a *solution* of the equation.

In this section we consider *solutions* of equations that include *two unknowns*. For example, the distance $d$ (in miles) and the time $t$ (in hours) of an automobile traveling at a constant rate of 55 miles per hour are related by the equation

$$d = 55t \tag{1}$$

For a given equation in two unknowns, such as Equation 1, we can prepare a table that pairs the values of the two unknowns. For example, we can prepare a table in which values of $d$ are paired with values of $t$ equal to 0, 1, 2, 3, 4, and 5, by first substituting these values of $t$ in Equation 1 and computing the corresponding values of $d$:

$$d = 55 \cdot 0 = 0 \qquad d = 55 \cdot 3 = 165$$
$$d = 55 \cdot 1 = 55 \qquad d = 55 \cdot 4 = 220$$
$$d = 55 \cdot 2 = 110 \qquad d = 55 \cdot 5 = 275$$

Now, we prepare a table of values, as follows:

| $t$ | 0 | 1 | 2 | 3 | 4 | 5 |
|---|---|---|---|---|---|---|
| $d$ | 0 | 55 | 110 | 165 | 220 | 275 |

If we substitute any pair of related numbers from this table for $t$ and for $d$ in Equation 1, we obtain a true statement, and we say that the numbers *satisfy* the equation. The pair of numbers is a **solution** of the equation. Thus the table of values is actually a list of some of the solutions of the equation.

**EXAMPLE 1**   Screws with diameter less than 0.25 in. are called *machine screws* (Figure 9.19). The size of a machine screw is indicated by assigning to it a number, $N$, from 0 to 12. The outside diameter $D$ (in inches) of a machine screw is related to $N$ by the equation

$$D = 0.013N + 0.060$$

Prepare a table of values of outside diameters $D$ for $N$ equal to 2, 4, 6, 8, 10, and 12.

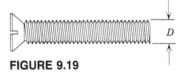

**FIGURE 9.19**

**Solution**   Substitute values for $N$ in the given equation, and compute each related $D$ value.

$$D = 0.013 \times 2 + 0.060 = 0.086 \qquad D = 0.013 \times 8 + 0.060 = 0.164$$
$$D = 0.013 \times 4 + 0.060 = 0.112 \qquad D = 0.013 \times 10 + 0.060 = 0.190$$
$$D = 0.013 \times 6 + 0.060 = 0.138 \qquad D = 0.013 \times 12 + 0.060 = 0.216$$

Prepare a table of values.

| N | 2 | 4 | 6 | 8 | 10 | 12 |
|---|---|---|---|---|----|----|
| D | 0.086 | 0.112 | 0.138 | 0.164 | 0.190 | 0.216 |

The computations needed to prepare a table sometimes include several operations. In such cases, it may be helpful to write a calculator sequence to use as a guide.

**EXAMPLE 2**  If a narrow steel bar is uniformly heated, its length $L$ changes according to the formula

$$L = 200(1 + 0.000013t)$$

where $L$ is the length (in centimeters) after heating, $t$ is the temperature in degrees Celsius, and the length of the bar at $0°$ C is 200 cm. Prepare a table of values of the length of the bar from $10°$ C to $50°$ C, using intervals of $10°$.

**Solution**  Substitute 10, 20, 30, 40, and 50 for $t$ in the given equation, and compute each related $L$ value. A suggested calculator sequence is

$$0.000013 \boxed{\times} t \boxed{+} 1 \boxed{\times} 200 \boxed{=}$$

Thus,

$$L = 200(1 + 0.000013 \times 10) = 200.026$$
$$L = 200(1 + 0.000013 \times 20) = 200.052$$
$$L = 200(1 + 0.000013 \times 30) = 200.078$$
$$L = 200(1 + 0.000013 \times 40) = 200.104$$
$$L = 200(1 + 0.000013 \times 50) = 200.130$$

Prepare a table of values.

| t | $10°$ | $20°$ | $30°$ | $40°$ | $50°$ |
|---|-------|-------|-------|-------|-------|
| L | 200.026 | 200.052 | 200.078 | 200.104 | 200.130 |

In the following examples we will no longer show individual computations.

Tables of values can also be prepared when negative as well as positive numbers are substituted for one of the unknowns.

**EXAMPLE 3**  Prepare a table of values for the equation

$$R = 7Q - 5$$

from  $Q = -2.0$  to  $Q = 2.0$,  using intervals of 0.5 unit.

**Solution**  Substitute $-2.0$, $-1.5$, $-1.0$, $-0.5$, 0, 0.5, 1.0, 1.5, and 2.0 for $Q$ in the given equation. A suggested calculator sequence for negative $Q$ values is

$$7 \boxed{\times} Q \boxed{+/_-} \boxed{-} 5 \boxed{=}$$

Prepare a table of values.

| Q | −2.0 | −1.5 | −1.0 | −0.5 | 0 | 0.5 | 1.0 | 1.5 | 2.0 |
|---|------|------|------|------|---|-----|-----|-----|-----|
| R | −19.0 | −15.5 | −12.0 | −8.5 | −5.0 | −1.5 | 2.0 | 5.5 | 9.0 |

**EXAMPLE 4**  Prepare a table of values for the equation

$$s = t^2 - 4$$

from $t = -1.0$ to $t = 3.0$, using intervals of 0.5 unit. Round off $s$ values to the nearest tenth.

**Solution** We can use either of the sequences

$$t \boxed{\times} t \boxed{-} 4 \boxed{=} \quad \text{or} \quad t \boxed{x^2} \boxed{-} 4 \boxed{=}$$

| $t$ | −1.0 | −0.5 | 0 | 0.5 | 1.0 | 1.5 | 2.0 | 2.5 | 3.0 |
|---|---|---|---|---|---|---|---|---|---|
| $s$ | −3.0 | −3.8 | −4.0 | −3.8 | −3.0 | −1.8 | 0 | 2.3 | 5.0 |

For computations involving a third (or greater) power, we can use the $\boxed{y^x}$ key. Since some calculators will not accept a negative base when using the $\boxed{y^x}$ key, we must use the rule introduced in Section 3.3 to raise a negative base to a power. For example, for $y$ values for an equation such as

$$y = \tfrac{1}{2}x^3$$

when $x$ is a negative number, say $-2.5$, we may use the sequence

$$2.5 \boxed{y^x} \boxed{3} \boxed{=} \boxed{\div} 2 \boxed{=} \longrightarrow 7.8125$$

and then precede 7.8125 by a minus sign because 3 is an odd number. That is,

$$y = \tfrac{1}{2}(-2.5)^3 = -7.8125$$

**EXAMPLE 5** Prepare a table of values for the equation

$$y = \tfrac{1}{2}x^3$$

from $x = -2.0$ to $x = 2.0$, using intervals of 0.5 unit. Round off $y$ values to the nearest tenth.

**Solution**

| $x$ | −2.0 | −1.5 | −1.0 | −0.5 | 0 | 0.5 | 1.0 | 1.5 | 2.0 |
|---|---|---|---|---|---|---|---|---|---|
| $y$ | −4.0 | −1.7 | −0.5 | −0.1 | 0 | 0.1 | 0.5 | 1.7 | 4.0 |

## ■ Equations Involving Square Root

The $\boxed{\sqrt{\phantom{x}}}$ key introduced in Section 1.4 can be used to make computations when preparing tables for equations that involve the square root symbol.

**EXAMPLE 6** Prepare a table of values for the equation

$$y = \sqrt{x - 2.5}$$

From $x = 2.5$ to $x = 6.0$, using intervals of 0.5 unit.

**Solution** Use the calculator sequence

$$x \boxed{-} 2.5 \boxed{=} \boxed{\sqrt{\phantom{x}}}$$

and round off results to the nearest tenth:

| $x$ | 2.5 | 3.0 | 3.5 | 4.0 | 4.5 | 5.0 | 5.5 | 6.0 |
|---|---|---|---|---|---|---|---|---|
| $y$ | 0 | 0.7 | 1.0 | 1.2 | 1.4 | 1.6 | 1.7 | 1.9 |

## Exercises for Section 9.3

*Prepare a table of values for each formula for the indicated values of one unknown.*

*See Examples 1 and 2 for Exercises 1 to 6.*

1. If an object is thrown downward with a speed of 20 ft per second, its speed $s$, after $t$ seconds, is given by

$$s = 32t + 20$$

where $s$ is measured in feet per second. Use $t = 0, 5, 10, 15, 20, 25$.

2. The length $L$ of a particular steel beam (Figure 9.20) changes with $w$, the load it supports, according to the formula

$$L = 50\left(\frac{w}{30,000} + 1\right)$$

where $w$ is in tons and $L$ is in feet. Use $w = 15, 20, 25, 30, 35$, and 40. Round off to the nearest hundredth of a foot.

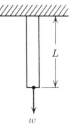

**FIGURE 9.20**

3. The number of feet, $L$, of $\frac{1}{2}$-inch rope that can be wound on the drum shown in Figure 9.21 is given by

$$L = 31.5(F + 24)$$

where the flange width, $F$, is in inches. Use $F = 6, 8, 10, 12, 14, 16$.

**FIGURE 9.21**

4. A certain dovetail machined in a block of steel is checked by using a round pin, as shown in Figure 9.22. The dimension $x$ is given by

$$x = 1.366d + 1$$

Use $d = 0.250, 0.3125, 0.375, 0.4375, 0.500, 0.625$. Round off to the nearest thousandth.

**FIGURE 9.22**

5. The pitch diameter $D$ of a gear having an outside diameter of 8 in. is given by

$$D = \frac{8N}{N + 2}$$

where $N$ is the number of teeth in the gear. Use $N = 50, 54, 58, 62, 66,$

and 70.   Round off to the nearest thousandth.

**6.** If an electric hand drill is connected in parallel with a second appliance with a resistance of $R_2$ ohms, the total resistance $R$ is given by

$$R = \frac{600R_2}{R_2 + 600}$$

Use   $R_2 = 100, 200, 300, 400, 500, 600.$   Round off to 10 ohms.

*For Exercises 7 to 26, prepare a table of values for each equation using the indicated interval (or using intervals of 1 unit).*

*See Example 3 for Exercises 7 to 12.*

**7.** $P = \dfrac{120}{V}$;   from   $V = 5$   to   $V = 40$,

using intervals of 5 units. Round off to the nearest tenth.

**8.** $s = \dfrac{3600}{t}$;   from   $t = 20$   to   $t = 60$,

using intervals of 10 units. Round off to the nearest whole number.

**9.** $P = \dfrac{-12}{D}$;   from   $D = 3$   to   $D = 12$.

Round off to the nearest tenth.

**10.** $r = \dfrac{-20}{t}$;   from   $t = 1$   to   $t = 10$.

Round off to the nearest tenth.

**11.** $B = \dfrac{12}{n + 5}$;   from   $n = -3$   to   $n = 7$.   Round off to the nearest hundredth.

**12.** $C = \dfrac{12}{d + 6} + 2$;   from   $d = -4$   to   $d = 4$.   Round off to the nearest hundredth.

*See Example 4 for Exercises 13 to 16.*

**13.** $P = 40I^2$;   from   $I = 10$   to   $I = 70$, using intervals of 10 units.

**14.** $A = 0.7854d^2$;   from   $d = 0.5$ to   $d = 3.5$,   using intervals of 0.5 unit. Round off to the nearest thousandth.

**15.** $R = \dfrac{16.888}{d^2}$;   from   $d = 2$   to $d = 7$,   using intervals of 0.5 unit. Round off to the nearest thousandth.

**16.** $m = \dfrac{B^2}{B^2 + 20.4}$;   from   $B = -15$ to   $B = 15$,   using intervals of 5 units. Round off to the nearest hundredth.

*See Example 5 for Exercises 17 to 22.*

**17.** $W = 2.879s^3$;   from   $s = -2$ to   $s = 2$,   using intervals of 0.5 unit. Round off to the nearest hundredth.

**18.** $V = \dfrac{4}{3}\pi r^3$;   from   $r = 2$   to   $r = 3$, using intervals of 0.2 unit. Round off to the nearest tenth.

**19.** $C = \dfrac{500}{s^3}$;   from   $s = 4$   to   $s = 8$, using intervals of 0.5 unit. Round off to the nearest tenth.

**20.** $y = \dfrac{10x}{x^3 + 3}$;   from   $x = -1$   to $x = 3$,   using intervals of 0.5 unit. Round off to the nearest hundredth.

**21.** $y = 2.513x^4$;   from   $x = -2$   to $x = 2$,   using intervals of 0.5 unit. Round off to the nearest tenth.

22. $y = 106.5x^5$; from $x = -1$ to
$x = 1$, using intervals of 0.2 unit.
Round off to the nearest hundredth.

*See Example 6 for Exercises 23 to 26.*

23. $d = \sqrt{1.273A}$; from $A = 1$
to $A = 4$, using intervals of 0.5 unit.
Round off to the nearest tenth.

24. $r = \sqrt{4.062S}$; from $S = 0.25$
to $S = 2.00$, using intervals of 0.25
unit. Round off to the nearest tenth.

25. $y = \sqrt{16 - x^2}$; from $x = -4$
to $x = 4$, using unit intervals.
Round off to the nearest tenth.

26. $y = \sqrt{x^2 - 2x}$; from $x = -2$
to $x = 0$, using intervals of 0.5 unit.
Round off to the nearest tenth.

27. For a particular paint, the number
of gallons $N$ required to cover a
spherical storage tank of radius $r$ is
given by

$$N = \frac{\pi r^2}{50}$$

Prepare a table showing, to the
nearest gallon, the number of
gallons required for tanks whose
diameters are 15 ft, 20 ft, 25 ft, 30
ft, 35 ft, 40 ft, and 45 ft.

28. The volume $V$ of the silo shown in
Figure 9.23, which is a hemisphere
on top of a cylinder, is given by

$$V = \frac{2\pi r^3}{3} + 50\pi r^2$$

Prepare a table showing, to the
nearest cubic meter, the volume of
silos whose diameters are 8 m, 10
m, 12 m, 14 m, 16 m, and 18 m.

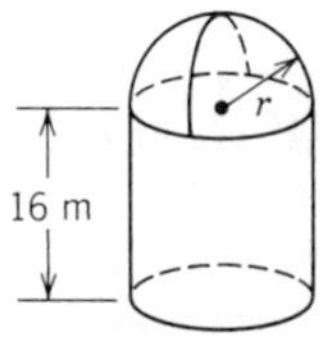

16 m

**FIGURE 9.23**

## 9.4  Graphing Linear Equations

In Section 9.2 we graphed tables of values that were not specifically related to equations. In this section we prepare tables from equations and then graph the resulting tables.

### ▪ Straight Line Graphs

Consider the equation

$$y = 2x + 3 \tag{1}$$

We can prepare the following table, showing solutions for values of $x$ from $-4$ to $4$.

| $x$ | $-4$ | $-3$ | $-2$ | $-1$ | 0 | 1 | 2 | 3 | 4 |
|---|---|---|---|---|---|---|---|---|---|
| $y$ | $-5$ | $-3$ | $-1$ | 1 | 3 | 5 | 7 | 9 | 11 |

Figure 9.24*a* shows the point that corresponds to each pair of related numbers in the table. Now, if we place a straightedge on these points, it appears that all the points are on one straight line, as in Figure 9.24*b*. In fact, that is the case. Equations such as Equation 1, for which the graphs are straight lines, are called *linear equations*.

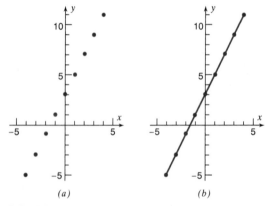

(a)                              (b)

**FIGURE 9.24**

**EXAMPLE 1**  Graph the linear equation  $y = 2x - 3$  from  $x = -2$  to $x = 4$,  using intervals of 1 unit.

**Solution**  Prepare a table of values.

| $x$ | $-2$ | $-1$ | 0 | 1 | 2 | 3 | 4 |
|---|---|---|---|---|---|---|---|
| $y$ | $-7$ | $-5$ | $-3$ | $-1$ | 1 | 3 | 5 |

Draw and label an $x$-axis and a $y$-axis. Graph the pairs of related numbers from the table, and draw a line through the resulting points, as shown in Figure 9.25.

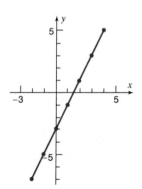

**FIGURE 9.25**

*In this section we consider only linear equations.* Thus, all graphs in this section will be straight lines. When we know that the graph of an equation is a straight line, it is only necessary to graph two pairs of related numbers that satisfy the equation in order to draw the line. However, it is good practice to graph three pairs of numbers, and if the points obtained are not all on the same line, the computations should be rechecked.

---

**To Graph a Linear Equation**
1. Compute a table of values, showing three pairs of numbers that satisfy the equation. (When possible, choose zero as one of the numbers.)
2. Draw and label a horizontal axis and a vertical axis.
3. Graph the three pairs of related numbers from the table.
4. Draw a straight line through the resulting points.

---

**EXAMPLE 2**  Graph the equation  $y = 3 - 2x$.  Use  $x = -3$,  $x = 0$,  and  $x = 4$  for the three points.

**Solution**
1. Prepare a table of values.

| $x$ | −3 | 0 | 4 |
|---|---|---|---|
| $y$ | 9 | 3 | −5 |

2. Draw and label an $x$-axis and a $y$-axis. Graph each of the three pairs of related numbers from the table.
3. Draw a line through the three resulting points, as shown in Figure 9.26.

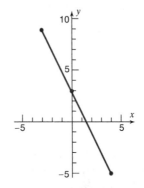

**FIGURE 9.26**

Although we use only three solutions of the equation to construct a straight-line graph, it is a fact that *every point* on the graph corresponds to a solution of the equation, and *every solution* of the equation corresponds to a point on the graph. In Example 2 you were given three values of $x$ to be used to prepare the tables of values. In fact, you could have selected any other values of $x$ and obtained the same line. For instance, if you chose $x$ equal to $-3$, 2, and 3, the line obtained by graphing the table

| $x$ | −3 | 2 | 3 |
|---|---|---|---|
| $y$ | 9 | −1 | −3 |

would be the same line as that obtained in the example, as you can verify by reading the coordinates from the graph.

In general, we try to choose numbers that will locate three points not too close to each other and yet will still be on the graph paper.

**EXAMPLE 3**  On a lathe, the cutting speed $C$ (in feet per minute) for machining a metal bar $\frac{3}{4}$ in. in diameter is related to $R$, the speed of the lathe (rpm), by the formula

$$C = \frac{\pi}{16} R$$

**(a)** Graph this equation for values of $R$ from 0 to 1000. Use a horizontal $R$-axis with one space equal to 100 units and a vertical $C$-axis with one space equal to 10 units. Round off $C$ values to the nearest whole number.
**(b)** From the graph, estimate the revolutions per minute of the lathe for a cutting speed of 150 ft per minute.

**Solutions**
**(a)** All $C$ values and $R$ values must be positive (or zero). Choose $R$ equal to 0, 500, and 1000, and prepare a table.

| $R$ | 0 | 500 | 1000 |
|---|---|---|---|
| $C$ | 0 | 98 | 196 |

Draw and label an *R*-axis and a *C*-axis, as shown in Figure 9.27 (there is no need for a negative *R*-axis or *C*-axis). Graph each pair of related numbers, and draw a line through the resulting points.

**(b)** From the point on the graph with *C* value 150, estimate 760 rpm as the speed of the lathe (*R* value).

Only integers appear in the tables of values in each of the preceding examples. In many practical situations, numbers in equations often involve decimals. When graphing such equations, we can still choose integers to substitute for one of the unknowns, but in general, values for the second unknown will not be integers. These values are usually rounded off, and the position of the graph of each such pair of numbers is *estimated*.

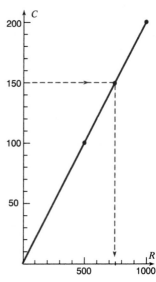

**FIGURE 9.27**

**EXAMPLE 4** If the diameter of a circular pin changes regularly from one end to the other, as shown in Figure 9.28, the pin is said to be *tapered*. A taper measurement *T* such as "2 centimeters per meter" means that the diameter of the pin changes by 2 cm for each meter of length. A relation between the taper *T* and the largest diameter *D*, the smallest diameter *d*, and the length of the pin *L* is given by the equation

$$T = \frac{D - d}{L}$$

**(a)** If $D = 8.5$ cm and $L = 20$ cm, write an equation that relates *T* and *d*.
**(b)** Graph the resulting equation from $d = 1$ cm to $d = 5$ cm. Use a horizontal *d*-axis with two spaces equal to 1 unit, and a vertical *T*-axis with one space equal to 0.025 unit.
**(c)** From the graph, estimate the taper for a diameter *d* equal to 1.5 cm.

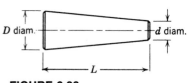

**FIGURE 9.28**

**Solutions**
**(a)** In the given equation, substitute 8.5 for *D* and 20 for *L* to obtain

$$T = \frac{8.5 - d}{20}$$

**(b)** Choose *d* equal to 1, 3, and 5, and prepare a table.

| *d* | 1 | 3 | 5 |
|---|---|---|---|
| *T* | 0.375 | 0.275 | 0.175 |

Draw and label a *d*-axis and a *T*-axis, as shown in Figure 9.29. Graph each pair of related numbers, and draw a line through the resulting points.
**(c)** From the point on the graph with *d* value 1.5, estimate 0.350 cm per meter as the taper *T*.

The computations needed to prepare tables can be done more efficiently when one of the unknowns appears by itself on one side of the equation, as in the above examples. Equations that are not in this form can be rewritten in this form by using the methods considered in Section 4.4.

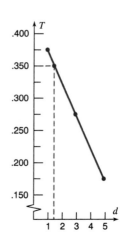

**FIGURE 9.29**

**EXAMPLE 5** Temperature scales in Fahrenheit (*F*) and Celsius (*C*) are related by the equation

$$5F - 9C = 160$$

**(a)** Solve the equation for *F*.

**(b)** Graph the equation from $C = -40°$ to $C = 40°$. Use a horizontal *C*-axis with one space equal to 10 degrees and a vertical *F*-axis with one space equal to 10 degrees.

### Solutions

**(a)** First, add $9C$ to each side of the equation.

$$5F - 9C + 9C = 160 + 9C$$
$$5F = 160 + 9C$$

Next, divide each side by 5 to obtain

$$F = \frac{160 + 9C}{5}$$

**(b)** Choose three values of *C*, say $-40°$, $0°$, and $40°$, and prepare a table of values. Use the calculator sequence

$$9 \;\boxed{\times}\; C \;\boxed{+}\; 160 \;\boxed{\div}\; 5 \;\boxed{=}$$

with the $\boxed{+/_-}$ key as needed.

| $C$ | $-40°$ | $0°$ | $40°$ |
|---|---|---|---|
| $F$ | $-40°$ | $32°$ | $104°$ |

Draw and label a *C*-axis and an *F*-axis as shown in Figure 9.30. Graph each pair of related numbers, and draw a line through the resulting points.

### Exercises for Section 9.4

*Graph each linear equation over the indicated interval, using intervals of one unit. See Example 1.*

**1.** $y = 2x + 2$;   $x = -3$   to   $x = 2$.

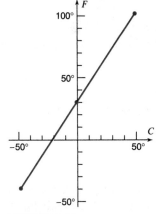

**FIGURE 9.30**

**2.** $y = 3x + 1$;   $x = -3$   to   $x = 2$.

**3.** $y = 1.4x + 3.8$;   $x = -3$   to   $x = 3$.

**4.** $y = 2.9x + 3.4$;   $x = -3$   to   $x = 2$.

**5.** $y = 3x - 9$;   $x = -1$   to   $x = 4$.

**6.** $y = 6x - 9$;   $x = 1$   to   $x = 4$.

**7.** $y = \dfrac{-0.3x - 2.1}{0.61}$;   $x = -5$   to   $x = 1$.

**8.** $y = \dfrac{-6.1x + 3.2}{0.92}$;  $x = -1$ to $x = 4$.

*Graph each linear equation using the given values of x. See Example 2.*

**9.** $y = x - 5$;  $x = 0$,  $x = 2$,  $x = 5$.

**10.** $y = x - 6$;  $x = 0$,  $x = 3$,  $x = 6$.

**11.** $y = 10 - 4x$;  $x = 1$,  $x = 3$,  $x = 4$.

**12.** $y = 8 - 5x$;  $x = 1$,  $x = 2$,  $x = 3$.

**13.** $y = \dfrac{2x + 6}{3}$;   $x = -3$,   $x = 0$,   $x = 3$.

**14.** $y = \dfrac{3x - 15}{5}$;   $x = -5$,   $x = 0$,   $x = 5$.

*For Exercises 15 to 18, see Example 3.*

**15.** The distance $d$ across the corners of
a square-head bolt (see Figure 9.31)
is given by the formula

$$d = 1.414f$$

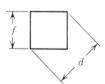

**FIGURE 9.31**

(a) Graph the equation for values
of $f$ from $\frac{1}{2}$ in. to 2 in., using
intervals of $\frac{1}{2}$ in. Take the $f$-axis
horizontal, with one space equal
to $\frac{1}{4}$ unit. On the $d$-axis, take
one space equal to 0.2 unit.
Round off $d$ values to the
nearest tenth.
(b) From the graph, estimate the
value of $d$ for $f = 1\frac{1}{4}$ in.

**16.** Repeat Exercise 15 using the
formula for a hexagon-head bolt
(see Figure 9.32):

$$d = 1.155f$$

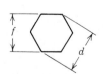

**FIGURE 9.32**

**17.** The thread depth, $h$, and pitch, $P$, of American National Standard threads (see Figure 9.33) are related by the formula

$$h = 0.6495P$$

(a) Graph the equation for values of $P$ from 0.030 to 0.070, using intervals of 0.010. Take the $P$-axis horizontal, with one space equal to 0.005 unit on both axes. Round off $h$ values to the nearest thousandth.

(b) From the graph, estimate the value of $h$ for $P = 0.042$ (the pitch for a No. 10-24 machine screw).

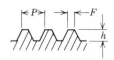

**FIGURE 9.33**

**18.** In Figure 9.33, the distance $F$ is given by

$$F = 0.125P$$

(a) Graph the equation for values of $P$ from 0.030 to 0.070, using intervals of 0.010. Take the $P$-axis horizontal, with one space equal to 0.005 unit. Take one space equal to 0.001 on the $F$-axis. Round off $F$ values to the nearest ten-thousandth.

(b) From the graph, estimate the value of $F$ for $P = 0.055$.

*In Exercises 19 to 22, use the formula from Example 4:*

$$T = \frac{D - d}{L}$$

*See Examples 4 and 5.*

**19.** (a) If $D = 6.4$ cm and $L = 15$ cm, write an equation relating $T$ and $d$.

(b) Graph the equation from $d = 1$ to $d = 6$. Use a horizontal $d$-axis with two spaces equal to 1 unit and a $T$-axis with one space equal to 0.05 unit.

(c) From the graph, estimate the taper of a pin with the smaller diameter 2.8 cm. (Answer to two significant digits.)

**20. (a)** If $D = 10$ cm   and   $L = 18.5$ cm,   write an equation relating $T$ and $d$.
**(b)** Graph the equation from $d = 2$  to  $d = 8$.  Use a horizontal $d$-axis, with two spaces equal to 1 unit and a $T$-axis with one space equal to 0.05 unit.
**(c)** From the graph, estimate the taper of a pin with a smaller diameter of 6.7 cm. (Answer to two significant digits.)

**21. (a)** If $T = 0.368$   and   $L = 15$ cm,   write an equation relating $D$ and $d$, and solve for $D$.
**(b)** Graph the equation from $d = 1$  to  $d = 5$.  Use a horizontal $d$-axis with two spaces equal to 1 unit on both axes.
**(c)** From the graph, estimate the larger diameter if the smaller diameter is 3.8 cm. (Answer to two significant digits.)

**22.** Follow the directions for Exercise 21 for $T = 0.407$   and   $L = 16.4$ cm.   (Answer to three significant digits.)

## 9.5  Graphing Nonlinear Equations

In Section 9.4 we considered techniques for graphing equations whose graphs were straight lines and noted that we only needed to obtain two points (a third point serves as a check) for such graphs. In this section we consider techniques for graphing equations whose graphs are *curved lines*, or simply, *curves*. For such graphs we usually need more than two or three points. Most curved-line graphs cannot be drawn completely; in the following examples we shall select values for the unknown that yield the important features of the graph, and we shall specify convenient scales to be used on the axes.

As an example, let us graph the equation

$$y = x^2 \tag{1}$$

from $x = -4$ to $x = 4$, using intervals of 1 unit. First, we prepare a table of values.

| $x$ | $-4$ | $-3$ | $-2$ | $-1$ | 0 | 1 | 2 | 3 | 4 |
|---|---|---|---|---|---|---|---|---|---|
| $y$ | 16 | 9 | 4 | 1 | 0 | 1 | 4 | 9 | 16 |

Next, we draw a horizontal $x$-axis with one space equal to 1 unit and a vertical $y$-axis with one space equal to 1 unit. Because the $y$-values are not less than zero, we show very little of the $y$-axis below the origin (Fig. 9.34a). Now, we graph each pair of related numbers from the table, as shown in Figure 9.34a, and draw a smooth curve through the resulting points, as shown in Figure 9.34b.

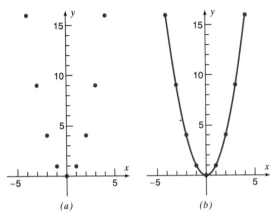

(a)                    (b)

**FIGURE 9.34**

Equations such as Equation 1, for which the graph is a curve (not a straight line), are called *nonlinear equations*.

**EXAMPLE 1**  Graph the equation $y = \frac{1}{2}x^3$ from $x = -3$ to $x = 3$, using intervals of 1 unit. Estimate the $y$ value paired with $x$ equal to 2.5.

**Solution**

**1.** Prepare a table of values; use the calculator sequence

$$x \boxed{y^x} 3 \boxed{\div} 2 \boxed{=}$$

(Remember that a negative number raised to the third power results in a negative number.)

| $x$ | $-3$ | $-2$ | $-1$ | 0 | 1 | 2 | 3 |
|---|---|---|---|---|---|---|---|
| $y$ | $-13.5$ | $-4$ | $-0.5$ | 0 | 0.5 | 4 | 13.5 |

**2.** Draw a horizontal $x$-axis with one space equal to 1 unit and a vertical $y$-axis with one space equal to 2 units.
**3.** Graph each pair of related numbers from the table, and draw a smooth curve through the resulting points, as shown in Figure 9.35.
**4.** From the point on the graph with $x$ value 2.5, estimate the $y$ value as 7.8.

**EXAMPLE 2**  Graph the equation $y = \sqrt{x} - 2$ from $x = 0$ to $x = 70$, using intervals of 10 units. (Round off $y$ values to the nearest tenth.) Estimate the $y$ value paired with $x$ equal to 45.

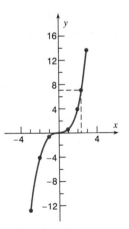

**FIGURE 9.35**

## Solution

**1.** Prepare a table of values; use the calculator sequence

$$x \boxed{\sqrt{\phantom{x}}} \boxed{-} 2 \boxed{=}$$

| $x$ | 0 | 10 | 20 | 30 | 40 | 50 | 60 | 70 |
|---|---|---|---|---|---|---|---|---|
| $y$ | −2 | 1.2 | 2.5 | 3.5 | 4.3 | 5.1 | 5.7 | 6.4 |

**2.** Draw a horizontal $x$-axis with one space equal to 5 units and a vertical $y$-axis with one space equal to 1 unit.

**3.** Graph each pair of related numbers from the table, and draw a smooth curve through the resulting points, as shown in Figure 9.36.

**4.** From the point on the graph with $x$ value 45, estimate the $y$ value as 4.7.

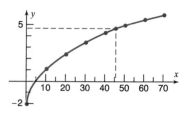

**FIGURE 9.36**

**EXAMPLE 3** The N.A.C.C. horsepower rating ($hp$) of a six-cylinder gasoline engine can be computed from the equation

$$hp = 2.4D^2$$

where $D$ is the bore (diameter) of each cylinder in inches. Graph this equation from $D = 1.0$ to $D = 4.0$, using intervals of 0.5 unit. Use a horizontal $D$-axis with two spaces equal to 1 unit and a vertical $hp$-axis with one space equal to 2 units. Estimate the $hp$ rating for a cylinder bore of 2.25 in.

## Solution

**1.** Prepare a table of values; use the calculator sequence

$$2.4 \boxed{\times} D \boxed{x^2} \boxed{=}$$

| $D$ | 1.0 | 1.5 | 2.0 | 2.5 | 3.0 | 3.5 | 4.0 |
|---|---|---|---|---|---|---|---|
| $hp$ | 2.4 | 5.4 | 9.6 | 15.0 | 21.6 | 29.4 | 38.4 |

**2.** Draw and label axes as shown in Figure 9.37.

**3.** Graph each pair of related numbers from the table, and draw a smooth curve through the resulting points.

**4.** From the point on the graph with $D$ value 2.25, estimate the $hp$ as 12.2.

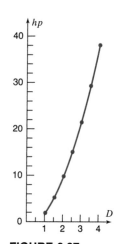

**FIGURE 9.37**

**EXAMPLE 4** In an electric circuit using 150 watts of power, the current $I$ (in amperes) is related to the resistance $R$ (in ohms) by the equation

$$RI^2 = 150$$

Graph the equation from $I = 4$ to $I = 10$, using intervals of 1 unit. Use a horizontal $I$-axis and a vertical $R$-axis, with one space equal to 1 unit on each axis. (Round off $R$ values to the nearest tenth). Estimate the current $I$ when the resistance is 3.5 ohms.

**Solution**  Because values of $R$ are to be computed, begin by solving the given equation for $R$. Divide each side by $I^2$ to obtain

$$R = \frac{150}{I^2}$$

**1.** Prepare a table of values. Use either of the sequences

$$150 \boxed{\div} I \boxed{\div} I \boxed{=}$$

or

$$150 \boxed{\div} \; I \; \boxed{x^2} \; \boxed{=}$$

(or square *I* mentally and do only the division on your calculator).

| *I* | 4 | 5 | 6 | 7 | 8 | 9 | 10 |
|---|---|---|---|---|---|---|---|
| *R* | 9.4 | 6.0 | 4.2 | 3.1 | 2.3 | 1.9 | 1.5 |

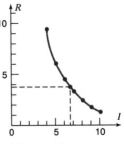

**FIGURE 9.38**

2. Draw and label axes, as shown in Figure 9.38.
3. Graph each pair of related numbers from the table, and draw a smooth curve through the resulting points.
4. From the point on the graph with *R* value 3.5, estimate 6.5 amp as the *I* value.

### Exercises for Section 9.5

*Graph each equation over the indicated interval.*

*See Example 1 for Exercises 1 to 6.*

1. $y = x^3$; from $x = -4$ to $x = 4$, using intervals of 1 unit. One space $= \frac{1}{2}$ unit on the *x*-axis; one space $= 5$ units on the *y*-axis. From the graph, estimate the *y* value for $x = 1.7$.

2. $y = -\frac{1}{2}x^3$; from $x = -4$ to $x = 4$, using intervals of 1 unit. One space $= \frac{1}{2}$ unit on the *x*-axis; one space $= 5$ units on the *y*-axis. From the graph, estimate the *y* value for $x = -2.3$.

3. $y = 4 - x^2$; from $x = -4$ to $x = 4$, using intervals of 1 unit. One space $= 1$ unit on both axes. From the graph, estimate the *y* value for $x = 0.8$.

4. $y = x^2 - 1$; from $x = -4$ to $x = 4$, using intervals of 1 unit. One space $= \frac{1}{2}$ unit on the *x*-axis; one space $= 2$ units on the *y*-axis. From the graph, estimate the *y* value for $x = -1.6$.

**5.** $x = y^2$; from $y = -4$ to $y = 4$, using intervals of 1 unit, $y$-axis vertical. One space $= \frac{1}{2}$ unit on the $y$-axis; one space $= 2$ units on the $x$-axis. From the graph, estimate the $x$ value for $y = 2.3$.

**6.** $x = 4 - y^2$; from $y = -4$ to $y = 4$, using intervals of 1 unit, $y$-axis vertical. One space $= 1$ unit on both axes. From the graph, estimate the $x$ value for $y = -0.7$.

*See Example 2 for Exercises 7 to 10.*

**7.** $y = \sqrt{x + 4}$; from $x = 0$ to $x = 40$, using intervals of 5 units. Round off $y$ values to the nearest tenth. One space $= 5$ units on the $x$-axis; one space $= 1$ unit on the $y$-axis. From the graph, estimate the $y$ value for $x = 28$.

**8.** $y = 4 - \sqrt{x}$; from $x = 0$ to $x = 40$, using intervals of 5 units. Round off $y$ values to the nearest tenth. One space $= 5$ units on the $x$-axis; one space $= \frac{1}{2}$ unit on the $y$-axis. From the graph, estimate the $y$ value for $x = 12$.

**9.** $y = \sqrt{x} + 4$; from $x = 0$ to $x = 70$, using intervals of 10 units. Round off $y$ values to the nearest tenth. One space $= 5$ units on the $x$-axis; one space $= 1$ unit on the $y$-axis. From the graph, estimate the $y$ value for $x = 55$.

**10.** $y = \sqrt{10 - x}$; from $x = -50$ to $x = 10$, using intervals of 10 units. Round off $y$ values to the nearest tenth. One space = 5 units on the *x*-axis; one space = 1 unit on the *y*-axis. From the graph, estimate the *y* value for   *x* = −25.

*See Examples 3 and 4 for Exercises 11 to 20.*

**11.** $P = \dfrac{120}{V}$; from *V* = 5 to *V* = 40,

using intervals of 5 units, *V*-axis horizontal. Round off *P* values to the nearest tenth. One space = 5 units on the *V*-axis; one space = 2 units on the *P*-axis. From the graph, estimate the *P* value for   *V* = 28 and the *V* value for   *P* = 17.

**12.** $n = \dfrac{120}{i + 5}$;  from  *i* = 10  to  *i* = 80,

using intervals of 10 units, *i*-axis horizontal. Round off *n* values to the nearest tenth. One space = 5 units on the *i*-axis; one space = 1 unit on the *n*-axis. From the graph, estimate the *n* value for   *i* = 55 and the *i* value for   *n* = 6.5.

**13.** A hole *d* cm in diameter is to be drilled lengthwise through a steel rod 4 cm in diameter (see Figure 9.39). The cross-sectional area, *A*, of the wall of the resulting tube is given by

$$A = 12.566 - 0.785d^2$$

Graph the equation from   *d* = 1.5  to  *d* = 2.9,  using intervals of 0.2 unit, *d*-axis horizontal. One space = 0.1 unit on the *d*-axis; one space = 1 unit on the *A*-axis. From the graph, estimate the area for *d* = 1.8.

*d* cm

←—4 cm—→

**FIGURE 9.39**

**14.** A concrete pipe 5 m long, with an
inside diameter of 2 m, is to be cast
with a wall thickness of $T$ meters
(see Figure 9.40). The volume $V$ in
cubic meters of concrete needed is
given by

$$V = 15.708T^2 + 31.416T$$

Graph the equation from $T =$
0.10 to $T = 0.40$, using intervals
of 0.05 unit, $T$-axis horizontal.
Round off $V$ values to the nearest
tenth. Two spaces = 0.05 unit on
the $T$-axis; one space = 1 unit on
the $V$-axis. From the graph, estimate
the volume of concrete needed
for $T = 0.32$.

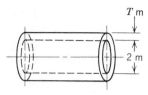

**FIGURE 9.40**

**15.** For a certain manufacturer, the cost,
$C$, in dollars of producing one
component of a stereo amplifier, if $n$
of them are produced in 1 hour, is

$$C = 1 + \frac{n}{3} + \frac{5}{n}$$

Graph the equation from $n = 2$
to $n = 8$, using the intervals of 1
unit, $n$-axis horizontal. Round off $C$
values to the nearest hundredth.
One space = 1 unit on the $n$-axis;
one space = 0.2 unit on the $C$-axis.
From the graph, estimate the
greatest whole number of
components that can be made per
hour if the cost for each component
is not to exceed $3.75.

**16.** The cost $C$ in dollars per mile of
running a certain miniature
locomotive at a speed of $s$ miles per
hour is given by

$$C = \frac{s}{12} + \frac{40}{s}$$

Graph the equation from $s = 10$
to $s = 40$, using intervals of 5
units, $s$-axis horizontal. Round off $C$
values to the nearest hundredth.
One space = 5 units on the $s$-axis;
four spaces = 1 unit on the $C$-axis.
From the graph, estimate the
greatest speed at which the engine
can be operated for $4 per mile.

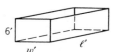

**FIGURE 9.41**

17. A storage bin is to have a volume of 1200 cu ft (see Figure 9.41). If it must be 6 ft high, the length $\ell$ and the width $w$ are related by the formula

$$\ell w = 200$$

Graph the equation from $w = 8$ to $w = 20$, using intervals of 2 units, $w$-axis horizontal. Round off $\ell$ values to the nearest tenth. One space = 2 units on both axes. From the graph, estimate the necessary length for a bin 15 ft wide.

18. If a steam engine is to develop 200 indicated horsepower at 300 power strokes per minute with an average pressure of 130 pounds per square inch, the length $L$ of the stroke, measured in feet, and the area $A$ of the piston, measured in square inches, are related by the formula

$$LA = 169.2$$

Graph the equation from $A = 40$ to $A = 80$, using intervals of 5 units, $A$-axis horizontal. Round off $L$ values to the nearest hundredth. One space = 5 units on the $A$-axis; one space = 0.2 unit on the $L$-axis. From the graph, estimate the length of stroke for a piston having an area of 52 sq in.

19. For a cylindrical storage tank that is to hold 5000 gal, the dimensions shown in Figure 9.42 are related by the formula

$$D^2 h = 851$$

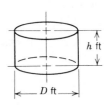

**FIGURE 9.42**

Graph the equation from $D = 4$ to $D = 16$, using intervals of 2 units, $D$-axis horizontal. Round off $h$ values to the nearest whole number. One space = 1 unit on the $D$-axis; one space = 5 units on the $h$-axis. From the graph, estimate the necessary height $h$ for a tank 13 ft in diameter.

**20.** If the formula in Exercise 19 is solved for $D$, we have

$$D = \frac{29.17}{\sqrt{h}}$$

Graph the equation from $h = 4$ to $h = 16$, using intervals of 2 units, $h$-axis horizontal. Round off $D$ values to the nearest tenth. One space = 1 unit on both axes. From the graph, estimate the diameter of the tank for a height of 9 ft.

## 9.6 More on Graphing

In the preceding sections we specified scales to be used on axes when drawing graphs. In actual practice, the scales used for any graph involve two choices:

**1.** The number of units to be associated with each space on an axis. This depends on the kind of numbers in the table of values.
**2.** The number of spaces to be included on an axis. This depends on the type of graph paper to be used. Axes that have 10 to 30 spaces are convenient.

There are no special rules for choosing scales. However, the following examples include suggestions that may be helpful.

**EXAMPLE 1**   Suggest a scale for each set of coordinates.
**(a)** $-8, -4, 0, 4, 8, 12$
**(b)** $-20, -15, -10, -5, 0, 5, 10, 15, 20, 25, 30, 35, 40$
**(c)** $-2.0, -1.6, -1.2, -0.8, -0.4, 0, 0.4, 0.8$

**Solutions**
**(a)** The difference of the largest and smallest numbers is $12 - (-8)$, or 20. A choice of one space equal to 1 unit will require at least 20 spaces on the axis, as shown in Figure 9.43.

**FIGURE 9.43**

**(b)** The difference of the largest and smallest numbers is $40 - (-20)$, or 60. Sixty spaces for an axis may be too long. Because each number is a multiple of 5, a convenient scale is one space equal to 5 units. Note that the axis will then require at least 12 spaces ($60 \div 5$), as shown in Figure 9.44.

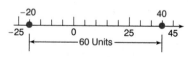

**FIGURE 9.44**

**(c)** The difference of the largest and smallest numbers is $0.8 - (-2.0)$, or 2.8, just under 3. In this case, one space equal to 1 unit is too "crowded." A convenient scale is one space equal to 0.2 unit. Note that the axis will then require at least 15 spaces ($3 \div 0.2$), as shown in Figure 9.45.

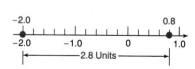

**FIGURE 9.45**

***EXAMPLE 2*** Graph the linear equation

$$s = \frac{3t - 160}{12}$$

from $t = -10$ to $t = 110$. Use a horizontal $t$-axis. Round off $s$ values to the nearest tenth.

### Solution

1. The graph of a linear equation is a straight line. Hence, choose any three $t$ values, say $-10$, $60$, and $110$, and compute a table of values.

| $t$ | $-10$ | $60$ | $110$ |
|---|---|---|---|
| $s$ | $-15.8$ | $1.7$ | $14.2$ |

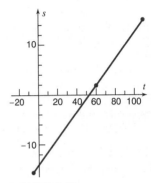

2. The $t$ values range from $-10$ to $110$, a total of 120 units. Because each $t$ value is a multiple of 10, use one space equal to 10 units. Note that the $t$-axis will then require at least 12 spaces ($120 \div 10$). Draw and label the $t$-axis, as shown in Figure 9.46.
3. The $s$ values range from $-15.8$ to $14.2$, a total of 30 units. Use one space equal to 2 units. Note that the $s$-axis will then require at least 15 spaces ($30 \div 2$). Draw and label an $s$-axis, as shown.
4. Graph each pair of related numbers and complete the graph, as in the figure.

**FIGURE 9.46**

***EXAMPLE 3*** Graph the nonlinear equation

$$V = r^3 - 2r^2 - 15r$$

from $r = -4$ to $r = 6$, using intervals of 1 unit. Use a horizontal $r$-axis.

### Solution

1. Prepare a table of values.

| $r$ | $-4$ | $-3$ | $-2$ | $-1$ | $0$ | $1$ | $2$ | $3$ | $4$ | $5$ | $6$ |
|---|---|---|---|---|---|---|---|---|---|---|---|
| $V$ | $-36$ | $0$ | $14$ | $12$ | $0$ | $-16$ | $-30$ | $-36$ | $-28$ | $0$ | $54$ |

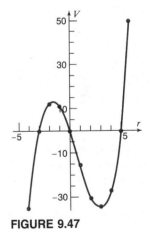

2. The $r$ values range from $-4$ to 6, a total of 10 units. Use one space equal to 1 unit. Draw and label an $r$-axis, as shown in Figure 9.47.
3. $V$ values range from $-36$ to 54, a total of 90 units. Use one space equal to 5 units. Note that the $V$-axis will then require at least 18 spaces ($90 \div 5$). Draw and label a $V$-axis, as shown.
4. Graph each pair of related numbers, and complete the graph, as in the figure.

**FIGURE 9.47**

## ■ Problem Solving by Graphical Methods

Coordinates of points on a graph can often be used to estimate solutions to certain kinds of problems.

***EXAMPLE 4*** A sheet of metal 30.6 cm wide is to be formed into a closed duct with a rectangular cross section (see Figure 9.48). The area $A$ of the cross section is related to the width $W$ of the duct by the equation

$$A = 15.3W - W^2$$

From the graph of the equation, estimate the value of $W$ (between 5 and 10) so that the cross-sectional area of the duct will be as great as possible. Use a horizontal $W$-axis; round off $A$ values to the nearest tenth.

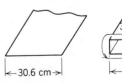

**FIGURE 9.48**

## Solution

1. Prepare a table, using intervals of 1 unit.

| W | 5 | 6 | 7 | 8 | 9 | 10 |
|---|---|---|---|---|---|----|
| A | 51.5 | 55.8 | 58.1 | 58.4 | 56.7 | 53.0 |

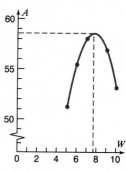

**FIGURE 9.49**

2. *W* values range from 5 to 10, a total of 5 units. Draw and label a *W*-axis, starting with zero and using one space equal to 1 unit, as shown.
3. *A* values range from 51.5 to 58.4, a total of just under 7 units. Use one space equal to 1 unit, but start at 50 and "compress" the *A*-axis as shown in Figure 9.49.
4. Graph each related pair of numbers, and complete the graph as shown in the figure.
5. Locate the "highest" point on the graph. Estimate the *A* value of that point as 58.5 sq cm and the related *W* value as 7.6 cm. Thus the duct will have the greatest cross-sectional area for the given sheet of metal if it is formed with a width of 7.6 cm.

**EXAMPLE 5**   Use the results of Example 4 to determine the height *h* of the cross section of the duct with greatest cross-sectional area.

**Solution**   Substitute 58.5 for *A* and 7.6 for *W* in the area formula

$$A = Wh$$

to obtain the height *h*:

$$58.5 = 7.6h$$

$$h = \frac{58.5}{7.6} = 7.6973684$$

Thus, the height of the cross section is 7.7 cm, to the nearest tenth. Note that the height and length are very nearly equal. In fact, the maximum cross-sectional area of the duct will be obtained if the cross section is *square*.

**EXAMPLE 6**   A right-circular cylindrical steel tank with an open top is to hold 64 cu m of oil (see Figure 9.50). The amount *A* of steel plate needed to form the tank is related to the radius *r* of the base of the tank by the equation

$$A = \pi r^2 + \frac{128}{r}$$

**FIGURE 9.50**

From the graph of the equation, estimate the radius of the base (between 2 and 4 m) so that the least amount of steel plate will be needed to form the tank. Use a horizontal *r*-axis; round off *A* values to the nearest tenth.

## Solution

1. Prepare a table of values, using intervals of 0.5 unit

| r | 2.0 | 2.5 | 3.0 | 3.5 | 4.0 |
|---|-----|-----|-----|-----|-----|
| A | 76.6 | 70.8 | 70.9 | 75.1 | 82.3 |

2. The *r* values range from 2.0 to 4.0, a total of 2 units. Use one space equal to 0.2 units, but start at 2 and compress the *r*-axis, as shown in Figure 9.51.
3. *A* values range from 70.8 to 82.3, just under 12 units. Use one space equal to 1 unit, but start at 70 and compress the *A*-axis, as shown.

**4.** Graph each related pair of numbers, and complete the graph, as shown in the figure.

**5.** Locate the "lowest" point in the graph. Estimate the *r* value for this point as 2.7 m.

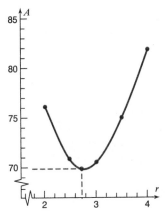

**FIGURE 9.51**

**EXAMPLE 7**  The height *h* of the steel tank of Example 6 is related to the radius *r* by the equation

$$h = \frac{64}{\pi r^2}$$

Use the result of Example 6 to compute the height (to the nearest tenth) of the tank formed from the least amount of steel.

**Solution**  Substitute 2.7 for *r*, and compute.

$$h = \frac{64}{\pi (2.7)^2} = 2.7944901$$

To the nearest tenth, the height of the tank is 2.8 m.

   Because there is a choice of scales when constructing a graph, different individuals may select different units for the same problem. As a consequence, in the following exercises, your graphs may be different from those shown in the answer section.

## Exercises for Section 9.6

*Graph each equation over the indicated interval. See Examples 2 and 3.*

**1.** $v = 32t + 20$;  $t = 0$  to  $t = 20$;  *t*-axis horizontal.

**2.** $s = 25v + 15$;  $v = 0$  to  $v = 30$;  *v*-axis horizontal.

**3.** $C = \dfrac{5F - 160}{9}$;  $F = -30$  to  $F = 100$;  *F*-axis horizontal.

**4.** $F = \dfrac{9C + 160}{5}$;   $C = -30$   to   $C = 100$;   $C$-axis horizontal.

**5.** $L = 100 - 20d$;   $d = -20$   to   $d = 20$;   $d$-axis horizontal.

**6.** $R = 80 - 40r$;   $r = -40$   to   $r = 40$;   $r$-axis horizontal.

**7.** $F = 314.16r^2$;   $r = 0$   to   $r = 20$;   $r$-axis horizontal.

**8.** $C = 62.75s^2$;   $s = 0$   to   $s = 30$;   $s$-axis horizontal.

**9.** $S = \dfrac{r^3}{12} - r;$ $r = -10$ to $r = 10;$ $r$-axis horizontal.

**10.** $V = d^3 - d^2 - 6d;$ $d = -3$ to $d = 4;$ $d$-axis horizontal.

**11.** $T = 0.01t^4 - t;$ $t = -5$ to $t = 5;$ $t$-axis horizontal.

**12.** $S = \dfrac{n^5}{215} - 0.1n^2;$ $n = -4$ to $n = 4;$ $n$-axis horizontal.

*For the following exercises, see Examples 4, 5, 6, and 7.*

**13.** An air duct with a rectangular cross section having an area of 144 sq in. is to be made from sheet metal (see Figure 9.52). The perimeter $P$ and the width $w$ are related by the equation

$$P = 2w + \frac{288}{w}$$

**(a)** To the nearest tenth, find the width (between 10 and 14 in.) that will give the smallest perimeter.

**(b)** Find the smallest perimeter.

144 sq in.

$w$

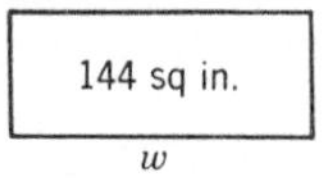

**FIGURE 9.52**

**14.** If the duct of Exercise 13 has a
cross section as shown in Figure
9.53, the perimeter $P$ and height $h$
are related by the equation

$$P = 1.571h + \frac{288}{h}$$

**FIGURE 9.53**

(a) To the nearest tenth, find the
height (between 12 and 15 in.)
that will give the smallest perimeter.
(b) Find the smallest perimeter.

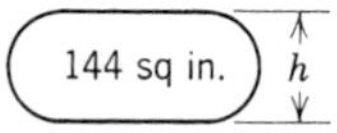

**15.** If an object is thrown upward with a
speed of 64 ft per second, its height
in feet above the ground after $t$
seconds is

$$s = -16t^2 + 64t$$

(a) To the nearest whole number,
find the time (between 0 and 4
seconds) at which the object
reaches its greatest height.
(b) Find its greatest height.
(c) Find the time at which it strikes
the ground.

**16.** An open-topped box is to be
formed by removing equal squares,
$x$ cm on a side, from each corner of
a piece of metal 30 cm square (see
Figure 9.54). The volume $V$ of the
box and the dimension $x$ are related
by the equation

$$V = 4x^3 - 120x^2 + 900x$$

**FIGURE 9.54**

To the nearest whole number, find
the dimension $x$ (between 0 and 10
cm) for which the volume of the
box will be as large as possible.

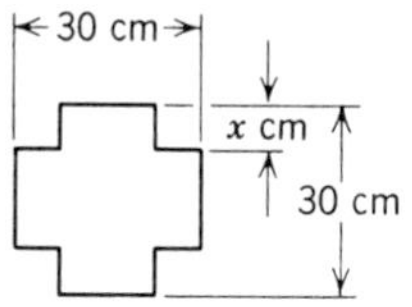

**17.** Two sources of heat, $A$ and $B$, are
10 cm apart (see Figure 9.55). The
intensity $I$ of the heat at a point $P$ is
given by

$$I = \frac{10}{x^2} + \frac{20}{(10 - x)^2}$$

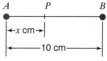

**FIGURE 9.55**

To the nearest tenth, locate the
point at which the intensity will be
the lowest.

18. For a certain type of screw, the efficiency $E$ and the pitch $p$ are related by the equation

$$E = \frac{p - p^2}{p + 1}$$

To the nearest tenth, find the pitch that will result in maximum efficiency. (Hint: Graph from $p = 0$ to $p = 1$, using intervals of 0.1.)

19. The cost $C$ in dollars per kilometer of operating a particular truck is related to the speed $s$, in kilometers per hour, by the equation

$$C = \frac{s}{15} + \frac{375}{s}$$

To the nearest whole number, find the speed (between 50 and 90 km per hour) at which the operating cost is the lowest possible.

20. The cost $C$ per unit, in dollars, of producing a particular refrigerator is related to the number $x$ of units produced in one day, by the equation

$$C = \frac{x}{12} + \frac{200}{x} + 127$$

To the nearest whole number, find the number of refrigerators (between 30 and 60) that should be made in one day in order that the cost per refrigerator be the lowest possible.

## 9.7  Sine and Cosine Curves

In earlier chapters we considered angles between 0° and 180°, and we obtained sine and cosine values for such angles. In mathematics we also consider angles greater than 180°, and we assign sine and cosine values to such angles. Figure 9.56 shows a 200° and a 345° angle. In this section we first consider how to find sines and cosines of such angles; then we use such values to construct sine and cosine graphs.

For graphing purposes, it is convenient to use $x$, rather than $A$, to name angles. For example, using such notation, in Figure 9.56a, $x = 200°$, and in Figure 9.56b, $x = 345°$.

The calculator sequences used to find $\sin x$ or $\cos x$ for angles between 0° and 180° can also be used for angles greater than 180°.

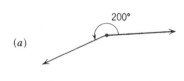

(a)

(b)

**FIGURE 9.56**

**EXAMPLE 1**

**(a)** Find sin 243°.      **(b)** Find cos 310.4°.

*Solutions*

**(a)** sin 243° = −0.8910,   using the calculator sequence

$$243 \; \sin \longrightarrow -0.8910065$$

**(b)** cos 310.4° = 0.6481,   using the calculator sequence

$$310.4 \; \cos \longrightarrow 0.6481199$$

Equations that involve sin $x$ and cos $x$ are not linear equations; curves that are graphs of such equations can be constructed by the methods introduced in Sections 9.3 and 9.5. For example, let us construct the graph of

$$y = \sin x$$

from 0° to 360°, using intervals of 30°. First, we prepare a table in which the $y$ values are rounded off to the nearest hundredth. (Use the appropriate sequence for your calculator.)

| $x$ | 0° | 30° | 60° | 90° | 120° | 150° | 180° |
|---|---|---|---|---|---|---|---|
| $y$ | 0 | 0.50 | 0.87 | 1.00 | 0.87 | 0.50 | 0 |

| $x$ | 210° | 240° | 270° | 300° | 330° | 360° |
|---|---|---|---|---|---|---|
| $y$ | −0.50 | −0.87 | −1.00 | −0.87 | −0.50 | 0 |

Next, because $x$ values range from 0° to 360° in 30° intervals, a convenient scale on the $x$-axis is one space equal to 15°. The $y$ values range from −1 to +1, a total of 2 units; a convenient scale on the $y$-axis is one space equal to 0.1 unit. We complete the graph by graphing each pair of related numbers from the table and drawing a smooth curve through the resulting points, as shown in Figure 9.57.

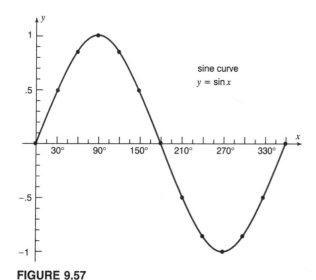

**FIGURE 9.57**

A similar approach can be used to graph the equation

$$y = \cos x$$

A table, using intervals of 30° and rounding off *y* values to the nearest hundredth, appears as

| x | 0° | 30° | 60° | 90° | 120° | 150° | 180° |
|---|----|-----|-----|-----|------|------|------|
| y | 1.00 | 0.87 | 0.50 | 0 | −0.50 | −0.87 | −1.00 |

| x | 210° | 240° | 270° | 300° | 330° | 360° |
|---|------|------|------|------|------|------|
| y | −0.87 | −0.50 | 0 | 0.50 | 0.87 | 1.00 |

We then graph each pair of related numbers. The smooth curve through the resulting points is shown in Figure 9.58.

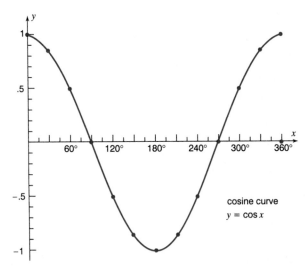

cosine curve
$y = \cos x$

**FIGURE 9.58**

Note that the graphs shown in Figures 9.57 and 9.58 reach a maximum value of +1 and a minimum value of −1. In general, for a given positive number *A*, graphs of equations in the form

$$y = A \sin x \quad \text{or} \quad y = A \cos x$$

reach a maximum value of *A* and a minimum value of −*A*.

**EXAMPLE 2**  Graph $y = 3 \sin x$ from $x = 0°$ to $x = 360°$, using intervals of 30°. Round off *y* values to the nearest tenth.

**Solution**
1. Prepare a table of values.

| x | 0° | 30° | 60° | 90° | 120° | 150° | 180° |
|---|----|-----|-----|-----|------|------|------|
| y | 0 | 1.5 | 2.6 | 3.0 | 2.6 | 1.5 | 0 |

| x | 210° | 240° | 270° | 300° | 330° | 360° |
|---|------|------|------|------|------|------|
| y | −1.5 | −2.6 | −3.0 | −2.6 | −1.5 | 0 |

2. Using one space equal to 15°, draw and label an *x*-axis.
3. The *y* values range from −3 to +3, a total of 6 units. A convenient scale is one space equal to 0.5 unit. Draw and label a *y*-axis.

**4.** Graph each pair of related numbers and draw a smooth curve through the resulting points, as shown in Figure 9.59.

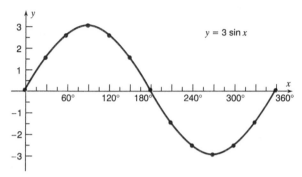

**FIGURE 9.59**

Note that each of the graphs shown above has a "wave" shape. Many technical applications lead to equations with "wave"-shaped graphs. For example, in a simple alternating current generator, a changing voltage $v$ is generated as a coil rotates between the two poles of a magnet. The voltage $v$ is related to the maximum generated voltage $V$ by the equation

$$v = V \sin x \qquad (1)$$

where $x$ is the angle through which the coil has rotated, $V$ is the maximum or peak voltage, and $v$ is the instantaneous voltage. (See Figure 9.60.) As a consequence

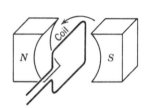 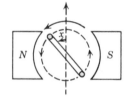

**FIGURE 9.60**

of the generated voltage, a changing current $i$ is established. The current $i$ is related to the maximum current $I$ by the equation

$$i = I \sin x$$

where, again, $x$ is the angle through which the coil has rotated.

**EXAMPLE 3**    If the maximum voltage $V$ generated by a simple alternating current generator is 10 volts, Equation 1 takes the form

$$v = 10 \sin x$$

Graph this equation for one complete revolution of the coil (0° to 360°). Use a horizontal $x$-axis with intervals of 30°. Round off $v$ values to the nearest tenth. From the graph, estimate the voltage in the coil when $x = 45°$ and when $x = 225°$.

*Solution*

**1.** Prepare a table of values.

| $x$ | 0° | 30° | 60° | 90° | 120° | 150° | 180° |
|---|---|---|---|---|---|---|---|
| $v$ | 0 | 5.0 | 8.7 | 10.0 | 8.7 | 5.0 | 0 |

| $x$ | 210° | 240° | 270° | 300° | 330° | 360° |
|---|---|---|---|---|---|---|
| $v$ | −5.0 | −8.7 | −10.0 | −8.7 | −5.0 | 0 |

**2.** The $x$ values range from 0° to 360° in 30° intervals. Draw and label an $x$ axis using one space equal to 15°.

**3.** The $v$ values range from −10 to +10, a total of 20 units. Draw and label a $v$-axis, using one space equal to 1 unit.

**4.** Graph each pair of related numbers, and draw a smooth curve through the resulting points, as shown in Figure 9.61.

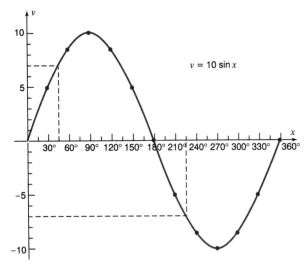

**FIGURE 9.61**

**5.** From the point with $x$ value 45°, estimate 7 volts as the $v$ value. From the point with $x$ value 225°, estimate −7 volts as the $v$ value.

## Exercises for Section 9.7

*Use your calculator to find the sine and cosine of each angle. See Example 1.*

**1.** 197°     **2.** 199°     **3.** 246.2°     **4.** 235.4°

**5.** 280°     **6.** 323°     **7.** 315.1°     **8.** 346.2°

*Graph each equation. See Example 2.*

**9.** $y = 2 \sin x$;   $x = 0°$   to   $x = 360°$,   intervals of 30°.

**10.** $y = 1.5 \sin x$;   $x = 0°$   to   $x = 360°$,   intervals of 30°.

**11.** $y = 2.5 \cos x$;   $x = 0°$   to   $x = 360°$,   intervals of 30°.

**12.** $y = 2 \cos x$;   $x = 0°$   to   $x = 360°$,   intervals of 30°.

**13.** $y = \sin 2x$;   $x = 0°$   to   $x = 180°$,   intervals of 15°.

**14.** $y = \cos 2x$;   $x = 0°$   to   $x = 180°$,   intervals of 15°.

*As given on page 366, the current i produced by a certain type of generator is*
$i = I \sin x$, *where I is the maximum current. Graph this equation for each value of I. See Example 3.*

**15.** $I = 5$, $x = 0°$ to $x = 360°$.
From the graph, estimate the value
of $i$ for $x = 60°$ and $x = 315°$.

**16.** $I = 8$, $x = 0°$ to $x = 360°$.
From the graph, estimate the value
of $i$ for $x = 135°$ and $x = 330°$.

**17.** For a pendulum 85 cm long (see
Figure 9.62), the displacement $x$ of
the pendulum bob from the center
is given by the equation

$$x = 85 \sin A$$

Graph the equation, with the $A$-axis
horizontal, from $A = 0°$ to $A = 90°$, using intervals of 10°. From
the graph, estimate the displacement
for $A = 25°$ and $A = 65°$.

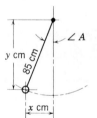

**FIGURE 9.62**

**18.** In Figure 9.62, the vertical
displacement is given by

$$y = 85 \cos A$$

Graph the equation, $A$-axis
horizontal, from $A = 0°$ to $A = 90°$, using intervals of 10°. From
the graph, estimate $y$ for $A = 25°$
and $A = 65°$.

**19.** One link in a mechanical linkage is 5 in. long (see Figure 9.63). The horizontal distance $x$ is given by the equation

$$x = 5 \cos B$$

Graph the equation, $B$-axis horizontal, from $B = 0°$ to $B = 90°$ using intervals of 10°. From the graph, estimate $x$ for $B = 15°$ and $B = 55°$.

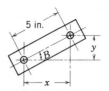

**FIGURE 9.63**

**20.** In Figure 9.63, the vertical displacement $y$ is given by

$$y = 5 \sin B$$

Graph the equation, $B$-axis horizontal, from $B = 0°$ to $B = 90°$, using intervals of 10°. From the graph, estimate $y$ for $B = 15°$ and $B = 55°$.

■ **Review Exercises**

**Section 9.1**

**1.** Construct a vertical bar graph for the following table showing the number of finished shafts produced by different machinists in one day.

| Machinist | A | B | C | D | E |
|---|---|---|---|---|---|
| Number of shafts | 12 | 15 | 10 | 16 | 11 |

**2.** Construct a horizontal bar graph for the following table showing the average wattage of various electric household appliances.

| Appliance | Oven | Sandwich grill | Toaster | Dishwasher | Broiler |
|---|---|---|---|---|---|
| Wattage | 1450 | 1161 | 1146 | 1201 | 1436 |

3. Construct a broken-line graph for the following table showing the commission earned by an air-conditioner salesperson for each of 5 consecutive months.

| Month | May | June | July | Aug | Sept |
|---|---|---|---|---|---|
| Commission | $1260 | $1580 | $2090 | $2215 | $1460 |

4. Construct a circle graph for the following table showing the percent of employees of a company belonging to various age groups.

| Age (years) | Under 20 | 20–25 | 26–30 | 31–35 | 36–40 | Over 40 |
|---|---|---|---|---|---|---|
| Percent | 8.3% | 20.5% | 28.2% | 19.4% | 8% | 15.6% |

## Section 9.2

*Graph each table of values, and connect the points with a smooth curve (or straight line). Use two spaces equal to 1 unit on each axis.*

5.

| $x$ | −2 | −1 | 0 | 1 | 2 |
|---|---|---|---|---|---|
| $y$ | −5 | −3 | −1 | 1 | 3 |

6.

| $x$ | −2 | −1 | 0 | 1 | 2 |
|---|---|---|---|---|---|
| $y$ | 2 | −1 | −2 | −1 | 2 |

7. From the graph of Exercise 5, estimate the value of $y$ for
   (*a*) $x = -1.5$, and (*b*) $x = 1.3$.

8. From the graph of Exercise 6, estimate the value of $y$ for
   (*a*) $x = -1.2$, and (*b*) $x = 0.7$.

**9.** Graph the table, and connect the points with a smooth curve. Compress the vertical *S*-axis.

| $t$ | 1.0 | 1.5 | 2.0 | 2.5 | 3.0 | 3.5 | 4.0 |
|---|---|---|---|---|---|---|---|
| $S$ | 36.71 | 36.50 | 36.41 | 36.39 | 36.42 | 36.55 | 37.00 |

*t*-axis horizontal; one space = 0.5
*S*-axis vertical; one space = 0.1

## Section 9.3

*Prepare a table of values for each equation over the indicated interval.*

**10.** $w = \dfrac{15.7}{h}$;   from   $h = 4$   to   $h = 12$,   using intervals of 2 units. Round off to the nearest tenth.

**11.** $S = 4.189r^3$;   from   $r = 2.0$   to   $r = 3.0$,   using intervals of 0.2 unit. Round off to the nearest tenth.

**12.** $y = \sqrt{36 - x^2}$;   from   $x = -6$   to   $x = 6$,   using intervals of 2 units. Round off to the nearest tenth.

## Section 9.4

**13.** Graph the equation   $y = 2x - 5$,   using   $x = -1$,   $x = 2$,   and $x = 4$.

**14.** Using the formula $R = \dfrac{r + a}{b}$ for the following:

(a) If   $a = 0.3$   and   $b = 2.1$, write an equation relating $R$ and $r$.

(b) Graph the equation from   $r = 1$   to   $r = 3$,   using intervals of 0.2 unit. Round off $R$ values to the nearest tenth. Use a horizontal $r$-axis with one space equal to 0.2 unit and an $R$-axis with one space equal to 0.1 unit.

(c) From the graph, estimate the value of $R$ for $r = 1.7$.

## Section 9.5

15. Graph the equation $y = 4x - x^2$ from $x = -1$ to $x = 5$, using intervals of 1 unit. One space equals 0.5 unit on the $x$-axis; one space equals 1 unit on the $y$-axis. From the graph, estimate the $y$ value for $x = 3.4$.

16. Graph the equation $y = 8 - \sqrt{x}$ from $x = 0$ to $x = 40$, using intervals of 5 units. One space equals 5 units on the $x$-axis; one space equals 0.5 unit on the $y$-axis. From the graph, estimate the $x$ value for $y = 5.5$.

17. The cost $C$, in dollars, of painting the outside of a spherical tank with a radius of $r$ feet is given by

$$C = 1.256r^2$$

Graph the equation from $r = 1$ to $r = 5$, using intervals of 0.5 unit; $r$-axis horizontal. One space equals 0.5 unit on the $x$-axis; one space equals 5 units on the $y$-axis. From the graph, estimate the cost, to the nearest dollar, of painting a sphere with a diameter of 8.6 ft.

## Section 9.6

18. Graph the equation $V = r^3 - r$ from $r = -2$ to $r = 2$. Take the $r$-axis horizontal; use intervals of 0.5 unit.

**19.** Under certain conditions, the power
$P$ developed in an electric circuit is
given by

$$P = 16.8I - 4.2I^2$$

where $I$ is the current in amperes.
Find the current (between 0 and 5
amp) that will give the maximum
power.

**Section 9.7**

**20.** $\cos 201° = ?$     **21.** $\sin 329.7° = ?$

**22.** Graph   $y = 4 \sin x$;   $x = 0°$   to
$x = 360°$,   intervals of 30°.

**23.** For a certain generator that will
produce a maximum current of $I$
amperes, the current $i$ produced is
given by

$$i = I \sin x$$

Graph the equation for   $I = 7.5$,
from   $x = 0°$   to   $x = 360°$.

# *Tables of Measurement*

## *U.S. Measurements*

**TABLE A.1**   Weight

| | |
|---|---|
| 16 ounces (oz) | = 1 pound (lb) |
| 2000 pounds | = 1 ton (T) |

**TABLE A.2**   Time

| | |
|---|---|
| 60 seconds (sec) | = 1 minute (min) |
| 60 minutes | = 1 hour (hr) |
| 24 hours | = 1 day (da) |
| 7 days | = 1 week (wk) |
| 30 days | = 1 month (mo) |
| 52 weeks | = 1 year (yr) |
| 12 months | = 1 year |
| 365 days | = 1 calendar year |
| 366 days | = 1 leap year |
| 360 days | = 1 business year |

**TABLE A.3**   Counting

| | |
|---|---|
| 20 units | = 1 score |
| 12 units | = 1 dozen (doz) |
| 12 dozen | = 1 gross (gro) |
| 12 gross | = 1 great gross |

**TABLE A.4**   Length

| | |
|---|---|
| 12 inches (in.) | = 1 foot (ft) |
| 3 feet = 36 inches | = 1 yard (yd) |
| 5280 feet = 1760 yards | = 1 mile (mi) |

**TABLE A.5**   Area

| | |
|---|---|
| 144 square inches (sq in.) | = 1 square foot (sq ft) |
| 9 square feet | = 1 square yard (sq yd) |
| 43,560 square feet | = 1 acre |
| 640 acres | = 1 square mile (sq mi) |

**TABLE A.6**   Volume

| | |
|---|---|
| 1728 cubic inches (cu in.) | = 1 cubic foot (cu ft) |
| 27 cubic feet | = 1 cubic yard (cu yd) |
| 231 cubic inches | = 1 gallon (gal) |
| 7.48 gallons | = 1 cu ft. |

**TABLE A.7**   Dry Capacity

| | |
|---|---|
| 2 pints (pt) | = 1 quart (qt) |
| 8 quarts | = 1 peck (pk) |
| 4 pecks | = 1 bushel (bu) |

**TABLE A.8**   Liquid Capacity

| | |
|---|---|
| 16 fluid ounces (fl oz) | = 1 pint (pt) |
| 2 pints | = 1 quart (qt) |
| 4 quarts | = 1 gallon (gal) |
| $31\frac{1}{2}$ gallons | = 1 barrel (bbl) |

## *Metric Measurements*

**TABLE A.9**   Length

| | |
|---|---|
| 10 millimeters (mm) | = 1 centimeter (cm) |
| 10 centimeters | = 1 decimeter (dm) |
| 10 decimeters | = 1 meter (m) |
| 10 meters | = 1 dekameter (dam) |
| 10 dekameters | = 1 hectometer (hm) |
| 10 hectometers | = 1 kilometer (km) |

**TABLE A.10**   Weight

| | |
|---|---|
| 10 milligrams (mg) | = 1 centigram (cg) |
| 10 centigrams | = 1 decigram (dg) |
| 10 decigrams | = 1 gram (g) |
| 10 grams | = 1 dekagram (dag) |
| 10 dekagrams | = 1 hectogram (hg) |
| 10 hectograms | = 1 kilogram (kg) |

**TABLE A.11**   Liquid Capacity

| | |
|---|---|
| 10 milliliters (ml) | = 1 centiliter (cl) |
| 10 centiliters | = 1 deciliter (dl) |
| 10 deciliters | = 1 liter ($\ell$) |
| 10 liters | = 1 dekaliter (dal) |
| 10 dekaliters | = 1 hectoliter (hl) |
| 10 hectoliters | = 1 kiloliter (kl) |

**TABLE A.12**   Area

| | | CONVERSION NUMBER |
|---|---|---|
| 100 square millimeters | = 1 square centimeter | $10^2$ ($n = 2$) |
| 10,000 square centimeters | = 1 square meter | $10^4$ ($n = 4$) |
| 1,000,000 square meters | = 1 square kilometer | $10^6$ ($n = 6$) |

**TABLE A.13**   Volume

| | | CONVERSION NUMBER |
|---|---|---|
| 1000 cubic millimeters | = 1 cubic centimeter | $10^3$ ($n = 3$) |
| 1,000,000 cubic centimeters | = 1 cubic meter | $10^6$ ($n = 6$) |

# Approximate Metric Equivalents of U.S. Units

**TABLE B.1**   Length

1 inch  = 2.540 centimeters
1 foot  = 30.48 centimeters = 0.305 meter
1 yard = 0.914 meter
1 mile = 1.609 kilometers

**TABLE B.2**   Weight

1 ounce  = 28.350 grams
1 pound = 453.592 grams = 0.454 kilogram

**TABLE B.3**   Area

1 square inch  = 6.452 square centimeters
1 square foot  = 0.093 square meter
1 square yard = 0.836 square meter
1 acre          = 0.004 square kilometer
1 square mile = 2.590 square kilometers

**TABLE B.4**   Volume

1 cubic inch  = 16.387 cubic centimeters
1 cubic foot  = 0.0283 cubic meter
1 cubic yard = 0.765 cubic meter
1 pint          = 0.473 liter
1 quart        = 0.946 liter
1 gallon       = 3.785 liters

## Exercises for Section 1.1

| | | | |
|---|---|---|---|
| **1.** 23 | **3.** 72.9713 | **5.** 76.84659 | **7.** 9 |
| **9.** 16.9409 | **11.** 9.8699 | **13.** 40 | **15.** 666.28808 |
| **17.** 126 | **19.** 0.593928 | **21.** 125.3259 | **23.** 8 |
| **25.** 16.974183 | **27.** 0.22495659 | **29.** 0.5 | **31.** 5.8823529 |
| **33.** 7.425 | **35.** 5.5288182 | **37.** 10.662921 | **39.** 1.9416996 |
| **41.** 0.1251745 | **43.** 1.3809615 | **45.** 1.64 inches | **47.** 0.719 inch |
| **49.** 4.266 inches | **51. (a)** 0.0625 **(b)** 0.03125 | | |

## Exercises for Section 1.2

**1.** 0.06886    **3.** 0.89416    **5.** 0.58140
**7. (a)** 10  **(b)** 14.8  **(c)** 14.77  **(d)** 14.774
**9. (a)** 80  **(b)** 76.3  **(c)** 76.28  **(d)** 76.283
**11. (a)** 70  **(b)** 69.9  **(c)** 69.90  **(d)** 69.899
**13. (a)** 50  **(b)** 45.9  **(c)** 45.91  **(d)** 45.910
**15. (a)** 70  **(b)** 71.0  **(c)** 70.96  **(d)** 70.960
**17. (a)** 20  **(b)** 20.0  **(c)** 19.95  **(d)** 19.951
**19. (a)** 1.9  **(b)** 1.91  **(c)** 1.907
**21. (a)** 0.9  **(b)** 0.92  **(c)** 0.920
**23. (a)** 0.1  **(b)** 0.10  **(c)** 0.099
**25. (a)** 6.2  **(b)** 6.17  **(c)** 6.170
**27.** $1.29    **29.** $8.40    **31.** $47.29    **33.** $58.00
**35.** $38    **37.** $510    **39.** $601    **41.** $1234
**43. (a)** 2.5  **(b)** 2.46          **45. (a)** 2.7  **(b)** 2.66
**47. (a)** 71.0  **(b)** 70.96        **49. (a)** 2.6  **(b)** 2.58
**51. (a)** 2,100,000  **(b)** 2,150,000    **53. (a)** 2,000,000  **(b)** 1,990,000
**55.** 2.37 inches                    **57.** 1.39 inches

## Exercises for Section 1.3

**1.** 3    **3.** 3    **5.** 2    **7.** 4
**9.** 4    **11.** 4    **13.** 2    **15.** 3
**17.** 4    **19. (a)** 4  **(b)** 4.0  **(c)** 3.98
**21. (a)** 6  **(b)** 5.7  **(c)** 5.70        **23. (a)** 2000  **(b)** 2500  **(c)** 2480
**25. (a)** 0.01  **(b)** 0.012  **(c)** 0.0124    **27. (a)** 20  **(b)** 15  **(c)** 15.0
**29. (a)** 100  **(b)** 100  **(c)** 105        **31.** 2.21 amperes

## Exercises for Section 1.4

**1.** 57.76      **3.** 0.6724      **5.** 1339.56      **7.** 3.3489
**9.** 0.001225      **11.** 1.014049      **13.** 343      **15.** 59,049
**17.** 0.0256      **19.** 0.000729      **21.** 2.012196      **23.** 192.4647
**25.** 2      **27.** 6      **29.** 8      **31.** 30
**33.** 3.146      **35.** 8.222      **37.** 22.361      **39.** 0.825
**41.** 0.301      **43.** 1.094

## Exercises for Section 1.5

**1.** 4,860,000      **3.** 2930      **5.** 65,500,000      **7.** 0.000000939
**9.** 0.000710      **11.** 0.009490      **13.** $1.048 \times 10^1$      **15.** $2.236 \times 10^2$
**17.** $4.21 \times 10^{-1}$      **19.** $3.70 \times 10^{-2}$      **21.** $7.2905 \times 10^4$      **23.** $1.74 \times 10^{-4}$
**25.** (a) $6.7 \times 10^3$  (b) 6700      **27.** (a) $2.13 \times 10^2$  (b) 213
**29.** (a) $3 \times 10^{-5}$  (b) 0.00003      **31.** 0.0000137 in.
**33.** 0.0000046 in.      **35.** 0.0000079 in.      **37.** 0.000001 in.
**39.** $5.98 \times 10^{27}$      **41.** $3.0 \times 10^8$      **43.** $1.5 \times 10^{-5}$
**45.** $1.4 \times 10^{10}$      **47.** $6.6 \times 10^{-7}$      **49.** $4.7 \times 10^{-6}$
**51.** $1.4 \times 10^7$

## Exercises for Section 1.6

**1.** multiplication; 47      **3.** addition; 40      **5.** subtraction; 4
**7.** division; 12      **9.** multiplication; 16      **11.** division; 16
**13.** $60 \div (5 + 7)$; addition; 5      **15.** $12 \div (6 - 2)$; subtraction; 3
**17.** $(19 - 3) \div 4$; subtraction; 4      **19.** $(8 + 12) \div 4$; addition; 5
**21.** $16 \div (12 - 8)$; subtraction; 4      **23.** $30 \div (6 + 9)$; addition; 2
**25.** 64      **27.** 32      **29.** 100      **31.** 20
**33.** 17      **35.** 7      **37.** 9      **39.** 150
**41.** 39      **43.** $3 + (6 \div 2)$; 6      **45.** $10 - (12 \div 4)$; 7
**47.** $(15 \div 3) + 8$; 13      **49.** $(16 \div 2) - 3$; 5      **51.** $[2 + (2 \cdot 5)] \div 3$; 4
**53.** $[20 - (2 \cdot 4)] \div 4$; 3      **55.** $[(5 \cdot 4) - 8] \div 3$; 4
**57.** $16 \div [2 + (3 \cdot 2)]$; 2      **59.** $18 \div [(5 \cdot 2) - 4]$; 3
**61.** 38.9      **63.** 6.8      **65.** 2.0      **67.** 3.5      **69.** 4.9
**71.** 10.8      **73.** 12.1      **75.** 15.9      **77.** 0.98

## Exercises for Section 1.7

**1.** 11      **2.** 14      **3.** 17      **4.** 23
**5.** 9      **6.** 8      **7.** 13      **8.** 6
**9.** 42      **10.** 40      **11.** 72      **12.** 81
**13.** 3      **14.** 8      **15.** 4      **16.** 3
**17.** 11      **18.** 14      **19.** 3      **20.** 3
**21.** 3      **22.** 4      **23.** 3      **24.** 7
**25.** 5      **26.** 6      **27.** 4      **28.** 9
**29.** 8      **30.** 10      **31.** 4      **32.** 36
**33.** 64      **34.** 81      **35.** 8      **36.** 27
**37.** 64      **38.** 1000      **39.** 1342      **40.** 762
**41.** 621      **42.** 124      **43.** 12.2      **44.** 62
**45.** 83.4; 834; 83400      **46.** 91.7; 917; 91700
**47.** 526.2; 5262; 526200      **48.** 731.1; 7311; 731100
**49.** 9.12; 91.2; 9120      **50.** 7.86; 78.6; 7860

**51.** 0.34;  3.4;  340
**52.** 0.62;  6.2;  620
**53.** 70.1;  7.01;  0.0701
**54.** 52;  5.2;  0.052
**55.** 0.2625;  0.02625;  0.0002625
**56.** 0.1137;  0.01137;  0.0001137
**57.** 0.0219;  0.00219;  0.0000219
**58.** 0.0687;  0.00687;  0.0000687
**59.** 0.0043;  0.00043;  0.0000043
**60.** 0.0026;  0.00026;  0.0000026

*Answers to Exercises 61–79 are estimates. Your answers may differ.*
**61.** 110　　**62.** 240　　**63.** 140　　**64.** 90
**65.** 2600　　**66.** 2000　　**67.** 200　　**68.** 200
**69.** 400　　**70.** 3800　　**71.** 500　　**72.** 80
**73.** 10,000　　**74.** 14000　　**75.** 280,000　　**76.** 3,150,000
**77.** 0.0048　　**78.** 0.035　　**79.** 700　　**80.** 0.0035

## Chapter 1 Review Exercises

**1.** 1.7857143　　**2.** 0.01785714　　**3.** 25.619　　**4.** 50.5025
**5.** 2072.325　　**6.** 14.882765　　**7.** 9.82407　　**8.** 0.43394156
**9.** 0.44660　　**10.** 3.9431
**11. (a)** 30　**(b)** 32.8　**(c)** 32.82　**(d)** 32.820　　**12. (a)** 50　**(b)** 45.3　**(c)** 45.27　**(d)** 45.268
**13.** \$2.37　　　　　　　　　　　　　**14.** \$3.12
**15. (a)** 3.7　**(b)** 3.66　　　　　　**16. (a)** 23.8　**(b)** 23.85
**17.** 1.7424　　**18.** 0.0729　　**19.** 0.59049　　**20.** 263.76683
**21.** 7　　**22.** 9　　**23.** 1.4387495　　**24.** 4.3931765
**25.** 84,200　　**26.** 0.00000145　　**27.** $1.23 \times 10^{-3}$　　**28.** $3.75 \times 10^4$
**29.** 2.01　　**30.** $8.2 \times 10^{-6}$　　**31.** 42　　**32.** $3.4 \times 10^{-4}$
**33.** multiplication;  31　　　　**34.** subtraction;  3
**35.** $40 \div (5 \times 4)$;  multiplication;  2　　**36.** $28 \div (5 + 9)$;  addition;  2
**37.** 43　　　　　　　　　　　　　**38.** 72
**39.** $[6 + (2 \cdot 6)] \div 9$;  2　　　　**40.** $24 \div [(5 \cdot 3) - 7]$;  3
**41.** 71　　**42.** 4.0　　**43.** 4.0　　**44.** 21
**45.** 77　　**46.** 65　　**47.** 490　　**48.** 0.14
**49.** 12　　**50.** 18　　**51.** 3　　**52.** 4
**53.** 4　　**54.** 1　　**55.** 25　　**56.** 169
**57.** 12　　**58.** 42　　**59.** 2　　**60.** 2
**61.** 5　　**62.** 6　　**63.** 5　　**64.** 13
**65.** 68;  680;  6800　　　　**66.** 122;  1220;  12200
**67.** 4.3;  43;  430　　　　**68.** 0.8;  8;  80
**69.** 43.9;  4.39;  0.439　　**70.** 2.8;  0.28;  0.028
**71.** 0.14;  0.014;  0.0014　　**72.** 0.081;  0.0081;  0.00081

## Exercises for Section 2.1

**1.** Yes;  $244 \div 4 = 61$　　**3.** No;  $142 \div 3 = 47.333$　　**5.** Yes;  $395 \div 79 = 5$
**7.** No;  $455 \div 21 = 21.666$　　**9.** No;  $752 \div 43 = 17.488$　　**11.** Yes;  $696 \div 29 = 24$
**13.** $2 \cdot 7$　　**15.** $3 \cdot 11$　　**17.** $2 \cdot 5$　　**19.** $3 \cdot 13$　　**21.** $2 \cdot 5 \cdot 7$　　**23.** $5 \cdot 13$
**25.** Divisible by 2, 3, and 5　　**27.** Divisible by 2 and 3　　**29.** Divisible by 3
**31.** Divisible by 3　　**33.** Divisible by 3 and 5　　**35.** Divisible by 2 and 3
**37.** $2^2$　　**39.** $3^3$　　**41.** $7^2$　　**43.** $2^2 \cdot 29$　　**45.** $2^2 \cdot 3^3 \cdot 5$　　**47.** $2^2 \cdot 3^3 \cdot 7$
**49.** $\sqrt{211} = 14.526$;  211 is not divisible by 2, 3, 5, 7, 11, or 13
**51.** $\sqrt{839} = 28.965$;  839 is not divisible by 2, 3, 5, 7, 11, 13, 17, 19, or 23
**53.** $2 \cdot 127$　　**55.** $2^2 \cdot 41$　　**57.** $2^2 \cdot 47$　　**59.** $3 \cdot 131$
**61.** $5 \cdot 83$　　**63.** $2 \cdot 373$　　**65.** 1409

## Exercises for Section 2.2

**1.** 0.875        **3.** 0.417        **5.** 0.594        **7.** 1.833
**9.** 1.762        **11.** 2.278       **13.** 3.563       **15.** 4.375
**17.** 5.833       **19.** 13.778      **21.** 11.713      **23.** 20.979
**25.** 45.813      **27.** 15.042      **29.** 109.969     **31.** 0.712
**33.** 0.562       **35.** 2.533       **37.** 98.81       **39.** 1022.63
**41.** 597.06      **43.** 12.39

## Exercises for Section 2.3

**1.** $5 \cdot 56 = 8 \cdot 35$;   equal
**3.** $17 \cdot 96 \neq 32 \cdot 52$;   not equal
**5.** $7 \cdot 507 \neq 11 \cdot 329$;   not equal

**7.** $\dfrac{4}{7}$        **9.** $\dfrac{15}{17}$       **11.** $\dfrac{5}{7}$       **13.** $\dfrac{7}{17}$

**15.** $\dfrac{5}{16}$      **17.** $\dfrac{2}{7}$        **19.** $\dfrac{3}{4}$       **21.** $\dfrac{1}{3}$

**23.** $\dfrac{4}{11}$      **25.** $\dfrac{3}{5}$        **27.** $\dfrac{1}{5}$       **29.** $\dfrac{3}{7}$

**31.** $\dfrac{12}{24}$     **33.** $\dfrac{21}{35}$      **35.** $\dfrac{25}{40}$     **37.** $\dfrac{21}{36}$

**39.** $\dfrac{55}{80}$     **41.** $\dfrac{81}{144}$     **43.** $\dfrac{48}{128}$    **45.** $\dfrac{30}{105}$

**47.** $\dfrac{30}{108}$    **49.** $\dfrac{15}{135}$     **51.** $\dfrac{12}{12}$     **53.** $\dfrac{80}{16}$

## Exercises for Section 2.4

**1.** $\dfrac{5}{7}$        **3.** $\dfrac{3}{8}$        **5.** $\dfrac{4}{7}$        **7.** $\dfrac{1}{16}$

**9.** $\dfrac{13}{16}$      **11.** $\dfrac{3}{32}$      **13.** $\dfrac{1}{2}$       **15.** $\dfrac{3}{8}, \dfrac{6}{8}$

**17.** $\dfrac{21}{48}, \dfrac{8}{48}$    **19.** $\dfrac{40}{96}, \dfrac{9}{96}$    **21.** $\dfrac{12}{48}, \dfrac{32}{48}, \dfrac{3}{48}$    **23.** $\dfrac{10}{36}, \dfrac{4}{36}, \dfrac{3}{36}$

**25.** $\dfrac{300}{60}, \dfrac{8}{60}, \dfrac{9}{60}$    **27.** $\dfrac{17}{12}$    **29.** $\dfrac{37}{40}$    **31.** $\dfrac{7}{30}$

**33.** $\dfrac{13}{40}$     **35.** $\dfrac{19}{16}$     **37.** $\dfrac{41}{64}$     **39.** $\dfrac{19}{32}$

**41.** $\dfrac{173}{48}$    **43.** $\dfrac{7}{8}$       **45.** $\dfrac{3}{8}, \dfrac{5}{8}$    **47.** $\dfrac{1}{8}, \dfrac{1}{4}, \dfrac{5}{16}$

**49.** $\dfrac{15}{32}, \dfrac{9}{16}, \dfrac{37}{64}, \dfrac{7}{8}$    **51.** $\dfrac{3}{8}$ in.;   $\dfrac{1}{2}$ in.;   $\dfrac{9}{16}$ in.    **53.** $\dfrac{3}{4}$ in.;   $\dfrac{7}{8}$ in.;   1 in.

**55.** $\dfrac{7}{16}$ in.;   $\dfrac{1}{2}$ in.;   $\dfrac{5}{8}$ inch    **57.** $\dfrac{15}{8}$ inches    **59.** $\dfrac{7}{16}$ inch

## Exercises for Section 2.5

**1.** $\dfrac{64}{5}$       **3.** $\dfrac{341}{16}$      **5.** $\dfrac{939}{31}$      **7.** $5\dfrac{3}{8}$

**9.** $6\frac{11}{16}$    **11.** $9\frac{20}{33}$    **13.** 12    **15.** $11\frac{1}{8}$

**17.** $8\frac{19}{40}$    **19.** $7\frac{7}{12}$    **21.** $3\frac{3}{8}$    **23.** $4\frac{7}{15}$

**25.** $7\frac{5}{8}$    **27.** $8\frac{11}{12}$    **29.** $2\frac{13}{20}$    **31.** $9\frac{21}{32}$

**33.** $6\frac{7}{8}$    **35.** $3\frac{7}{12}$    **37.** $4\frac{3}{4}$    **39.** 12 feet

**41.** 12 feet    **43.** $26\frac{7}{12}$ feet    **45.** $1\frac{7}{12}$ feet    **47.** $1\frac{13}{16}$ inches

## Exercises for Section 2.6

**1.** $\frac{1}{12}$    **3.** $\frac{2}{15}$    **5.** $\frac{5}{14}$    **7.** $\frac{11}{96}$

**9.** 40    **11.** 36    **13.** $28\frac{1}{2}$    **15.** $16\frac{1}{2}$

**17.** $534\frac{5}{8}$    **19.** 9    **21.** $\frac{1}{5}$    **23.** $\frac{5}{3}$

**25.** $\frac{2}{7}$    **27.** $1\frac{1}{5}$    **29.** $\frac{1}{15}$    **31.** $19\frac{1}{2}$

**33.** 2    **35.** $2\frac{1}{2}$    **37.** $7\frac{1}{3}$    **39.** $1\frac{4}{5}$

**41.** **(a)** $\frac{3}{8}$ inch    **(b)** $\frac{5}{8}$ inch    **(c)** $\frac{15}{16}$ inch    **(d)** $\frac{7}{24}$ inch

**43.** **(a)** 6 amperes   **(b)** 8 amperes    **(c)** $1\frac{1}{4}$ amperes    **(d)** 17 amperes

## Exercises for Section 2.7

**1.** $2^2$    **2.** $2^2 \cdot 3$    **3.** $2 \cdot 3^2$    **4.** $3 \cdot 5$
**5.** $2^4$    **6.** $3^3$    **7.** $2 \cdot 3 \cdot 5$    **8.** $2 \cdot 19$

**9.** $\frac{1}{2}$    **10.** $\frac{1}{2}$    **11.** $\frac{2}{3}$    **12.** $\frac{2}{3}$

**13.** $\frac{2}{3}$    **14.** $\frac{3}{4}$    **15.** $\frac{3}{5}$    **16.** $\frac{2}{3}$

**17.** 8    **18.** 18    **19.** 18    **20.** 15
**21.** 28    **22.** 20    **23.** 6    **24.** 12

**25.** 18    **26.** 16    **27.** $\frac{1}{2}$    **28.** 1

**29.** $1\frac{1}{4}$    **30.** 2    **31.** $5\frac{1}{2}$    **32.** $5\frac{1}{2}$

**33.** $7\frac{1}{4}$    **34.** $8\frac{1}{2}$    **35.** $5\frac{15}{16}$    **36.** $\frac{8}{9}$

**37.** $\frac{5}{8}$    **38.** $\frac{15}{16}$    **39.** $\frac{5}{4}$    **40.** $\frac{5}{2}$

**41.** $\frac{10}{3}$    **42.** $\frac{21}{8}$    **43.** $\frac{22}{5}$    **44.** $\frac{23}{4}$

**45.** $\frac{38}{7}$    **46.** $\frac{39}{16}$    **47.** $\frac{93}{10}$    **48.** $\frac{37}{10}$

**49.** $\frac{517}{100}$    **50.** $\frac{393}{100}$    **51.** $1\frac{1}{2}$    **52.** $4\frac{3}{4}$

**53.** $2\frac{3}{8}$     **54.** $2\frac{1}{3}$     **55.** $2\frac{7}{10}$     **56.** $2\frac{1}{4}$

**57.** $5\frac{2}{5}$     **58.** $4\frac{2}{3}$     **59.** $\frac{1}{2}$     **60.** $\frac{1}{3}$

**61.** $\frac{5}{7}$     **62.** $1\frac{3}{4}$     **63.** $1\frac{5}{8}$     **64.** $\frac{1}{16}$

**65.** $\frac{1}{9}$     **66.** $\frac{3}{16}$     **67.** $\frac{7}{20}$     **68.** $\frac{3}{4}$

**69.** 1     **70.** 0     **71.** $\frac{1}{4}$     **72.** $\frac{1}{3}$

**73.** 1     **74.** 6     **75.** 10     **76.** 9

**77.** 2     **78.** 1     **79.** 1     **80.** $\frac{5}{4}$

**81.** $\frac{1}{4}$     **82.** 5     **83.** $\frac{2}{3}$     **84.** 1

**85.** $\frac{1}{5}$     **86.** $\frac{1}{2}$     **87.** 12     **88.** 18

## Chapter 2 Review Exercises

**1.** Yes;   $936 \div 26 = 36$     **2.** No;   $884 \div 48 = 18.417$

**3.** $2 \cdot 3 \cdot 7$     **4.** $2 \cdot 3 \cdot 11$     **5.** $11 \cdot 37$     **6.** $2^2 \cdot 107$

**7.** 0.34375     **8.** 1.1578947     **9.** 2.3125     **10.** 93.375

**11.** 13.854     **12.** 11.156     **13.** 36.875     **14.** 0.651

**15.** $5 \cdot 80 = 16 \cdot 25$;   equal     **16.** $15 \cdot 64 \neq 32 \cdot 31$;   not equal

**17.** $\frac{11}{36}$     **18.** $\frac{2}{3}$     **19.** $\frac{5}{7}$     **20.** $\frac{9}{16}$

**21.** $\frac{12}{32}$     **22.** $\frac{75}{180}$     **23.** $\frac{14}{35}, \frac{15}{35}$     **24.** $\frac{9}{12}, \frac{5}{12}$

**25.** $\frac{20}{30}, \frac{25}{30}, \frac{4}{30}$     **26.** $\frac{288}{72}, \frac{20}{72}, \frac{21}{72}$

**27.** $\frac{5}{13}, \frac{6}{13}, \frac{8}{13}$     **28.** $\frac{1}{7}, \frac{9}{25}, \frac{2}{5}$

**29.** $\frac{1}{2}$     **30.** $\frac{11}{15}$     **31.** $\frac{7}{36}$     **32.** $\frac{7}{32}$

**33.** $\frac{51}{8}$     **34.** $9\frac{2}{15}$     **35.** $9\frac{5}{7}$     **36.** $22\frac{7}{16}$

**37.** $7\frac{5}{8}$     **38.** $12\frac{7}{12}$     **39.** $11\frac{1}{6}$     **40.** $4\frac{7}{10}$

**41.** $\frac{3}{10}$     **42.** $\frac{1}{10}$     **43.** $\frac{1}{4}$     **44.** 52

**45.** $2\frac{1}{3}$     **46.** $\frac{5}{13}$     **47.** $1\frac{1}{4}$     **48.** $\frac{1}{16}$

**49.** 2     **50.** $\frac{7}{8}$

## Exercises for Section 3.1

**1.**

**3.**

**5.**

-10   -5   0   5   10

-10   -5   0   5   10

**7.** $-4$      **9.** $-5$

**11.** $+6$     **13.** $+10$     **15.** $-7$     **17.** $+8$
**19.** $a < b$     **21.** $c < d$     **23.** $d > e$     **25.** $-15 < +8$
**27.** $+8 < +17$     **29.** $-0.9 > -1.2$     **31.** $-10 < -7$     **33.** $+1.1 > +0.3$
**35.** $+1 > -10$     **37.** $3.2$     **39.** $2.15$     **41.** $6.4$
**43.** $-6$     **45.** $+4$     **47.** $-2.4$

## Exercises for Section 3.2

**1.** $8 + 6$;   $14$     **3.** $3 + 3$;   $6$     **5.** $-6 + (-7)$;   $-13$
**7.** $-8 + (-1)$;   $-9$     **9.** $7 + (-4)$;   $3$     **11.** $4 + (-8)$;   $-4$
**13.** $-5 + 9$;   $4$     **15.** $-4 + 6$;   $2$     **17.** $-6 + 6$;   $0$
**19.** $-5 - 2$;   $-7$     **21.** $-7 + 4$;   $-3$     **23.** $-9 + 4$;   $-5$
**25.** $9 + 2$;   $11$     **27.** $-10 - 3$;   $-13$     **29.** $-6 + 4$;   $-2$
**31.** $-26.2$     **33.** $4.2$     **35.** $0.86$
**37.** $7.50$     **39.** $45.43$     **41.** $-0.57$
**43.** $20.07$     **45.** $-8.54$     **47.** $2.0°$
**49.** $0.8°$     **51.** $-1.6°$

## Exercises for Section 3.3

**1.** $-12$     **3.** $-10$     **5.** $15$     **7.** $-6$
**9.** $4$     **11.** $-2$     **13.** $399.9$     **15.** $0.706$
**17.** $-5.52$     **19.** $0.743$     **21.** $-0.040$     **23.** $-0.600$
**25.** $-3.31$     **27.** $-75.5$     **29.** $136$     **31.** $2.48$
**33.** $-4.76$     **35.** $25.2$     **37.** $-32$     **39.** $4.5949728$
**41.** $16304.736$     **43.** $0.017576$     **45.** $6.13$     **47.** $-36.63$
**49.** $70.81$     **51.** $-106.844$     **53.** $-15.13$

## Exercises for Section 3.4

**1.** $\dfrac{1}{2^5}$     **3.** $\dfrac{1}{5^2}$     **5.** $2^5$     **7.** $5^2$

**9.** $\dfrac{1}{4^{-6}}$     **11.** $\dfrac{1}{8^{-3}}$     **13.** $4^{-6}$     **15.** $8^{-3}$

**17.** $0.0313$     **19.** $0.0625$     **21.** $0.579$     **23.** $3.32$
**25.** $0.0370$     **27.** $0.000772$     **29.** $0.269$     **31.** $41.0$
**33.** $5^3$;   $125$     **35.** $4^3$;   $64$     **37.** $3^{-3}$;   $0.03703704$     **39.** $(1.2)^{-2}$;   $0.69444444$
**41.** $6^3$;   $216$     **43.** $(0.45)^4$;   $0.04100625$     **45.** $2^5$;   $32$     **47.** $5^{-3}$;   $0.008$
**49.** $(6.2)^{-4}$;   $0.00067675$     **51.** $3^1$;   $3$     **53.** $(1.3)^2$;   $1.44$     **55.** $5^{-1}$;   $0.2$
**57.** $10^3$;   $1000$     **59.** $0.01234568$     **61.** $-0.1859344$     **63.** $-0.08112675$
**65.** $6.574622$

## Exercises for Section 3.5

**1.** $+200$     **2.** $+100$     **3.** $+30$     **4.** $-30$
**5.** $+30$     **6.** $+70$     **7.** $+30$     **8.** $-230$
**9.** $-20$     **10.** $-40$     **11.** $-300$     **12.** $-100$
**13.** $+10$     **14.** $-10$     **15.** $+250$     **16.** $+400$
**17.** $-70$     **18.** $+800$     **19.** $-150$     **20.** $-240$
**21.** $+200$     **22.** $-160$     **23.** $+100$     **24.** $+10$

**25.** +20      **26.** +20      **27.** $\dfrac{1}{9}$      **28.** $\dfrac{1}{8}$

**29.** 6      **30.** 2      **31.** 2      **32.** 3

**33.** 10      **34.** 100      **35.** 1000      **36.** 0.001

**37.** 10      **38.** 10      **39.** 100      **40.** 10

## Chapter 3 Review Exercises

**1.**

**2.** −3;   +6      **3.** $a > b$

**4.** $d < c$      **5.** +3 > −4      **6.** +3 < +4      **7.** −3 > −4

**8.** 3.7      **9.** 4.2      **10.** +2      **11.** −3

**12.** +4.1      **13.** −6 + 2;   −4      **14.** 4 + (−2);   +2      **15.** −9 + 2;   −7

**16.** −7 + (−4);   −11      **17.** 12 − 2 + 6;   +16      **18.** 16 + 4 − 1 − 2;   +17      **19.** −10.5

**20.** 36.3      **21.** −36.3      **22.** 10.5      **23.** −34.13

**24.** 25.56      **25.** −30      **26.** 15      **27.** −4

**28.** 5      **29.** 2.5971892      **30.** −16.718972      **31.** 81

**32.** −343      **33.** 1.2538462      **34.** −3.5352941      **35.** 39.1801

**36.** −146.4134      **37.** $\dfrac{1}{5^3}$      **38.** $4^2$      **39.** $3^{-4}$

**40.** $\dfrac{1}{6^{-3}}$      **41.** 0.1080      **42.** 1.731      **43.** 0.4823

**44.** −0.006717      **45.** $3^{-2}$;   0.111      **46.** $(0.87)^3$;   0.659      **47.** $(0.15)^{-3}$;   296.296

**48.** $3^{-5}$;   0.004      **49.** +30      **50.** −70      **51.** +90

**52.** +20      **53.** −100      **54.** +60      **55.** +6

**56.** 0      **57.** $10^1$ or 10      **58.** $10^{-2}$ or 0.01      **59.** 10

**60.** 100

## Exercises for Section 4.1

**1.** 23      **3.** 1      **5.** −1.2

**7.** 12.93      **9.** 12      **11.** 3.13

**13.** (a) 0.051   (b) 0.05      **15.** (a) 3.8982036   (b) 3.90

**17.** (a) 0.35925926   (b) 0.36      **19.** (a) 6.6   (b) 6.60

**21.** (a) 11.664   (b) 11.66      **23.** (a) 2.4255952   (b) 2.43

**25.** (a) 1.875   (b) 1.9      **27.** (a) 0.95301205   (b) 0.95

**29.** (a) −0.57142851   (b) −0.57      **31.** (a) 16.372549   (b) 16

**33.** (a) −1.425   (b) −1.4      **35.** (a) 17.490566   (b) 17

**37.** (a) −0.26210526   (b) −0.26      **39.** (a) 4.7390411   (b) 4.7

**41.** 57.5      **43.** 6.93      **45.** −11.8      **47.** 212

**49.** 5      **51.** 0.750      **53.** 0.938      **55.** 8

## Exercises for Section 4.2

**1.** 0.9      **3.** 3.77      **5.** 2.13      **7.** −2.12

**9.** 4.32      **11.** 10.4      **13.** 13.0      **15.** 2.9

**17.** −3.1      **19.** 0.39      **21.** −0.59      **23.** 0.69

**25.** 0.700      **27.** 0.201      **29.** −0.450      **31.** 19.5

**33.** 9.22      **35.** 0.211      **37.** 113.6      **39.** 2436.3

**41.** 69,242.0      **43.** 11.18      **45.** 31.83      **47.** 1.17

**49.** 2.26      **51.** 1.21      **53.** 7.34      **55.** 10.09

## Exercises for Section 4.3

**1.** 489.3 miles    **3.** 89.7 miles per hour    **5.** 4.8 minutes    **7.** 1044.8
**9.** 77,181    **11.** 1.3    **13.** 1.0    **15.** 7.6
**17.** 149    **19.** 122.4    **21.** −34.6    **23.** −51.7
**25.** 85,753.7    **27.** 768.5    **29.** 3.3    **31.** 115.0
**33.** 2.0    **35.** 290.1    **37.** 0.776    **39.** 5.579
**41.** 0.551    **43.** 1.0    **45.** 22.0    **47.** 3.1

## Exercises for Section 4.4

**1.** $l = 16 - d$    **3.** $g = f - 5$    **5.** $x = 7 - y$    **7.** $t = \dfrac{500}{r}$

**9.** $V = \dfrac{37.5}{P}$    **11.** $R = \dfrac{E^2}{P}$    **13.** $R = \dfrac{P}{I^2}$    **15.** $R = \dfrac{E}{I}$

**17.** $A = \dfrac{10L}{R}$    **19.** $I = \dfrac{1374H}{E}$    **21.** $R_x = \dfrac{R_1 R_3}{R_2}$    **23.** $P = \dfrac{I}{rt}$

**25.** $w = \dfrac{V}{lh}$    **27.** $P_f = \dfrac{P}{IE}$    **29.** $I = \dfrac{746H}{EE_p}$    **31.** $d = \dfrac{l - 16}{n}$

**33.** $I = \dfrac{E - 220}{R}$    **35.** $I = \dfrac{7.8 - E}{R}$    **37.** $I = \sqrt{\dfrac{P}{R}}$    **39.** $D = \sqrt{\dfrac{2.5H}{N}}$

**41.** $N = \sqrt{\dfrac{LI}{4\pi A}}$    **43.** $a = \sqrt{c^2 - b^2}$    **45.** $y = \sqrt{x^2 - z^2}$    **47.** $x = \sqrt{\dfrac{36 - 9y^2}{4}}$

## Exercises for Section 4.5

**1.** 15 inches    **3.** 120.5 kilograms    **5.** 1.52 centimeters    **7.** 1.48 centimeters
**9.** 0.75 inch    **11.** 2.466 inches    **13.** 4.76 inches    **15.** 283 tons

**17.** $\dfrac{5}{8}$ inch    **19.** $\dfrac{7}{8}$ inch    **21.** Sum not zero; −12 volts

**23.** Sum not zero; −15 volts    **25.** 62 volts    **27.** −22 volts

## Exercises for Section 4.6

**1.** **(a)** $m = 23 \times 19$   **(b)** 437 miles    **3.** **(a)** $p = 45 \times 58$   **(b)** 2610 parts

**5.** **(a)** $d = \dfrac{4625}{37}$   **(b)** \$125    **7.** **(a)** $m = \dfrac{768}{48}$   **(b)** 16 miles per gallon

**9.** **(a)** $l = 35 \times 125$   **(b)** 4375 lines    **11.** **(a)** $l = \dfrac{3300}{25}$   **(b)** 132 lines per minute

**13.** **(a)** $d = 12 \times 175$   **(b)** \$2100    **15.** **(a)** $y = \dfrac{704}{16}$   **(b)** 44 yards

**17.** M-1: \$18,237.80;   M-2: \$13,688.40;   M-3: \$18,048.15;   M-4: \$36,331.25
**19.** A134: \$957.76;   A270: \$2691.54;   B985: \$4531.68;   Grand total: \$8180.98
**21.** **(a)** 441.4   **(b)** 417.4   **(c)** 829.3   **(d)** 490
**23.** **(a)** 56   **(b)** 67   **(c)** 53   **(d)** 62   Average: 60
**25.** \$207.92
**27.** *A*: 18.99;   *B*: 22.50;   *C*: 13.10;   *D*: 27.30;   Average: 20.47
**29.** **(a)** −34.5°   **(b)** −14.7°    **31.** **(a)** 99.97 inches   **(b)** 100.04 inches

## Exercises for Section 4.7

**1.** $\frac{5}{6}$;  5 and 6          **3.** $\frac{1.2}{4.3}$;  1.2 and 4.3

**5.** $\frac{13.2}{2.12}$;  13.2 and 2.12          **7.** $\frac{9.7}{3.5}$;  9.7 and 3.5

**9. (a)** Extremes: 6 and 4.375  **(b)** Means: 7 and 3.75  **(c)** $6 \times 4.375 = 7 \times 3.75 = 26.25$
**11. (a)** Extremes: 3.5 and 4.6875  **(b)** Means: 7.5 and 2.1875
  **(c)** $3.5 \times 4.6875 = 7.5 \times 2.1875 = 16.40625$
**13. (a)** Extremes: 42.7 and 3.3  **(b)** Means: 13.2 and 10.675
  **(c)** $42.7 \times 3.3 = 13.2 \times 10.675 = 140.91$
**15.** $n = 14.91$     **17.** $d = 3.94$     **19.** $t = 30.28$     **21.** $s = 0.68$
**23.** $r = 6.54$     **25.** $a = 1.42$     **27.** \$231.73     **29.** \$20.97
**31.** 18 inches     **33.** \$338     **35.** 165 gallons     **37.** 32 inches
**39. (a)** 128 miles  **(b)** 160 miles  **(c)** 256 miles  **(d)** 230 miles  **(e)** 376 miles
**41.** 75 teeth     **43.** 300 horsepower

## Exercises for Section 4.8

**1.** $x = 20$     **2.** $x = 35$     **3.** $t = -20$     **4.** $t = -10$
**5.** $r = 30$     **6.** $r = 25$     **7.** $s = 50$     **8.** $s = 50$
**9.** $p = 4$     **10.** $p = 3$     **11.** $n = 3$     **12.** $n = 4$
**13.** $x = 12$     **14.** $x = 30$     **15.** $t = 4$     **16.** $t = 3$
**17.** $t = 4$     **18.** $t = 9$     **19.** 3500 feet     **20.** 2000 pieces
**21.** \$16.00     **22.** 18,000 pieces     **23.** $V = 400$     **24.** $h = 4$

## Chapter 4 Review Exercises

**1.** $r = 10.29$     **2.** $s = -2.9$     **3.** $t = 5.56$          **4.** $d = 3.21$
**5. (a)** 6.5644171  **(b)** 6.564  **(c)** 6.56
**6. (a)** 4.8462222  **(b)** 4.846  **(c)** 4.85
**7. (a)** 6.3692308  **(b)** 6.369  **(c)** 6.37
**8. (a)** 8.9068965  **(b)** 8.907  **(c)** 8.91
**9.** 89.0          **10.** 1.8          **11.** $-2.9$          **12.** 0.2
**13.** 12.0          **14.** 7.4          **15.** 26.303          **16.** 29.336

**17.** 19.161          **18.** 225.960          **19.** $d = l - 12$     **20.** $m = \dfrac{E}{c^2}$

**21.** $l = \dfrac{V}{wh}$          **22.** $c = \sqrt{\dfrac{E}{m}}$          **23.** 1827 feet          **24.** 0.8525 inch

**25.** 0.931 inch     **26.** 0.433
**27. (a)** $h = 52 \times 20$  **(b)** 1040 hours     **28. (a)** $g = 12 \times 141.3$  **(b)** 1695.6 grams

**29. (a)** $p = \dfrac{27{,}800}{8}$  **(b)** 3475 pieces     **30. (a)** $l = \dfrac{84}{6}$  **(b)** 14 inches

**31.** $\frac{5}{7}$; 5 and 7     **32. (a)** Extremes: 30.6 and 2.7  **(b)** Means: 16.2 and 5.1
                    **(c)** $30.6 \times 2.7 = 16.2 \times 5.1 = 82.62$

**33.** 4.41          **34.** 46.3     **35.** \$3.97     **36.** 15,306 pieces
**37.** 12          **38.** 12     **39.** 4          **40.** 6
**41.** 24,000 pieces     **42.** 240

## Exercises for Section 5.1

**1.** 33%  **3.** 70%  **5.** 8%  **7.** 50.4%
**9.** 78.7%  **11.** 2.1%  **13.** 0.8%  **15.** 550%
**17.** 0.15  **19.** 0.0027  **21.** 0.068  **23.** 0.027
**25.** 1.19  **27.** 2.59  **29.** 0.0325  **31.** 0.075
**33.** 0.02375  **35.** 0.084  **37.** 0.16333333  **39.** 0.15666667
**41.** 0.005  **43.** 0.00166667  **45.** 75%  **47.** 37.5%
**49.** 60%  **51.** 225%  **53.** 240%  **55.** 0.4%
**57. (a)** 28.6%  **(b)** 28.57%  **59. (a)** 41.7%  **(b)** 41.67%
**61. (a)** 45.8%  **(b)** 45.83%  **63. (a)** 83.3%  **(b)** 83.33%
**65. (a)** 216.7%  **(b)** 216.67%  **67. (a)** 4.2%  **(b)** 4.17%

**69.** $\dfrac{1}{4}$  **71.** $\dfrac{3}{4}$  **73.** $\dfrac{5}{4}$  **75.** $\dfrac{9}{1000}$

**77.** $\dfrac{3}{8}$  **79.** $\dfrac{61}{400}$  **81.** $\dfrac{1}{3}$  **83.** $\dfrac{5}{6}$

## Exercises for Section 5.2

**1.** 24.06  **3.** 49.04  **5.** 61.87  **7.** 3.88
**9.** 234.49  **11.** 152.68  **13.** 0.07  **15.** 0.04
**17.** 0.32  **19.** 1142.93  **21.** 37.23  **23.** 46.84
**25.** 3.05  **27.** 18.98  **29.** 30.84  **31.** 19.1 amperes
**33.** 26.4 amperes  **35.** 232.7 volts  **37.** 282.0 volts  **39.** 6.05 kilograms
**41.** 9.51 kilograms  **43.** 43.44 kilograms  **45.** $1895.20  **47.** 2200 rpm
**49.** 2.296 kilograms

## Exercises for Section 5.3

**1. (a)** $8 = P \times 25$  **(b)** 32.0%  **3. (a)** $16\frac{1}{2} = P \times 64$  **(b)** 25.8%
**5. (a)** $102 = P \times 42$  **(b)** 242.9%  **7. (a)** $1.56 = P \times 63$  **(b)** 2.5%
**9.** 71.2%  **11.** 57.0%  **13.** 20.6%  **15.** 11.4%
**17.** 13.9%  **19.** 0.4%  **21.** 4.1%
**23. (a)** $68 = 0.27 \times n$  **(b)** 251.9  **25. (a)** $80 = 0.6 \times n$  **(b)** 133.3
**27. (a)** $212 = 0.0775 \times n$  **(b)** 2735.5  **29. (a)** $443 = 0.0605 \times n$  **(b)** 7322.3
**31. (a)** $555.069 = 1.06 \times n$  **(b)** 523.7
**33.** 57.0 amperes  **35.** 4680  **37.** 118.9%  **39.** 5.69
**41.** 141.52  **43.** 83.6 liters  **45.** 31.3%

## Exercises for Section 5.4

**1.** $67.63  **3.** $123.00
**5.** Chain: $25.92;  camshaft: $74.79;  water pump: $29.56;  fuel pump: $22.46  **7.** $125
**9.** $54.41  **11.** $283.42  **13.** $443.16
**15.** First supplier: $452.83;  second supplier: $447.24;  second is better
**17.** $468.59  **19.** $404.18  **21.** $7477.74
**23.** Axe: $10.87;  box: $105.05;  vise: $46.67;  saw: $13.08
**25. (a)** $888.22  **(b)** $799.40  **27. (a)** $2029.01  **(b)** $1680.22
**29.** $122,262  **31.** $49.49  **33.** $397.98  **35.** $2128.50
**37.** $86.10  **39.** $32.01  **41.** $9.96

**43. (a)** $14.44  **(b)** $1596.67
**45. (a)** $73.33  **(b)** $7623.33  **(c)** $69.88  **(d)** $7243.21

## Exercises for Section 5.5

**1.** 6.72       **2.** 7.18       **3.** 8.92       **4.** 3.35
**5.** 20        **6.** 300       **7.** 200       **8.** 3000
**9.** 10        **10.** 100      **11.** 50%      **12.** 25%
**13.** $66\frac{2}{3}\%$   **14.** $33\frac{1}{3}\%$   **15.** 75%      **16.** 30%
**17.** 160      **18.** 50       **19.** 80       **20.** 48
**21.** 180      **22.** 120      **23.** 5        **24.** 30
**25.** 2        **26.** 6        **27.** 16       **28.** 15
**29.** $16\frac{2}{3}\%$   **30.** $83\frac{1}{3}\%$

*Answers to Exercises 31–38 are estimates. Your answers may differ.*
**31.** 9        **32.** 20       **33.** 22       **34.** 20
**35.** $33\frac{1}{3}\%$   **36.** 20%      **37.** 10%      **38.** 50%

## Chapter 5 Review Exercises

**1.** 35%         **2.** 8.7%          **3.** 500%        **4.** 0.16
**5.** 2.15        **6.** 0.003         **7.** 0.0225      **8.** 0.0075
**9.** 0.21666667  **10.** 40%          **11.** 0.25%
**12. (a)** 57.1%  **(b)** 57.14%       **13. (a)** 136.4%  **(b)** 136.36%
**14.** $\frac{3}{4}$       **15.** $\frac{7}{8}$          **16.** 41.91       **17.** 2.76
**18.** 24.92      **19.** 26.32        **20.** 54.82       **21.** 77.43
**22.** 2.30       **23.** 61.08        **24.** 68.27       **25.** 63.78
**26. (a)** $17 = P \times 32$  **(b)** 53.1%   **27. (a)** $16.3 = P \times 70$  **(b)** 23.3%
**28.** 67.9%                          **29. (a)** $58.2 = 0.17 \times n$  **(b)** 342.4
**30. (a)** $9.18 = 0.0725 \times n$  **(b)** 127
**31.** 33.1 pounds  **32.** 1293.1 gallons  **33.** $71.31      **34.** $31.11
**35.** $71.35      **36.** $412.57       **37.** $23,352     **38.** $2201.60
**39.** $196.33     **40.** $653.11       **41.** $1866.80    **42.** 14.56
**43.** 713.59      **44.** 31.9%         **45.** 14.35       **46.** 158.38
**47.** 29.3%       **48.** 1.67          **49.** 10          **50.** 25%
**51.** 4           **52.** 12            **53.** 1%          **54.** 30
**55.** 40          **56.** 4             **57.** 20          **58.** $33\frac{1}{3}\%$
**59.** $66\frac{2}{3}\%$     **60.** 75%

## Exercises for Section 6.1

**1.** 14 ft          **3.** 9 yd          **5.** 40,000 lb       **7.** 30 pt
**9.** 900 min        **11.** 126,720 ft    **13.** 126,720 in     **15.** 80 gal
**17.** 20,160 min    **19.** 272 pt        **21.** 768 qt         **23.** 21,600 min
**25.** 158 oz        **27.** 53 ft         **29.** 532 min        **31.** 23 qt
**33.** 6 ft 5 in.    **35.** 14 gal 3 qt   **37.** 2 mi 1440 ft   **39.** 8 lb 2 oz
**41.** 4.2 pt        **43.** 7.9 lb        **45.** 3.2 yr         **47.** 17.1 ft
**49.** 1.8 mi        **51.** 1.9 da        **53.** 4.58 ft        **55.** 5.13 mi
**57.** 6.69 lb       **59.** 5.75 gal      **61.** 25,200 ft;  4.8 mi     **63.** 1331 yd;  3993 ft
**65.** 1.2 mi        **67.** 2.66 in.

## Exercises for Section 6.2

**1. (a)** 22 ft 5 in.   **(b)** 22.42 ft
**5. (a)** 49 ft 2 in.   **(b)** 49.17 ft
**9. (a)** 50 yd 1 ft   **(b)** 50.33 yd
**13. (a)** 8 da 3 hr   **(b)** 8.13 da
**17. (a)** 18 lb 13 oz   **(b)** 18.81 lb
**21.** 1147 ft          **23.** 785 ft 4 in.
**29.** 19 hr 4 min      **31.** 11 lb 2 oz
**37.** 94              **39.** 60
**45.** 40              **47.** 6 ft 4 in.

**3. (a)** 36 lb 7 oz   **(b)** 36.44 lb
**7. (a)** 19 hr 20 min   **(b)** 19.33 hr
**11.** 557 mi
**15. (a)** 54 ft 5 in.   **(b)** 54.42 ft
**19. (a)** 179 gal 3 qt   **(b)** 179.95 gal
**25.** 116 gal        **27.** 4 da
**33.** 7 in.          **35.** 43
**41.** 4 ft 11 in.    **43.** 3 ft 3 in.
**49.** 10 ft 2 in.

## Exercises for Section 6.3

**1.** One-thousandth of a meter
**5.** 1000 meters        **7.** 750 mm
**13.** 1200 g           **15.** 3.9 kg
**21.** 11.9 cm          **23.** 4.4 m
**29.** 30.5 ft          **31.** 108.2 yd
**37.** 14.3 lb          **39.** 327.9 ft
**45.** 7.8 km per liter **47.** 6.2 km per liter
**53.** 1.4 in.          **55.** 10.4 mm

**3.** One-hundredth of a gram
**9.** 380 mg           **11.** 0.148 l
**17.** 2449.4 g        **19.** 258.0 g
**25.** 43.5 m          **27.** 3.4 in
**33.** 347.4 m         **35.** 0.5 oz
**41.** 0.4 gal         **43.** 25.6 gal
**49.** 20.5 mpg        **51.** 18.6 mpg
**57.** 4.1 mm

## Exercises for Section 6.4

**1.** 1845 ft-lb        **3.** 1902 ft-lb
**9.** 375.7 ft-lb/sec   **11.** 2217.1 ft-lb/min
**17.** 1.2 hp          **19.** 10.7 hp
**25.** 41.5 hp         **27.** 136.4 bhp
**33.** 6880.1 hp       **35.** 257.0 ihp
**41.** 150 lb          **43.** 46 kg

**5.** 240 ft-lb        **7.** 21,156 ft-lb
**13.** 18,177 ft-lb    **15.** 1357.2 ft-lb
**21.** 23.1 hp         **23.** 43.4 hp
**29.** 228.7 bhp       **31.** 5285.1 hp
**37.** 147.8 ihp       **39.** 5.3 in.

## Exercises for Section 6.5

**1.** 3.7 amp        **3.** 2435.3 volts
**7.** 4.2 amp        **9.** 3630.3 volts
**13. (a)** 127.8 ohms   **(b)** 31.6 ohms
**17. (a)** 194.2 ohms   **(b)** 21.2 ohms
**21. (a)** 174.3 ohms   **(b)** 5.5 ohms
**25.** 1.5 kw        **27.** 3.6 kw
**33.** 1.2 hp        **35.** 120.1 volts
**41.** 1.9 watts     **43.** 0.170 watt

**5.** 11.9 ohms
**11.** 10.6 ohms
**15. (a)** 41.2 ohms   **(b)** 5.5 ohms
**19. (a)** 29.2 ohms   **(b)** 1.9 ohms
**23. (a)** 70.8 ohms   **(b)** 1.9 ohms
**29.** 9.8 kw        **31.** 25.9 amp
**37.** 9.2 kwh       **39.** 7.6 kwh

## Exercises for Section 6.6

**1.** 48 lb          **3.** 64 lb
**9.** 21.4 in.       **11.** 12.7 in.
**13.** $50 \cdot 6 = 3 \cdot 100$;  in balance
**17. (a)** 72.4 lb   **(b)** 73 lb
**21.** Clockwise;  80 rpm
**25.** Counterclockwise;  30 teeth
**29.** Clockwise

**5.** 525 in.-lb       **7.** 612 in.-lb

**15.** $175 \cdot 4.5 \neq 8.5 \cdot 93$;   not in balance
**19.** 1369 lb
**23.** Clockwise;   128 rpm
**27.** Counterclockwise;   20 teeth
**31.** Clockwise

**33.** Counterclockwise          **35.** Clockwise;   125 rpm
**37.** Clockwise;   180 rpm
**39.** 14.8 in.      **41.** 22.9 in.          **43.** 140 rpm      **45.** 56 rpm
**47. (a)** Clockwise;   90 rpm   **(b)** Counterclockwise;   90 rpm
      **(c)** Idler does not affect speed of *B*   **(d)** Direction of *B* changes

## Exercises for Section 6.7

**1.** 10 ft              **2.** 20 qt              **3.** 2 lb              **4.** 90 ft
**5.** 30 in.            **6.** 50 qt              **7.** 1 lb 9 oz         **8.** 5.3 ft
**9.** 8 ft 5 in.        **10.** 10 da            **11.** 2 gal 3 qt       **12.** 4 lb 10 oz
**13.** 8 ft 8 in.       **14.** 26 ft            **15.** 10 cm            **16.** 10 m
**17.** 10 km            **18.** 1000 mm          **19.** 10,000 cm        **20.** 100,000 m
**21.** 10 cg            **22.** 10 g             **23.** 10 kg            **24.** 1000 mg
**25.** 10,000 cg        **26.** 100,000 g        **27.** 10 cl            **28.** 10 $\ell$
**29.** 10 kl            **30.** 1000 ml          **31.** 10,000 cl        **32.** 100,000 $\ell$
**33.** 2 cm             **34.** 2 m             **35.** 1 m              **36.** 0.6 m
**37.** 1300 kg          **38.** 14 g             **39.** 0.25 $\ell$      **40.** 68 $\ell$

## Chapter 6 Review Exercises

**1.** 27 ft              **2.** 3200 fl oz        **3.** 26 qt
**4.** 13 lb 13 oz       **5.** 2.4 mi            **6.** 8.6 ft
**7. (a)** 24 ft 7 in.   **(b)** 24.58 ft        **8. (a)** 12 hr 39 min   **(b)** 12.65 hr
**9. (a)** 20 da 20 hr   **(b)** 20.83 da        **10.** 768 ft
**11.** 14 gal           **12.** 86              **13. (a)** One-thousandth of a gram   **(b)** 1000 g
**14.** 1130 mm          **15.** 7500 mg          **16.** 74,900 m         **17.** 3650 ml
**18.** 140,000 mg       **19.** 18.8 cm          **20.** 91.4 m           **21.** 8.1 $\ell$
**22.** 12.1 ft          **23.** 12.3 lb          **24.** 7.9 gal          **25.** 6164 ft-lb
**26.** 322.3 ft-lb/sec  **27.** 14,288 ft-lb     **28.** 0.4 hp           **29.** 42.3 hp
**30.** 248.2 hp         **31.** 2.3 amp          **32.** 3.0 ohms         **33. (a)** 136.4 ohms   **(b)** 11.0 ohms
**34.** 8.5 kw           **35.** 15.5 amp         **36.** 219.4 volts      **37.** 68 lb
**38.** 96 in.           **39.** 90 rpm           **40.** 327 rpm          **41.** 4 ft
**42.** 10 lb            **43.** 6 gal            **44.** 14 ft            **45.** 3 km
**46.** 200 cm           **47.** 5000 g           **48.** 100 ml

## Exercises for Section 7.1

**1.** *A* is vertex; *AB* and *AE* are sides     **3.** *C* is vertex; *CB* and *CD* are sides
**5.** *E* is vertex; *AG* and *ED* are sides     **7.** 56°
**9.** 53°            **11.** 33°           **13.** 104°          **15.** 40.38°
**17.** 23.07°        **19.** 16.63°        **21.** 123.35°       **23.** 90°43'
**25.** 89°37'        **27.** 115°40'       **29.** 173°7'        **31.** Acute
**33.** Right         **35.** Obtuse        **37.** Straight      **39.** 122°
**41.** 120°          **43.** 150°          **45.** 162°

## Exercises for Section 7.2

**1.** Scalene       **3.** Equilateral     **5.** Isosceles
**7.** Obtuse        **9.** Acute
**11.** Right;   *FG* is hypotenuse        **13.** Altitude: 4.6 in.;   base: 10 in.
**15.** Altitude: 6 ft;   base: 8 ft       **17.** Altitude: 7.5 m;   base: 8 m

**19.** 100°     **21.** 70°     **23.** 149°10′     **25.** 95.5°
**27.** 70°     **29.** 70°40′     **31.** 16.3°
**33.** ∠D = 70°10′;  ∠F = 39°40′     **35.** ∠A = ∠D = 79°40′
**37.** 6.4 cm     **39.** 6.2 km     **41.** 81.6 mi     **43.** 0.94 in.
**45.** 0.92 in.     **47.** 16 ft 11 in.     **49.** 2.33 in.

## Exercises for Section 7.3

**1.** 13 ft     **3.** 10 ft     **5.** 27 ft 6 in.     **7.** 41.0 cm
**9.** 35 in.     **11.** 35.8 in.     **13.** 241.2 cm     **15.** 16 ft 4 in.
**17.** $w = 5.7$ m     **19.** $\ell = 90.2$ km
**21.** **(a)** 17 ft 3 in.  **(b)** 23 ft  **(c)** 34 ft 6 in.  **(d)** 46 ft
**23.** **(a)** 26.7 km  **(b)** 35.6 km  **(c)** 53.4 km  **(d)** 71.2 km
**25.** **(a)** 23.2 km  **(b)** 17.4 km  **(c)** 11.6 km  **(d)** 8.7 km
**27.** **(a)** 9.5 in.  **(b)** 7.1 in.  **(c)** 4.7 in.  **(d)** 3.6 in.
**29.** 18.2 yd.     **31.** 52.8 m     **33.** 263.9 km     **35.** 1.352 ft
**37.** 10.30 m     **39.** 122.6 mm     **41.** 14.4 in.     **43.** 57.4 cm
**45.** 224.2°     **47.** 215.1°     **49.** 357.08 m     **51.** 13.49 in.
**53.** 31.97 cm     **55.** 15 ft 8 in.     **57.** $56.78     **59.** $889.20

## Exercises for Section 7.4

**1.** 162 sq in.     **3.** 245 sq km     **5.** 67 sq ft     **7.** 810 sq cm
**9.** **(a)** 86.1 sq mm  **(b)** 198.8 sq mm  **(c)** 516.5 sq mm  **(d)** 959.9 sq mm
**11.** **(a)** 23.1 sq ft  **(b)** 53.3 sq ft  **(c)** 138.5 sq ft  **(d)** 257.3 sq ft
**13.** 78.5 sq in.     **15.** 422.7 sq m     **17.** 25.4 sq m     **19.** 7290 sq ft
**21.** 73.2 sq in.     **23.** 889 sq cm     **25.** 10.19 cm     **27.** 26.14 in.
**29.** 4.3 m     **31.** 5.5 km     **33.** 23.4 in.     **35.** 76.9 m
**37.** 1839 sq in.     **39.** 36.89 sq cm     **41.** 9.678 sq m     **43.** 6960 sq m
**45.** 7.43 sq in.     **47.** 62.9 sq cm     **49.** **(a)** 3753 sq ft  **(b)** $713.07
**51.** **(a)** 5888  **(b)** $2885.12     **53.** **(a)** 12  **(b)** $238.08
**55.** 0.49 in.     **57.** 0.75 in.

## Exercises for Section 7.5

**1.** 63.2 cu cm     **3.** 3938.5 cu ft     **5.** 232.7 cu cm     **7.** 100.9 cu in.
**9.** 250 cu m     **11.** 550 cu cm     **13.** 22,871 gal     **15.** 2543 gal
**17.** 512 cu ft     **19.** 227 cu m     **21.** 0.0741 cu in.     **23.** 38,100 cu in.
**25.** 91,000 cu cm     **27.** 62.4 cu cm     **29.** 843.7 cu in.     **31.** 209 cu m
**33.** 179 cu ft     **35.** 9030 cu in.     **37.** 6.9 cm     **39.** 12.5 in.
**41.** 1472 lb     **43.** **(a)** 170 cu cm  **(b)** 439 g
**45.** **(a)** 594 cu ft  **(b)** $647.46     **47.** 9.4 min

## Exercises for Section 7.6

**1.** 103.42 cu in.     **3.** 15,935.39 cu cm     **5.** 8.93 in.     **7.** 9.12 m
**9.** 706 cu in.     **11.** 4200 cu cm     **13.** 10.19 ft     **15.** 7.78 cm
**17.** 1560.85 cu ft     **19.** 16.36 cu m     **21.** 448.9 cu ft     **23.** 249,200 cu cm
**25.** 3172 gal     **27.** **(a)** 2606.0 cu ft  **(b)** 19,493 gal
**29.** 40,527 g     **31.** 23.6 cu in.     **33.** 412.3 cu cm     **35.** 7.6 lb
**37.** 3686.0 g     **39.** 131 sec     **41.** 25.9 cu in.

## Exercises for Section 7.7

**1.** 406.7 sq cm          **3.** 588.8 sq ft          **5.** 170 sq cm          **7.** 180 sq in.
**9.** 2460 sq ft          **11.** 2820 sq cm          **13.** 519 sq m          **15.** 330 sq in.
**17.** **(a)** 153.63 sq in.  **(b)** 234.16 sq in.          **19.** **(a)** 292.40 sq m  **(b)** 732.58 sq m
**21.** **(a)** 99.1 sq ft  **(b)** 141 sq ft          **23.** **(a)** 8820 sq cm  **(b)** 11,300 sq cm
**25.** **(a)** 1225.77 sq ft  **(b)** 1861.06 sq ft          **27.** **(a)** 244.48 sq m  **(b)** 380.49 sq m
**29.** 4584 sq mm          **31.** 1056 sq ft          **33.** 8.5 in.          **35.** 2.8 m
**37.** 3 gal          **39.** 332 sq in.          **41.** 233 sq in.

## Chapter 7 Review Exercises

**1.** $B$ is vertex;  $BC$ and $BE$ are sides          **2.** $C$ is vertex;  $CD$ and $CF$ are sides
**3.** 24.75°          **4.** 56°24′          **5.** **(a)** Acute  **(b)** Obtuse  **(c)** Right
**6.** 100°          **7.** Scalene;  right triangle          **8.** Altitude = $GK$;  base = $BF$
**9.** 110.3°          **10.** 67°;  67°          **11.** 10.4 mm
**12.** 48 ft          **13.** 7.1 m
**14.** **(a)** 98.7 cm  **(b)** 131.6 cm  **(c)** 197.4 cm  **(d)** 263.2 cm
**15.** 206.1 sq in.          **16.** 15.0 in.          **17.** 30.6 sq mi          **18.** 705 sq km
**19.** **(a)** 88.5 sq in.  **(b)** 204 sq in.  **(c)** 531 sq in.  **(d)** 987 sq in.
**20.** 81.71 sq m          **21.** 30.1 sq m          **22.** 24.06 sq in.          **23.** 9.299 cm
**24.** 14.3 sq in.          **25.** 2.6 sq cm          **26.** 280.6 cu in.          **27.** 45,957 gal
**28.** 3375 cu m          **29.** 17,150 cu ft          **30.** 381 cu ft          **31.** 1700 cu cm
**32.** 180 cu in.          **33.** 515 cu m          **34.** 9200 cu ft          **35.** 319 sq cm
**36.** 304 sq ft          **37.** 91 sq m          **38.** 363 sq ft          **39.** 109 sq cm
**40.** 749 sq in.

## Exercises for Section 8.1

**1.** Hypotenuse = $AC$;  side opposite $\angle A = BC$;  side adjacent to $\angle A = AB$;  side opposite $\angle C = AB$;  side adjacent to $\angle C = BC$

**3.** Hypotenuse = $DE$;  side opposite $\angle D = CE$;  side adjacent to $\angle D = CD$;  side opposite $\angle E = CD$;  side adjacent to $\angle E = CE$

**5.** Hypotenuse = $EG$;  side opposite $\angle E = FG$;  side adjacent to $\angle E = EF$;  side opposite $\angle G = EF$;  side adjacent to $\angle G = FG$

**7.** In $\triangle GKL$, hypotenuse = 8 in.;  side opposite $\angle G$ = 4.8 in.;  side adjacent to $\angle G$ = 6.4 in.;  side opposite $\angle K$ = 6.4 in.; side adjacent to $\angle K$ = 4.8 in.
In $\triangle HKL$, hypotenuse = 6 in.;  side opposite $\angle H$ = 4.8 in.;  side adjacent to $\angle H$ = 3.6 in.;  side opposite $\angle K$ = 3.6 in.; side adjacent to $\angle K$ = 4.8 in.
In $\triangle GHK$, hypotenuse = 10 in.;  side opposite $\angle G$ = 6 in.;  side adjacent to $\angle G$ = 8 in.;  side opposite $\angle H$ = 8 in.; side adjacent to $\angle H$ = 6 in.

**9.** In $\triangle LMN$, hypotenuse = 7.1 cm;  side opposite $\angle L$ = 5 cm;  side adjacent to $\angle L$ = 5 cm;  side opposite $\angle N$ = 5 cm; side adjacent to $\angle N$ = 5 cm
In $\triangle KMN$, hypotenuse = 11.2 cm;  side opposite $\angle K$ = 5 cm;  side adjacent to $\angle K$ = 10 cm;  side opposite $\angle N$ = 10 cm; side adjacent to $\angle N$ = 5 cm

**11.** In $\triangle MNR$, hypotenuse = 12.8 in.;  side opposite $\angle N$ = 8 in.;  side adjacent to $\angle N$ = 10 in.;  side opposite $\angle R$ = 10 in.; side adjacent to $\angle R$ = 8 in.
In $\triangle NPQ$, hypotenuse = 6.4 in.;  side opposite $\angle N$ = 4 in.;  side adjacent to $\angle N$ = 5 in.;  side opposite $\angle Q$ = 5 in.; side adjacent to $\angle Q$ = 4 in.

**13.** $\sin B = 0.38462$;  $\cos B = 0.92308$;  $\tan B = 0.41667$

**15.** sin $G$ = 0.88235;  cos $G$ = 0.47059;  tan $G$ = 1.87500
**17.** sin $H$ = 0.51449;  cos $H$ = 0.85749;  tan $H$ = 0.60000
**19.** sin $DBC$ = 0.6000;  cos $DBC$ = 0.8000;  tan $DBC$ = 0.75000
**21.** sin $A$ = 0.39391;  cos $A$ = 0.91912;  tan $A$ = 0.42857
**23.** sin $F$ = 0.80000;  cos $F$ = 0.60000;  tan $A$ = 1.33333
**25.** sin $GKH$ = 0.80000;  cos $GKH$ = 0.60000;  tan $GKH$ = 1.33333
**27.** sin 68° = 0.92718;  cos 68° = 0.37461;  tan 68° = 2.47509
**29.** sin 49° = 0.75471;  cos 49° = 0.65606;  tan 49° = 1.15037
**31.** sin 44.9° = 0.70587;  cos 44.9° = 0.70834;  tan 44.9° = 0.99652
**33.** sin 26.2° = 0.44151;  cos 26.2° = 0.89726;  tan 26.2° = 0.49206
**35.** sin 27°43′ = 0.46510;  cos 27°43′ = 0.88526;  tan 27°43′ = 0.52538
**37.** sin 10°13′ = 0.17737;  cos 10°13′ = 0.98414;  tan 10°13′ = 0.18023
**39.** 22.5°  **41.** 57.8°  **43.** 20.7°  **45.** 81.6°
**47.** 78.1°  **49.** 80.4°  **51.** 28.1°  **53.** 22.6°

## Exercises for Section 8.2

**1.** 11.3  **3.** 31.9  **5.** 0.221  **7.** 10.0
**9.** 1090  **11.** 3.95  **13.** 0.1572  **15.** 1.016
**17.** 0.8676  **19.** 104.4  **21.** 56.0°  **23.** 35.1°
**25.** 41.6°  **27.** 70.9°  **29.** 29.4°  **31.** 6.0
**33.** 6.8

## Exercises for Section 8.3

**1.** $\angle A$ = 29.7°;  $\angle B$ = 60.3°  **3.** $\angle A$ = 48.1°;  $\angle B$ = 41.9°
**5.** $\angle A$ = 14.7°;  $\angle B$ = 75.3°  **7.** $\angle A$ = 78.5°;  $\angle B$ = 11.5°
**9.** $\angle B$ = 33.8°;  $a$ = 16.0 m;  $b$ = 10.7 m  **11.** $\angle A$ = 55.7°;  $a$ = 11.3 yd;  $c$ = 13.7 yd
**13.** $\angle A$ = 25.0°;  $\angle B$ = 65.0°;  $c$ = 24.6 cm  **15.** $\angle A$ = 51.5°;  $\angle B$ = 38.5°;  $b$ = 61.9 mi
**17.** 9.8°  **19.** $a$ = 1.1 in.;  $b$ = 1.7 in.  **21.** 75.4°
**23.** 2.75 in.  **25.** 6.7 m  **27.** 69.2 ft

## Exercises for Section 8.4

**1.** 0.97030  **3.** 0.97815  **5.** 0.26387  **7.** 0.61704
**9.** 0.82692  **11.** 0.99803  **13.** 154.9°  **15.** 135.7°
**17.** 160.0°  **19.** $\angle C$ = 82°;  $a$ = 5.42 in.;  $b$ = 12.8 in.
**21.** $\angle B$ = 27.3°;  $a$ = 55.0 km;  $b$ = 54.7 km
**23.** $\angle C$ = 88°10′;  $a$ = 146 m;  $c$ = 153 m
**25.** 64.7°  **27.** 126.8°  **29.** 322 m  **31. (a)** 798.7 yd  **(b)** 582.2 yd
**33. (a)** 4.22 cm  **(b)** 3.65 cm  **(c)** 2.11 cm

## Exercises for Section 8.5

**1.** −0.95106  **3.** −0.58779  **5.** −0.95052  **7.** −0.83292
**9.** −0.67043  **11.** −0.10742  **13.** 118.0°  **15.** 131.4°
**17.** 86.6°  **19.** 157.7°  **21.** 59.7°  **23.** 135.2°
**25.** 24.9 in.  **27.** 104 m  **29.** $\angle A$ = 64.5°;  $\angle B$ = 83.5°

**31.** $\angle B = 18.8°$;   $\angle C = 43.8°$    **33.** $\angle C = 97.6°$;   $\angle A = 54.1°$;   $\angle B = 28.3°$
**35.** $\angle B = 136.8°$;   $\angle A = 23.0°$;   $\angle C = 20.2°$
**37.** 7.4 in.          **39.** 15.5 ft        **41. (a)** 38.6°   **(b)** 7.5 ft

## Exercises for Section 8.6

**1.** 103.1°          **3.** 90.0°          **5.** 229.8°          **7.** 1.71 radians
**9.** 0.51 radian     **11.** 3.44 radians  **13.** 0.51550        **15.** −0.58850
**17.** −1.70985       **19.** 0.57735       **21.** 0.86603        **23.** −0.50000
**25.** 0.60 radian    **27.** 2.00 radians  **29.** 1.28 radians
**31. (a)** 1.9 cm   **(b)** 2.7 sq cm       **33. (a)** 80.7 ft   **(b)** 1416.8 sq ft
**35.** 72.3 volts     **37.** 7.00 cm/sec   **39.** −0.185 cm

## Chapter 8 Review Exercises

**1.** In $\triangle ABD$, hypotenuse = 3 in.;   side opposite $\angle A$ = 2.4 in.;   side adjacent to $\angle A$ = 1.8 in.;   side opposite $\angle D$ = 1.8 in.;   side adjacent to $\angle D$ = 2.4 in.
In $\triangle BCD$, hypotenuse = 4 in.;   side opposite $\angle C$ = 2.4 in.;   side adjacent to $\angle C$ = 3.2 in.;   side opposite $\angle D$ = 3.2 in.;   side adjacent to $\angle D$ = 2.4 in.
In $\triangle ACD$, hypotenuse = 5.0 in.;   side opposite $\angle A$ = 4 in.;   side adjacent to $\angle A$ = 3 in.;   side opposite $\angle C$ = 3 in.;   side adjacent to $\angle C$ = 4 in.
**2.** sin $F$ = 0.81395;   cos $F$ = 0.58140;   tan $F$ = 1.40000
**3.** sin 16.4° = 0.28234;   cos 16.4° = 0.95931;   tan 16.4° = 0.29432
**4.** 38.9°               **5.** 56.6°          **6.** 79.2°        **7.** 10.0
**8.** 4.44               **9.** 0.146          **10.** 2.72        **11.** 56.6°
**12.** 78.5°             **13.** $\angle A$ = 28.1°;   $\angle B$ = 61.9°
**14.** $\angle A$ = 39.7°;   $\angle B$ = 50.3°;   $c$ = 9.9 in.
**15.** 0.91140           **16.** 172.2°        **17.** $\angle C$ = 110°;   $b$ = 8.46 in.;   $c$ = 27.2 in.
**18.** 31.0°             **19.** −0.17937      **20.** 93.8°
**21.** $e$ = 6.4 in.;   $\angle D$ = 70.9°;   $\angle F$ = 38.2°
**22.** $\angle C$ = 119.7°;   $\angle A$ = 38.0°;   $\angle B$ = 22.3°
**23. (a)** 99.1°   **(b)** 0.86        **24. (a)** 0.86281   **(b)** 1.05$^R$
**25.** 138.7 volts       **26.** −0.267

## Exercises for Section 9.1

**1.**
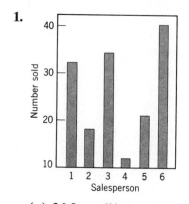
(*a*) 26.2      (*b*) 25.4%

**3.**
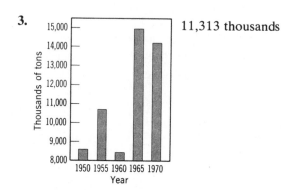
11,313 thousands

**5.**

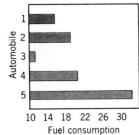

(*a*) 19.5     (*b*) 60.6%

**7.**

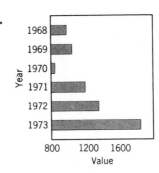

1174 millions

**9.** 1300 thousands     **11.** 2000 thousands     **13.** 1425 thousands

**15.**

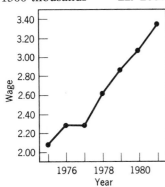

**17.**

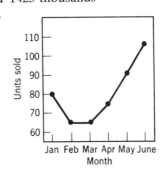

**19.** $29.00 millions     **21.** 1970 and 1971

Milling: 130°;   drilling: 133°;   grinding: 68°;   bench work: 29°

**23.**

**25.**

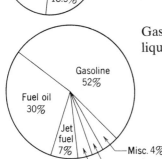

Gasoline: 188°;   fuel oil: 108°;   jet fuel: 25°;   asphalt: 13°;
liquefied gas: 10°;   miscellaneous: 16°

## Exercises for Section 9.2

**1.**

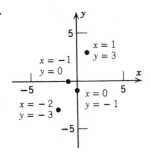

**3.**

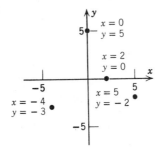

**5.**

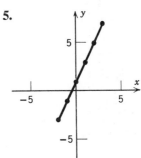

**7.**

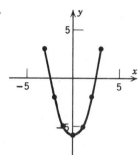

**9.**

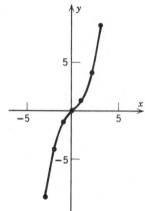

**11.** −0.4

**13.** 5.1

**15.** 4.1

**17.** −4.7

**19.** (a)

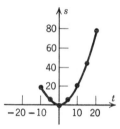

(b)

| $t$ | −7 | 12 | 12 | 18 |
|---|---|---|---|---|
| $s$ | 10 | 29 | 30 | 65 |

**21.** (a)

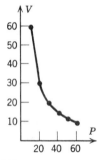

(b)

| $P$ | 18 | 43 | 21 | 11 |
|---|---|---|---|---|
| $V$ | 38 | 13 | 27 | 54 |

**23.**

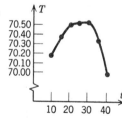

**25.**

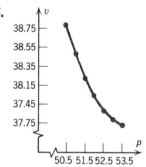

---

## Exercises for Section 9.3

**1.**

| $t$ | 0 | 5 | 10 | 15 | 20 | 25 |
|---|---|---|---|---|---|---|
| $s$ | 20 | 180 | 340 | 500 | 660 | 820 |

**3.**

| $F$ | 6 | 8 | 10 | 12 | 14 | 16 |
|---|---|---|---|---|---|---|
| $L$ | 945 | 1008 | 1071 | 1134 | 1197 | 1260 |

**5.**

| N | 50 | 54 | 58 | 62 | 66 | 70 |
|---|---|---|---|---|---|---|
| D | 7.692 | 7.714 | 7.733 | 7.750 | 7.765 | 7.778 |

**7.**

| V | 5 | 10 | 15 | 20 | 25 | 30 | 35 | 40 |
|---|---|---|---|---|---|---|---|---|
| P | 24.0 | 12.0 | 8.0 | 6.0 | 4.8 | 4.0 | 3.4 | 3.0 |

**9.**

| D | 3 | 4 | 5 | 6 | 7 | 8 | 9 | 10 | 11 | 12 |
|---|---|---|---|---|---|---|---|---|---|---|
| P | −4.0 | −3.0 | −2.4 | −2.0 | −1.7 | −1.5 | −1.3 | −1.2 | −1.1 | −1.0 |

**11.**

| n | −3 | −2 | −1 | 0 | 1 | 2 | 3 | 4 | 5 | 6 | 7 |
|---|---|---|---|---|---|---|---|---|---|---|---|
| B | 6.00 | 4.00 | 3.00 | 2.40 | 2.00 | 1.71 | 1.50 | 1.33 | 1.20 | 1.09 | 1.00 |

**13.**

| I | 10 | 20 | 30 | 40 | 50 | 60 | 70 |
|---|---|---|---|---|---|---|---|
| P | 4,000 | 16,000 | 36,000 | 64,000 | 100,000 | 144,000 | 196,000 |

**15.**

| d | 2 | 2.5 | 3 | 3.5 | 4 | 4.5 | 5 | 5.5 | 6 | 6.5 | 7 |
|---|---|---|---|---|---|---|---|---|---|---|---|
| R | 4.222 | 2.702 | 1.876 | 1.379 | 1.056 | 0.834 | 0.676 | 0.558 | 0.469 | 0.400 | 0.345 |

**17.**

| s | −2 | −1.5 | −1 | −5 | 0 | 0.5 | 1 | 1.5 | 2 |
|---|---|---|---|---|---|---|---|---|---|
| W | −23.03 | −9.72 | −2.88 | −0.36 | 0 | 0.36 | 2.88 | 9.72 | 23.03 |

**19.**

| s | 4 | 4.5 | 5 | 5.5 | 6 | 6.5 | 7 | 7.5 | 8 |
|---|---|---|---|---|---|---|---|---|---|
| C | 7.8 | 5.5 | 4.0 | 3.0 | 2.3 | 1.8 | 1.5 | 1.2 | 1.0 |

**21.**

| x | −2 | −1.5 | −1 | −0.5 | 0 | 0.5 | 1 | 1.5 | 2 |
|---|---|---|---|---|---|---|---|---|---|
| y | 40.2 | 12.7 | 2.5 | 0.2 | 0 | 0.2 | 2.5 | 12.7 | 40.2 |

**23.**

| A | 1 | 1.5 | 2 | 2.5 | 3 | 3.5 | 4 |
|---|---|---|---|---|---|---|---|
| d | 1.1 | 1.4 | 1.6 | 1.8 | 2.0 | 2.1 | 2.3 |

**25.**

| x | −4 | −3 | −2 | −1 | 0 | 1 | 2 | 3 | 4 |
|---|---|---|---|---|---|---|---|---|---|
| y | 0 | 2.6 | 3.5 | 3.9 | 4.0 | 3.9 | 3.5 | 2.6 | 0 |

**27.**

| r | 7.5 | 10 | 12.5 | 15 | 17.5 | 20 | 22.5 | ft |
|---|---|---|---|---|---|---|---|---|
| N | 4 | 6 | 10 | 14 | 19 | 25 | 32 | gal |

## Exercises for Section 9.4

**1.**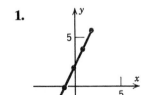

**3.**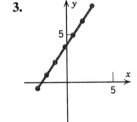

**5.**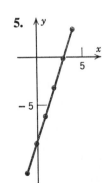

**7.**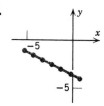

**9.**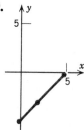

**11.**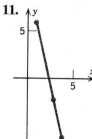

**13.**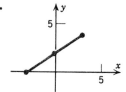

**15.** (*a*) 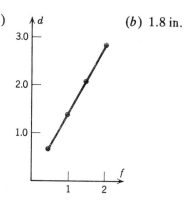   (*b*) 1.8 in.

**17.** (*a*) 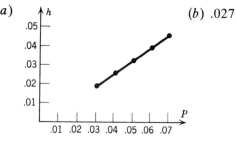   (*b*) .027

**19.** (*a*) $T = \dfrac{6.4 - d}{15}$   (*b*) 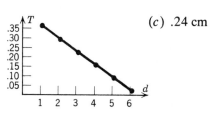   (*c*) .24 cm

**21.** (*a*) $D = d + 5.52$   (*b*) 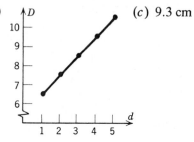    (*c*) 9.3 cm

## Exercises for Section 9.5

**1.** 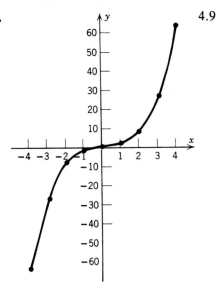 4.9

**3.** 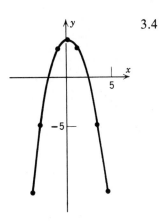 3.4

**7.** 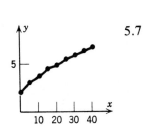 5.7

**5.** 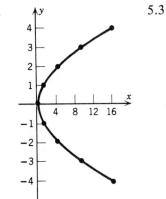 5.3

**9.** 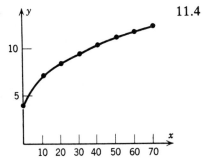 11.4

**11.** 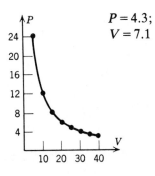 $P = 4.3;$ $V = 7.1$

**13.** 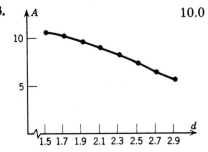 10.0

**15.** 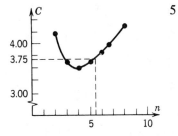 5

**17.** 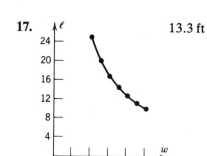                                    13.3 ft

**19.** 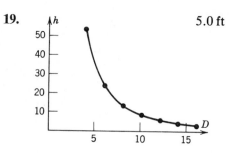                                    5.0 ft

---

## Exercises for Section 9.6

**1.**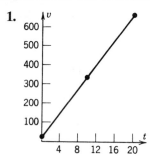

**3.**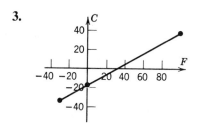

**5.**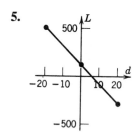

**7.**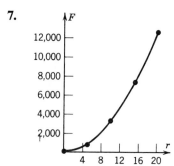

**9.**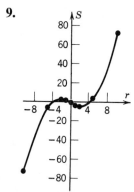

**11.**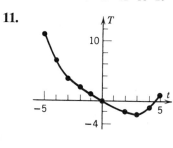

**13. (a)** 12 in.    **(b)** 48 in.     **15. (a)** 2 sec    **(b)** 64 ft    **(c)** 4 sec
**17.** 4.5 cm                           **19.** 75 km/hr

---

## Exercises for Section 9.7

**1.** sine 197° = −0.29237;   cosine 197° = −0.95630
**3.** sine 246.2° = −0.91496;   cosine 246.2° = −0.40355
**5.** sine 280° = −0.98481;   cosine 280° = 0.17365
**7.** sine 315.1° = −0.70587;   cosine 315.1° = 0.70834

**9.**

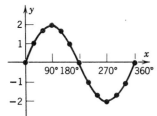

**11.**

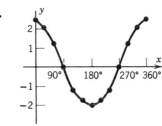

**13.**

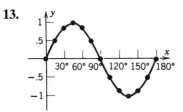

**15.**

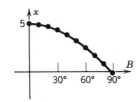

For $x = 60°$, $i = 4.3$; for $x = 315°$, $i = -3.5$

**17.**

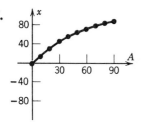

**19.**

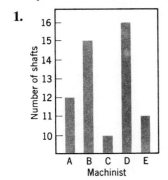

For $A = 25°$, $x = 35.9$; for $A = 65°$, $x = 77.0$

For $B = 15°$, $x = 4.8$; for $B = 55°$, $x = 2.9$

## *Chapter 9 Review Exercises*

**1.**

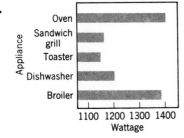

**2.**

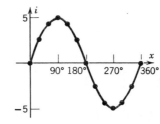

**3.**

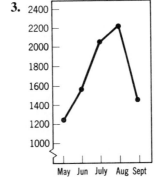

**4.**

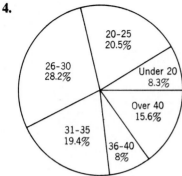

20-25
20.5%

26-30
28.2%

Under 20
8.3%

Over 40
15.6%

31-35
19.4%

36-40
8%

**5.**

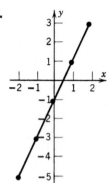

**6.**

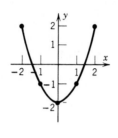

**7. (a)** $-4.0$  **(b)** $1.6$

**8. (a)** $-0.6$  **(b)** $-1.5$

**9.**

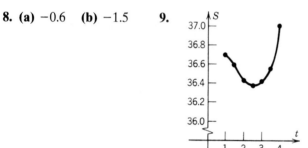

**10.**

| $h$ | 4 | 6 | 8 | 10 | 12 |
|---|---|---|---|---|---|
| $w$ | 3.9 | 2.6 | 2.0 | 1.6 | 1.3 |

**11.**

| $r$ | 2.0 | 2.2 | 2.4 | 2.6 | 2.8 | 3.0 |
|---|---|---|---|---|---|---|
| $S$ | 33.5 | 44.6 | 57.9 | 73.6 | 92.0 | 113.1 |

**12.**

| $x$ | $-6$ | $-4$ | $-2$ | 0 | 2 | 4 | 6 |
|---|---|---|---|---|---|---|---|
| $y$ | 0 | 4.5 | 5.7 | 6 | 5.7 | 4.5 | 0 |

**13.**

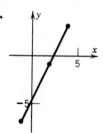

**14. (a)** $R = \dfrac{r + 0.3}{0.21}$ **(b)** **(c)** 0.95

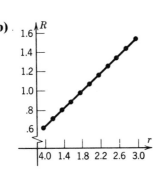

**15.** 2.0

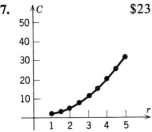

**16.** 6.3

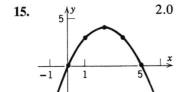

**17.** $23

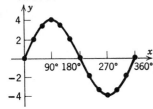

**18.**

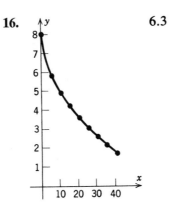

**19.** 2 amp
**20.** −0.93358
**22.**

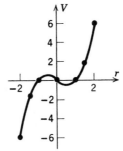

**21.** −0.50453
**23.**

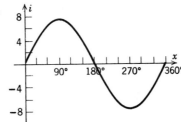

# Formulas for Geometric Figures

## Section 7.3  Perimeter and Circumference Formulas

| | | |
|---|---|---|
| Rectangle | $P = (l + w) \times 2$ | |
| Circle | $C = 2\pi r, \quad C = \pi d$ | |
| Arc | $L = \dfrac{\pi D°r}{180} = r\theta^R$ | |

## Section 7.4  Area Formulas

| | | |
|---|---|---|
| Triangle | $A = \dfrac{bh}{2}$ | |
| Equilateral triangle | $A = \dfrac{s^2\sqrt{3}}{4}$ | |
| Rectangle | $A = lw$ | |
| Square | $A = s^2$ | |
| Parallelogram | $A = bh$ | |
| Trapezoid | $A = \dfrac{(b_1 + b_2)h}{2}$ | |
| Circle | $A = \pi r^2, \quad A = \dfrac{\pi d^2}{4}$ | |
| Sector | $A = \dfrac{\pi D°r^2}{360} = \dfrac{r^2\theta^R}{2}$ | |

## Sections 7.5–7.6    Volume Formulas

| | | |
|---|---|---|
| Prism | $V = $ **area of base** $\times$ **altitude** | |
| Rectangular prism | $V = lwh$ | |
| Pyramid | $V = \frac{1}{3} \times$ **area of base** $\times$ **altitude** | |
| Frustum of pyramid | $V = \frac{1}{3}h(B + b + \sqrt{Bb})$ | |
| Circular cylinder | $V = \pi r^2 h$ | |
| Circular cone | $V = \frac{1}{3}\pi r^2 h$ | |
| Frustum of cone | $V = \frac{1}{3}\pi h(R^2 + r^2 + Rr)$ | |
| Sphere | $V = \frac{4}{3}\pi r^3$ | |